T0145244

Advances in Information Security

Volume 54

Series Editor

Sushil Jajodia, George Mason University, Fairfax, VA, USA

The purpose of the *Advances in Information Security* book series is to establish the state of the art and set the course for future research in information security. The scope of this series includes not only all aspects of computer, network security, and cryptography, but related areas, such as fault tolerance and software assurance. The series serves as a central source of reference for information security research and developments. The series aims to publish thorough and cohesive overviews on specific topics in Information Security, as well as works that are larger in scope than survey articles and that will contain more detailed background information. The series also provides a single point of coverage of advanced and timely topics and a forum for topics that may not have reached a level of maturity to warrant a comprehensive textbook.

Mark Stamp • Corrado Aaron Visaggio
Francesco Mercaldo • Fabio Di Troia
Editors

Artificial Intelligence for Cybersecurity

 Springer

Editors
Mark Stamp
San Jose, CA, USA

Corrado Aaron Visaggio
Benevento, Italy

Francesco Mercaldo
Campobasso, Italy

Fabio Di Troia (iD)
San Jose, CA, USA

ISSN 1568-2633   ISSN 2512-2193   (electronic)
Advances in Information Security
ISBN 978-3-030-97086-4   ISBN 978-3-030-97087-1   (eBook)
https://doi.org/10.1007/978-3-030-97087-1

This Springer imprint is published by the registered company Springer Nature Switzerland AG
The registered company address is: Gewerbestrasse 11, 6330 Cham, Switzerland

Preface

We are on the cusp of a revolution in artificial intelligence (AI). Today, AI plays a significant role in daily life, and the impact of AI is sure to increase dramatically over the coming years. Perhaps surprisingly, the net effect of this AI revolution on cybersecurity is, at present, unclear, as both the "good guys" and the "bad guys" can employ such technology. If cybersecurity is to reap major benefits from AI, the technology itself must be better understood—black boxes are inherently the enemy of security.

Models used in AI are notoriously opaque, which creates numerous potential problems. From a cybersecurity perspective, one of the greatest of these problems is the threat of adversarial attacks. It follows that "explainable AI," for example, is of fundamental importance in information security.

This book includes chapters that attempt to illuminate various aspects of the AI black boxes that have come to dominate cybersecurity. The topics of explainable AI and adversarial attacks—as well as the closely related issue of model robustness—are considered. Most of the chapters explore these and similar topics in the context of specific security threats. The security domains considered include such diverse areas as malware, biometrics, and side-channel attacks, among others. We have strived to make the material accessible to the widest possible audience of researchers and practitioners.

We are confident that this book will prove valuable to practitioners working in the field and to researchers in both academia and industry. The chapters include insights that should help to illuminate some of the darkest corners of popular AI models that are used in cybersecurity.

San Jose, CA, USA   Mark Stamp
Benevento, Italy   Corrado Aaron Visaggio
Campobasso, Italy   Francesco Mercaldo
San Jose, CA, USA   Fabio Di Troia
December 2021

Contents

Part I   Malware-Related Topics

**Generation of Adversarial Malware and Benign Examples Using
Reinforcement Learning**.. 3
Matouš Kozák, Martin Jureček, and Róbert Lórencz
1   Introduction 3
2   Background 5
     2.1   Adversarial Machine Learning 5
     2.2   Reinforcement Learning.. 6
     2.3   Portable Executable File Format 8
3   Implementation 8
     3.1   Overview 9
     3.2   Dataset 9
     3.3   PE File Modifications 9
     3.4   Target Classifier.. 10
     3.5   Agent and Its Environment 11
4   Evaluation 11
     4.1   Adversarial Malware Examples 12
     4.2   Adversarial Benign Examples.. 17
5   Related Work... 20
     5.1   Gradient-Based Attacks 20
     5.2   Reinforcement Learning-Based Attacks.............................. 20
     5.3   Other Methods 21
6   Conclusion 22
     6.1   Future Work... 23
References 23

Auxiliary-Classifier GAN for Malware Analysis 27
Rakesh Nagaraju and Mark Stamp
1   Introduction 27
2   Related Work... 28

3   Methodology 30
     3.1   Data... 30
     3.2   AC-GAN 31
     3.3   Evaluation Plan 33
     3.4   Accuracy 35
4   Implementation 36
     4.1   Dataset Analysis and Conversion 37
     4.2   AC-GAN Implementation 38
     4.3   Evaluation Models.. 40
5   Experimental Results 42
     5.1   AC-GAN Experiments 42
     5.2   CNN and ELM Experiments 48
6   Conclusion and Future Work 65
References 66

**Assessing the Robustness of an Image-Based Malware Classifier
with Smali Level Perturbations Techniques**................................. 69
Giacomo Iadarola, Fabio Martinelli, Antonella Santone, and Francesco
Mercaldo
1   Introduction 69
2   Background and Related Works...................................... 71
     2.1   Static Malware Analysis....................................... 71
     2.2   Convolutional Neural Network 72
     2.3   Dalvik VM and Dalvik EXecutable............................. 74
     2.4   Image-Based Malware Classification 75
3   Methodology 76
     3.1   Untargeted Misclassification 78
4   Implementation and Experiments 80
5   Conclusion and Future Work.. 82
References 82

Detecting Botnets Through Deep Learning and Network Flow Analysis .. 85
Ji An Lee and Fabio Di Troia
1   Introduction 85
2   Background 86
     2.1   Introduction to Botnets 87
     2.2   Autocorrelation Analysis...................................... 88
     2.3   Deep Neural Networks 89
3   Related Work... 90
4   Dataset... 91
     4.1   CTU-13 Dataset Features 92
5   Proposed Methodology 92
     5.1   Data Preprocessing Phase 94
     5.2   Deep Learning Phase 99

6   Results 101
7   Conclusions 103
References 103

**Interpretability of Machine Learning-Based Results of Malware
Detection Using a Set of Rules** 107
Jan Dolejš and Martin Jureček
1   Introduction 107
2   Related Works.. 109
3   Rule-Based Classification 110
      3.1   From Trees to Rules 112
      3.2   Rule-Learning Algorithms 113
4   Implementation of Rule-Based Classifiers................................ 115
      4.1   Decision List 115
      4.2   I-REP 116
      4.3   RIPPER 117
5   Experiments 118
      5.1   Dataset Description.. 118
      5.2   Data Splitting 119
      5.3   Feature Transformation and Selection 120
      5.4   Evaluation Metrics... 122
      5.5   Interpretability of Machine Learning Models 123
      5.6   Measuring Performance of RBCs on ML Predictions 124
      5.7   Interpreting ML Results Using RBCs 126
      5.8   Pruning and Metrics 128
      5.9   Does Order of the Rules Matter?................................. 130
6   Conclusion and Future Work 133
References 135

Mobile Malware Detection Using Consortium Blockchain 137
George Martin, Dona Spencer, Aditya Hair, Deepa K, Sonia Laudanna,
Vinod P, and Corrado Aaron Visaggio
1   Introduction 138
2   Use Case... 139
3   Android Application Components 140
      3.1   Activities 140
      3.2   Services 141
      3.3   Broadcast Receivers 141
      3.4   Content Providers.. 141
4   Role in Malware Detection 141
5   The Blockchain Network 142
6   Related Works.. 143
7   Methodology 144
      7.1   APK Files 145
      7.2   Trusted Server 145
      7.3   Adding a Record 145

7.4   Members of the Consortium 146
7.5   Blockchain Ledger.. 146
7.6   Final Response... 146
7.7   Technology Behind Blockchain Network 147
8   Implementation Details... 149
8.1   Scenario 1 149
8.2   Scenario 2 151
8.3   Initializing Block for Unknown apk 152
8.4   Updating Block with Vote and Features 153
8.5   Setting the State of the apk After Counting All the Votes 154
9   Feature Extraction and Model Training.................................. 155
10   Dataset and Experimentation 156
11   Results... 157
12   Conclusion 158
References 159

BERT for Malware Classification 161
Joel Alvares and Fabio Di Troia
1   Introduction 161
2   Related Work... 162
3   Background 163
3.1   NLP Models 164
3.2   Classifiers 168
4   Experiments and Results 171
4.1   Dataset 172
4.2   Methodology 173
4.3   Classifier Parameters 173
4.4   Logistic Regression Results 174
4.5   SVM Results 174
4.6   Random Forest Results 175
4.7   MLP Results 176
4.8   Further Analysis 176
4.9   Summary 178
5   Conclusions and Future Work.. 179
References 180

Machine Learning for Malware Evolution Detection 183
Lolitha Sresta Tupadha and Mark Stamp
1   Introduction 183
2   Background 185
2.1   Malware 185
2.2   Related Work 186
2.3   Dataset 187
2.4   Learning Techniques 189

3   Experiments and Results 196
    3.1   Logistic Regression Experiments 196
    3.2   Hidden Markov Model Experiments................................ 197
    3.3   HMM2Vec Experiments.. 198
    3.4   Word2Vec Experiments 199
    3.5   Discussion... 199
4   Conclusion and Future Work 202
Appendix... 202
References 212

Part II   Other Security Topics

**Gambling for Success: The Lottery Ticket Hypothesis in Deep
Learning-Based Side-Channel Analysis** 217
Guilherme Perin, Lichao Wu, and Stjepan Picek
1   Introduction 218
2   Background 220
    2.1   Notation 220
    2.2   Supervised Machine Learning in Profiling SCA 220
    2.3   Leakage Models and Datasets....................................... 221
3   Related Works... 222
4   The Lottery Ticket Hypothesis (LTH) 223
    4.1   Pruning Strategy 224
    4.2   Winning Tickets in Profiling SCA 225
5   Experimental Results 226
    5.1   Baseline Neural Networks.. 226
    5.2   ASCAD with a Fixed Key.. 229
    5.3   ASCAD with Random Keys 232
    5.4   CHES CTF 2018.. 235
    5.5   General Observations... 238
6   Conclusions and Future Work.. 238
References 239

**Evaluating Deep Learning Models and Adversarial Attacks on
Accelerometer-Based Gesture Authentication** 243
Elliu Huang, Fabio Di Troia, and Mark Stamp
1   Introduction 243
2   Related Work.. 244
3   Background 245
    3.1   Support Vector Machines... 246
    3.2   1D Convolutional Neural Networks 246
    3.3   Adversarial Strategy 247
4   Dataset.. 248
    4.1   Data Collection 248
    4.2   Data Preprocesssing 250

5   Implementation 251
     5.1   DC-GAN Structure 251
     5.2   Adversarial Attack....................................... 253
6   Experiments and Results 253
     6.1   SVM Results.. 254
     6.2   1D-CNN Results... 255
     6.3   Adversarial Results 256
7   Conclusion and Future Work 257
References 258

Clickbait Detection for YouTube Videos............................... 261
Ruchira Gothankar, Fabio Di Troia, and Mark Stamp
1   Introduction 261
2   Background 263
     2.1   Related Work 263
     2.2   Natural Language Processing 265
     2.3   Learning Techniques 268
3   Implementation 269
     3.1   Hardware and Software 270
     3.2   Approach... 271
     3.3   Features 271
     3.4   Dataset 272
     3.5   Experiments.. 272
4   Results.. 274
5   Conclusion and Future Works 278
Appendix: Model Architectures 279
References 282

**Survivability Using Artificial Intelligence Assisted Cyber Risk
Warning** 285
Nikolaos Doukas, Peter Stavroulakis, Vyacheslav Kharchenko, Nikolaos
Bardis, Dimitrios Irakleous, Oleg Ivanchenko, and Olga Morozova
1   Introduction 286
2   Related Work.. 288
3   Security Infringement Detection 289
     3.1   Static Analysis of Code.................................. 289
     3.2   Methodology 291
     3.3   Results... 294
     3.4   Evaluation 295
4   Digital Twin Cyber Resilience Decision Support 295
     4.1   Landscape Model Development 298
5   Semi-Markov Cloud Availability Model........................... 300
6   Future Work 305
7   Conclusions 306
References 307

Machine Learning and Deep Learning for Fixed-Text Keystroke Dynamics 309
Han-Chih Chang, Jianwei Li, Ching-Seh Wu, and Mark Stamp
1   Introduction 309
2   Background 311
      2.1   Keystroke Dynamics.. 311
      2.2   Learning Techniques 311
3   Previous Work.. 314
4   Dataset... 316
5   Experiments and Results 316
      5.1   Data Exploration... 317
      5.2   Classification Results 319
      5.3   Summary and Discussion 325
6   Conclusion and Future Work 327
References 327

Machine Learning-Based Analysis of Free-Text Keystroke Dynamics 331
Han-Chih Chang, Jianwei Li, and Mark Stamp
1   Introduction 331
2   Background 332
      2.1   Keystroke Dynamics.. 333
      2.2   Previous Work 333
3   Implementation 334
      3.1   Dataset 334
      3.2   Techniques Considered 335
4   Free-Text Experiments... 336
      4.1   Text-Based Classification 336
      4.2   Keystroke Dynamics Models....................................... 337
5   Conclusion and Future Work 354
References 355

Free-Text Keystroke Dynamics for User Authentication 357
Jianwei Li, Han-Chih Chang, and Mark Stamp
1   Introduction 357
2   Background 359
      2.1   Related Work 360
      2.2   Datasets 361
      2.3   Deep Leaning Algorithms 362
3   Feature Engineering 363
      3.1   Features 364
      3.2   Length of Keystroke Sequence..................................... 365
      3.3   Keystroke Dynamics Image 365
      3.4   Keystroke Dynamics Sequence 366
      3.5   Cutout Regularization 367

4   Architecture 367
     4.1   Multiclass vs Binary Classification 367
     4.2   Hyperparameter Tuning 368
     4.3   Implementations 368
5   Experiment and Result.. 369
     5.1   Metrics 369
     5.2   Result of Free-Text Experiments...................................... 370
     5.3   Discussion... 375
6   Conclusion 377
Appendix.. 378
References 379

Correction to: Artificial Intelligence for Cybersecurity C1

About the Editors

Mark Stamp has extensive experience in information security and machine learning, having worked in these fields within academic, industrial, and government environments. After completing his PhD research in cryptography at Texas Tech University, he spent more than 7 years as a cryptanalyst with the United States National Security Agency (NSA), followed by 2 years developing a security product for a Silicon Valley start-up company. Since early in the present century, Dr. Stamp has been employed as a professor in the Department of Computer Science at San Jose State University, where he teaches courses in machine learning and information security. To date, he has published more than 150 research articles, most of which deal with problems at the interface between machine learning and information security. Dr. Stamp served as a co-editor of the *Handbook of Information and Communication Security* (Springer, 2010) and *Malware Analysis Using Artificial Intelligence and Deep Learning* (Springer 2020), and he is the author of multiple textbooks, including *Information Security: Principles and Practice* (Wiley, third edition, 2021) and *Introduction to Machine Learning with Applications in Information Security* (Chapman and Hall/CRC, second edition, 2022).

Corrado Aaron Visaggio is an associate professor in the Department of Engineering at the University of Sannio, where he teaches "Security of Networks and Software Systems" in the MSc in computer engineering program. Currently, he is also chief scientific officer at Defence Tech, a company operating in cybersecurity, aerospace, and military engineering. He obtained his MSc in electronic engineering (2001) from Politecnico di Bari and his PhD in information engineering (2005) from the University of Sannio. His main research interests are malware analysis, data protection, and threat intelligence. Dr. Visaggio teaches in the master's program in cybersecurity at the University of Rome "Tor Vergata," and at the International School Against Organized Crime established by the Italian Ministry of the Interior for education of international law enforcement agencies, and he has been instructor in the Department of Intelligence at the Italian Ministry of the Interior. He is director of the Unisannio Chapter of the CINI Cybersecurity National Lab. He is in the Organizing Board of CINI Cybersecurity National Lab. Dr. Visaggio leads the Cybersecurity Lab in the Department of Engineering at the University of Sannio. He is the scientific leader of several research projects in cybersecurity, funded by private and public organizations. He collaborates with several universities (ETH Zurich, the University of San Jose, the University of Castilla-La-Mancha, the University of Lugano, University College Dublin, The University of Delft, Cochin University of Science & Technology, and SCMS School of Engineering & Technology). Dr. Visaggio has authored more than 100 scientific papers and he serves in the editorial boards of international journals and program committees of international conferences. He is one of the founders of the SER&Practice software house, and SLIMER software house.

Francesco Mercaldo received his master's degree in computer engineering from the University of Sannio (Benevento, Italy), with a thesis in software testing. He obtained his PhD in 2015 with a dissertation on malware analysis using machine learning techniques. The research areas of Francesco are software testing, verification, and validation, with emphasis on the application of empirical methods. Currently, he is working as a researcher at the University of Molise (Italy). He has written approximately 70 papers for international journals and conferences.

Fabio Di Troia is an assistant professor in the Computer Science Department at San Jose State University, where he teaches information security and machine learning courses. He completed his PhD in computer science at Kingston University, London, researching applications of machine learning in the field of cybersecurity. His areas of focus are malware detection, malware design, cryptology, biometrics, and access control. In collaboration with colleagues sharing similar academic background, he co-founded the Silicon Valley Cybersecurity Institute (SVCSI) in 2019, a non-profit organization that aims to increase awareness in the cybersecurity domain for high-school, undergraduate, and graduate students, with particular emphasis in the underrepresented community. Within this organization, he holds the role of program director in software security, and he is also the program committee chair for the Silicon Valley Cybersecurity Conference (SVCC).

The original version of the book has been revised. A correction to this book can be found at https://doi.org/10.1007/978-3-030-97087-1_16

Part I
Malware-Related Topics

Generation of Adversarial Malware and Benign Examples Using Reinforcement Learning

Matouš Kozák, Martin Jureček, and Róbert Lórencz

Abstract Machine learning is becoming increasingly popular among antivirus developers as a key factor in defence against malware. While machine learning is achieving state-of-the-art results in many areas, it also has drawbacks exploited by many with white-box attacks. Although the white-box scenario is possible in malware detection, the detailed structure of antivirus is often unknown. Consequently, we focused on a pure black-box setup where no information apart from the predicted label is known to the attacker, not even the feature space or predicted score. We implemented our exploratory integrity attack using a reinforcement learning approach on a dataset of portable executable binaries. We tested multiple agent configurations while targeting LightGBM and MalConv classifiers. We achieved an evasion rate of 68.64% and 13.32% against LightGBM and MalConv classifiers, respectively. Besides traditional modelling of malware adversarial samples, we present a setup for creating benign files that can increase the targeted classifier's false positive rate. This problem was considerably more challenging for our reinforcement learning agents, with an evasion rate of 3.45% and 36.62% against LightGBM and MalConv classifier, respectively. To understand how these attacks transfer from classifiers based purely on machine learning to real-world anti-malware software, we tested the same modified files against seven well-known antiviruses. We achieved an evasion rate of up to 47.09% in malware and 14.29% in benign adversarial attacks.

1 Introduction

Malware detection is one of the most important problems in information security since the detection of malware in advance allows us to block it. Malware detection is a binary classification problem of distinguishing between malware and benign

M. Kozák · M. Jureček (✉) · R. Lórencz
Faculty of Information Technology, Czech Technical University in Prague, Prague, Czechia
e-mail: kozakmat@fit.cvut.cz; martin.jurecek@fit.cvut.cz

© The Author(s), under exclusive license to Springer Nature Switzerland AG 2022
M. Stamp et al. (eds.), *Artificial Intelligence for Cybersecurity*, Advances in
Information Security 54, https://doi.org/10.1007/978-3-030-97087-1_1

files [3]. One of the main problems of malware detection systems is insufficient accuracy while keeping the false positive rate at an acceptable level. There is a need to build a machine learning framework suited for real-life practical use that generically detects as many malware samples as possible, with a very low false positives rate. The significant problem to be solved is how to detect malware that has never been seen before.

To defend against malware, users typically rely on antivirus (AV) products to detect a threat before it can damage their systems. Antivirus vendors rely mainly on a database of sequences of bytes (signatures) that uniquely identify the suspect files and are unlikely to be found in benign programs [36]. The major weakness of signature detection is that malware writers can easily modify their code, thereby changing their program's signature and evading virus scanners. The signature detection technique is unable to detect obfuscated and zero-day malware. Encryption, polymorphism, metamorphism, and other code obfuscation techniques are widely used by malware authors to evade signature detection techniques. For this reason, malware researchers are investigating novel detection strategies.

Nowadays, antivirus vendors face several problems concerning malware detection. The concept of employing machine learning to malware detection provides promising solutions [31]. Moreover, since malware developers create more and more sophisticated techniques, it is necessary to use the latest techniques from machine learning to keep the error rate and false positive rate as low as possible. This game may someday converge to the point when artificial intelligence of attackers will fight against the artificial intelligence of malware researchers.

Machine learning models are vulnerable to adversarial attacks that can fool the models [9]. For instance, an adversary can craft malware that has a similar feature vector to some benign file's feature vector. As a result, the training set may have different statistical distribution than the distribution of the testing set. Therefore, it is necessary to create defence techniques in order for machine learning algorithms can resist such adversarial attacks.

The goal of this paper is to implement a black-box exploratory integrity attack using reinforcement learning. We implement executability preserving modifications and train reinforcement learning agents to alter Windows portable executable binaries with an aim to avoid detection by a targeted machine learning classifier. These evasion techniques are later tested on real-world antivirus software. In comparison with other works, we do not only focus on malware adversarial samples, but we also deal with an inverted scenario of benign adversarial examples.

This paper is organized in the following form. Section 2 gives the necessary background to our work with a brief description of adversarial machine learning, reinforcement learning, portable executable file format, etc. In Sect. 3, we present our implementation and dataset description. Next, Sect. 4 contains all information about experiments and achieved results. We summarize related work in Sect. 5. The conclusion to this paper and ideas for future work can be found in Sect. 6.

2   Background

In this section, we provide the minimal necessary background to understand our paper. We firstly introduce adversarial and reinforcement machine learning. Then we briefly describe portable executable file format and finish with the definition of used evaluation metrics.

2.1   Adversarial Machine Learning

Machine learning outperforms human capabilities in many ways, yet we are reserved in trusting its decisions in areas such as self-less car driving or disease diagnostics. One of the reasons is the insufficient interpretability of the decisions and the resulting possible weakness or bias of the system [15]. *Adversarial machine learning* is a research area specializing in strengthening machine learning (ML) systems to be resistant against attacks both from the inside (data poisoning) and outside (evasion attacks).

In common terminology, an action to bypass or mislead a ML system is an adversarial attack. An attacker is called an adversary but both are acceptable and used interchangeably. In this section, we summarize the taxonomy of adversarial attacks and describe some prevalent adversarial attack strategies, focusing mainly on malware detection domain.

2.1.1   Taxonomy

In this part, we will closely follow the taxonomy laid down by Huang et al. in [18] as it is one of the most complete overviews of this topic we have found. They identify three main ways how to break down adversarial attacks based on these three properties: *influence, security violation, specificity*.

Influence   The first property is the way we can look at how adversary influences targeted ML model. There are two main categories. The first is *exploratory* attacks that do not alter the model itself but try to circumvent the model to achieve attacker's goal—usually, misclassification of a group of malicious files.
The second group is called *causative* attacks where the attacker impacts the model itself, in particular the training phase. It can be in form of wrongly labelled malware samples in training dataset.

Security violation   The second property is characterized by the objective of adversary. There are three groups. *Integrity* attacks cause an increase in false-negative rate. In the domain of malware detection, a false negative is malware sample classified as benign.

The second group is *availability* attacks where the adversary does not focus on a single class to be misclassified but targets the model's accuracy as a whole resulting in the model becoming completely unusable.

When an adversary tries to steal information from the model itself, e.g., what training data the model was learned on, it is called a *privacy* attack.

Specificity   The last property describes how large the attacker's target set is. In *targeted* attacks, there is a small set of samples that are supposed to be misclassified, i.e., an author of malicious software wants his particular program to be installed on the victim's device.

Whereas in *indiscriminate* attacks adversary does not specify which but rather how many samples should be mislabelled. This attack can be used as a one of the proves that given AV is not secure.

Individual categories can be combined together, e.g., a causative targeted attack can be when an adversary inserts malware binaries labelled as benign into a classifiers training dataset to prevent the classifier from correctly predicting for a particular malicious file.

Attacks can also be classified based on the available knowledge of the targeted model to attackers. If the attacker does not have any information about the model and is left with only the model's output, it is called a *black-box* attack. An opposite attack is called a *white-box* where all information about the model is known. In between these two is a *grey-box* attack where partial knowledge is accessible, e.g., feature space of targeted classifier.

2.2   Reinforcement Learning

The key idea of *reinforcement learning* (*RL*) is a simple one. An *agent* (a learning system) wants to achieve some goal. To achieve its objective, the agent must adapt its behaviour. To learn which actions are good and bad, the agent needs a stimulus in the form of signals from an *environment* where the agent works. This section is based on [29] from which we adopted the following notions.

Reinforcement learning is a triplet of an agent policy, a reward function and a value function. In some cases a model of environment is used as well.

The *agent policy* represents the agent's behaviour at a given time and environment state. It is a function which maps pairs of states and action to the probability of taking individual action at given state. The policy can be in the form of a simple automaton, lookup table, but also a sophisticated algorithm.

The *reward function* is an immediate response from the environment to the agent's action. The return value is a number that grades the action taken by the agent based on the goal. Formally, this number defines the agent's purpose, and under no circumstances can the agent change this function, i.e., change the goal it is facing.

The *value function* is a look to the future on what is an expected cumulative reward from the current environment state. This function can be understood as a heuristic function, and it is a critical part of any RL model. While reward function can be usually calculated easily, the value function must be recalculated over and over based on past observations. It is clear that agent looks for states with the highest value since these will maximize future rewards.

The *model of environment* simulates the environment and allows the agent to predict future states and rewards. This part is optional, and not all RL systems contain it.

Formally reinforcement learning is defined as repeated interactions between agent and environment. In our space of modifications of binary files these interaction take place at discrete time steps $t = 0, 1, 2, \ldots$ At time step t the environment is at state $S_t \in S$ where S is the set of all observable states. The agent receives state S_t and, based on its policy π, chooses action $A_t \in A(S_t)$ where $A(S_t)$ is a set of all possible actions at state S_t. When the environment receives response from agent in the form of action A_t it computes and sends back a reward $R_{t+1} \in R \subset \mathbb{R}$ and changes its state: $S_t \xrightarrow{A_t} S_{t+1}$. In this notation, R_{t+1} represents the reward for the state S_{t+1}.

As stated above, the agent looks for states that maximize future rewards. To formalize future rewards, we denote $G_t = R_{t+1} + R_{t+2} + \cdots + R_T$ a sum of all rewards after time step t, where T denotes the final time step. If $T \neq \infty$ we can call the S_T a *terminal state* and the entire process of states, actions and rewards from time step $t = 0$ to $t = T$ an *episode*. After the end of each episode the environment is reset to initial state and new independent episode begins. This repeats until terminal condition is met. Many problems will not have any terminal state. These problems are called *continuous*, and they are not covered in this work.

In real situations, the agent does not know the exact value of G_t at time step t. The value of G_t is approximated by the value function. In the computation, it is common to use *discounted* future rewards,

$$G_t \approx R_{t+1} + \gamma R_{t+2} + \gamma^2 R_{t+3} + \cdots + \gamma^{T-t-1} R_T = \sum_{i=0}^{T-t-1} = \gamma^i R_{t+i+1}$$

where $0 \leq \gamma \leq 1$ is called *discount rate*. This helps regulate importance of looking far into the future. Further we outline an *action-value function for policy* π, $q_\pi(s, a)$, that defines value of action a when in state s if following policy π. The value is in form of expected future reward G_t, formally,

$$q_\pi(s, a) = \mathbb{E}_\pi[G_t | S_t = s, A_t = a]$$

where $\mathbb{E}_\pi[\cdot]$ is expected value of random variable for policy π.

The rest of this section describes two algorithms which we later used in Sect. 3. Note that there are numerous other algorithms for reinforcement learning.

Q-learning is an algorithm introduced by Watkins [34] to learn the action-value function. It works by iteratively updating the learned action-value function, Q, in the following manner,

$$Q^{new}(S_t, A_t) = Q(S_t, A_t) + \alpha[R_{t+1} + \gamma \max_a \{Q(S_{t+1}, a)\} - Q(S_t, A_t)]$$

where α is learning rate, $a \in A(S_{t+1})$. Therefore, $\max_a\{Q(S_{t+1}, a)\}$ is an estimation of the best future value. Learned values of function Q are stored in so-called Q-table. In our work we used extension of this algorithm called *deep Q-learning* or *deep Q network* (DQN) where the Q-table is replaced with neural network [24].

Second algorithm we describe is *proximal policy optimization (PPO)* [27]. It is based on *policy gradient* methods. Informally, these methods use *gradient ascent* algorithm to approximate policy weight vector $\theta \in \mathbb{R}^n$. The policy of taking action a is then conditioned not only by the current state S_t but also weight vector θ. PPO improves the approximation by vanilla policy gradient method with multiple epochs of the gradient ascent before updating the policy vector.

2.3   Portable Executable File Format

Portable executable (PE) format is a file format commonly found on Windows operating system. Executable files (EXE), object codes and dynamically link libraries used on 32-bit and 64-bit systems adhere to this format. The structure of the PE file is as follows. The program starts with *MS-DOS header* and *stub* which are nowadays almost unused. Following is the signature and *file header* which contains information such as a target machine or size of section table. Next is *optional header* that contains, among other things, the necessary data directories. Finishing the program is a section table with corresponding section data. Precise specification is available in [19].

3   Implementation

Based on the taxonomy introduced in Sect. 2.1.1 our main approach belongs to the category of exploratory integrity attack. In other words, we are trying to mislead the classifier (antivirus) to predict malicious files as benign incorrectly. We also experimented with an idea to make an exploratory integrity attack where we interject classes. Therefore our goal would be to mislead the classifier to incorrectly predict benign files as malicious.

We utilized the existing `gym-malware` [2] framework. This framework provides a setup for deploying a custom RL agent to generate adversarial samples against a malware classifier. Both the agent and the classifier can be easily changed with even remote classifier supported. The environment is in `OpenAI Gym` [5] format and binary manipulations are implemented in `LIEF` [30] library. However, we have found that their implementation is not ideal. In particular, modifications of

PE files using `LIEF` do not preserve functionality for most of their implemented modifications according to our testing procedure and make unnecessary changes to original files. We also disagree with their approach to using the target classifier's feature space as an observation space for the RL agent as it can give an unfair advantage to the RL agent as opposed to a pure black-box setup where no information apart from the result label is known. For the reasons mentioned above, we have modified the existing implementation [2] and our implementation contain the following differences: minimizing unneeded changes to PE files, deploying different RL agents, implementing pure black-box setup.

3.1 Overview

We propose a complete setup consisting of custom gym class in `OpenAI Gym` format, manipulator of PE files preserving their executability and reinforcement learning agent which learns to maximize evasion rate against targeted classifier while minimizing the number of PE file modifications. *Evasion rate* is a key metric that is used in adversarial machine learning on malware detection domain. It represents the proportion of files that were misclassified by the target classifier,

$$evasion\ rate = \frac{misclassified}{total}$$

3.2 Dataset

We use two separate datasets of PE executables. The first, malware dataset consists of 5000 malware files from the VirusShare [32] repository. The second, benign dataset was gathered from fresh Windows 10 installation and Windows university computers and contains 1592 files. Both datasets contain only executables. Dynamically linked libraries and object files were not included in the datasets.

3.3 PE File Modifications

We implemented most of the modifications in `pefile` [6] library by extending existing implementation of PE file modifications by `MAB-Malware` [28] and `PEsidious` [7]. `MAB-Malware` implementation is described in Related Work under paper [28] and `PEsidious` is an adversarial malware generator built on the top of `gym-malware` framework. We opted to rather create more than fewer modifications and let the RL agents choose which are the most valuable. Be aware that we focus on black-box attacks where we do not have any information about what features the targeted model might use so we cannot design modifications tailored against specific malware detector.

All modifications were tested on a fresh Windows 10 virtual box environment. We selected randomly 50 benign binaries from our dataset as a test set and considered modification successful if the given program executed and showed itself in the `tasklist` command. This protocol has its limitation which we discuss in Sect. 6. We accepted the modification if at least 45 out of 50 binaries pass the test. We ensured that a set of 50 benign binaries passed the test in the first place. In total we have 9 working modifications:

Rename Section: Chooses a section at random and renames it to one of common benign section name.

Add Section: Adds new section with benign content (if enough space for a new entry in Section table).

Remove Certificate: Removes certificate table.

Remove Debug: Removes debug data.

Break Checksum: Zeros out CheckSum in Optional header.

Append Overlay: Adds benign content at the end of file.

Increase TimeDateStamp: Increases TimeDateStamp by 500 days.[1]

Decrease TimeDateStamp: Decreases TimeDateStamp by 500 days.

Append Imports: Selects library from a list of common imported libraries. Adds a new section with library name and its typical functions.

We had experimented with other modifications as well, e.g., shuffling section headers. However, we did not achieve an acceptable execution ratio. Comparison of number of executing files after modifications can be seen in Table 1. We compared `gym-malware`, `PEsidious` and our modifications extending `MAB-Malware` and using `pefile` library. We can see that our set of PE file modifications (pefile column) achieves higher execution rates than other implementations on most of the tested modifications. All nine operations from the pefile column were later used as the RL agent action space.

3.4   Target Classifier

We studied two primary scenarios. In the first one, the target classifier is the LightGBM model, and in the second one, it is MalConv, both trained by authors of EMBER dataset [1]. LightGBM model is a gradient boosted method and is trained on PE files transformed to feature space of 2,381 float numbers. On the other hand, the MalConv classifier is deep convolutional neural network and represents binary files with their first 200,000 raw bytes.

---

[1] 500 days were chosen to represent a substantial period of time and not a multiple of one year.

Table 1 Numbers of files executed successfully after modification from total of 50 binaries. The symbol × denotes that given operation was not implemented

Action	Gym-malware	PEsidious	pefile
Break checksum	4	×	47
Create new entry point	14	×	×
Append new import	42	48	48
Overlay append	50	47	46
Remove debug	5	×	50
Remove certificate	22	×	49
Add new section	4	48	46
Append to section	8	×	×
Rename section	5	4	49
upx pack	46	×	×
upx unpack	49	×	×
Increase TimeDateStamp	×	×	49
Decrease TimeDataStamp	×	×	49

3.5 Agent and Its Environment

We implemented multiple environment setups, all adhering to the OpenAI Gym structure. Key methods which must be implemented are reset and step methods. The reset method resets the environment to the initial state to get ready for the next episode. The return value of reset method is an *observation*. The **Observation** is a representation of the environment state which is presented to the agent. We implemented different observations based on targeted classifier. We either used raw bytes from the beginning of the binary or extracted features from PE files such as bytes histogram, imports and sections info or printable string. The step method performs the given agent's action on the environment. It is responsible for changing the environment state, tracking episode length and calculating reward for the agent's action. We limited the length of the episode to 10 calls of step method. It returns quadruplet of observation, reward, done (flag if an episode has ended), info (debugging information). The *reward* is either 0 if the action does not cause misclassification or (maximum episode length - number of taken actions) × 10 + 100. By taking the number of taken actions into account, we tried to force the agent to prioritize minimal modifications to the binaries.

4 Evaluation

In this section, we describe all the experiments we performed and present our results. In total, we have two main experiments with multiple evaluation phases and initial setups.

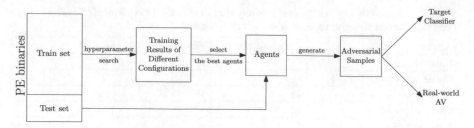

Fig. 1 Overview of our experiment workflow

All experiments follow this protocol. We start with a dataset of PE binaries and split it to train set and test set in 80:20 ratio. We create a training gym environment and use `Ray Tune` [23] to find optimal hyperparameters. `Ray Tune` is a Python library providing easy to use hyperparameter tuning interface. We explore the space of several hyperparameters. We perform a grid search on: agent type (DQN vs PPO), learning rate (lr) and discount rate (gamma γ) with other parameters left to default values as set by the authors of `Ray Tune`.

After identifying suitable hyperparameters values, we train the best performing (the one with the highest mean reward in a single training iteration, in the subsequent tables marked as episode reward) agents again up to 15,000 episodes. In the end, we test the agents on the test set. In the test results, we highlighted the highest evasion rate values in bold. In both the training and testing procedure, we discard any files that are already misclassified before adversarial modification. After evaluation on our test set against target classifier, we test the same modifications made by the best performing agent on real-world AVs from cybersecurity companies Avast, Cylance, Symantec (NortonLifeLock), ESET, Kaspersky, McAfee and Microsoft using VirusTotal [33] website. In our presentation of results, we anonymize the names of AVs to minimize the potential misuse of this work. Overview of our experiment workflow is pictured in Fig. 1.

4.1 Adversarial Malware Examples

In the first experiment, we focused on generating adversarial malware samples, i.e., we modified malware binaries to evade detection by the target classifier. This procedure is called an exploratory integrity attack. For this task, we defined multiple environment setups. In the first one, we use the LightGBM classifier as a target model, and we use the first 1024 (4096, 8192) bytes of PE binary as observation space for the agent. We labelled this setup *M-1*.

Table 2 presents a comparison of agents with different hyperparameters values with regard to their mean reward and episode length. It is clear that DQN agents significantly outperform PPO agents in all environment and hyperparameters setups.

Table 2 Training results of the search for hyperparameters for the M-1 setup. Each table represents different observation space (1024B, 4096B, 8192B). Tables are sorted by episode reward in descending order

Agent	Gamma	lr	Episode reward	Episode length
DQN	0.5	0.001	125.8	5.0
DQN	0.75	0.001	119.47	5.4
DQN	0.999	0.001	118.17	5.4
DQN	0.999	0.01	115.46	5.74
DQN	0.75	0.01	110.97	5.84
DQN	0.5	0.0001	110.4	5.65
DQN	0.5	0.01	109.51	5.88
DQN	0.75	0.0001	107.44	5.86
DQN	0.999	0.0001	105.91	5.96
PPO	0.5	0.001	92.09	6.82
PPO	0.999	0.001	86.86	7.1
PPO	0.75	0.001	86.47	7.1
PPO	0.5	0.0001	83.83	7.09
PPO	0.5	0.01	83.04	7.26
PPO	0.75	0.0001	82.22	7.19
PPO	0.999	0.01	81.62	7.35
PPO	0.999	0.0001	80.3	7.38
PPO	0.75	0.01	74.95	7.64

(a) 1024B

Agent	Gamma	lr	Episode reward	Episode length
DQN	0.75	0.01	110.99	5.85
DQN	0.5	0.0001	110.49	5.92
DQN	0.5	0.01	110.04	5.93
DQN	0.999	0.001	109.96	6.02
DQN	0.75	0.0001	108.25	5.9
DQN	0.999	0.0001	107.62	6.09
DQN	0.5	0.001	106.08	6.15
DQN	0.75	0.001	105.45	6.17
DQN	0.999	0.01	100.91	6.34
PPO	0.75	0.001	90.97	6.9
PPO	0.5	0.001	90.07	6.89
PPO	0.75	0.01	86.97	7.08
PPO	0.75	0.0001	86.37	7.04
PPO	0.5	0.0001	86.16	7.09
PPO	0.999	0.0001	84.34	7.22
PPO	0.999	0.001	82.1	7.34
PPO	0.999	0.01	66.63	8.09
PPO	0.5	0.01	65.72	7.93

(b) 4096B

Agent	Gamma	lr	Episode reward	Episode length
DQN	0.5	0.0001	110.64	5.91
DQN	0.999	0.0001	110.48	5.95
DQN	0.75	0.0001	109.88	5.97
DQN	0.75	0.01	109.23	6.02
DQN	0.5	0.01	107.96	6.14
DQN	0.75	0.001	105.94	6.15
DQN	0.5	0.001	105.94	6.24
DQN	0.999	0.01	105.91	6.14
DQN	0.999	0.001	104.41	6.24
PPO	0.5	0.0001	91.32	6.86
PPO	0.5	0.001	89.74	6.92
PPO	0.999	0.0001	88.32	7.06
PPO	0.75	0.001	87.14	7.06
PPO	0.75	0.0001	86.83	7.11
PPO	0.75	0.01	86.45	7.21
PPO	0.999	0.001	80.48	7.47
PPO	0.5	0.01	61.25	8.32
PPO	0.999	0.01	60.79	8.32

(c) 8192B

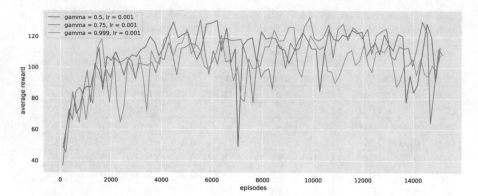

Fig. 2 Training runs of the three most promising agents from the M-1 setup

Table 3 Test results of the three most promising agents from the M-1 setup

Agent	Gamma	lr	Evasion rate [%]
DQN	0.5	0.001	**68.64**
DQN	0.75	0.001	65.21
DQN	0.999	0.001	60.22

Based on the results from Table 2 the environment with an observation space of 1024 bytes performed the best.

We took the first three best configurations (based on the mean episode reward) from Table 2 and tested them up to 15,000 episodes. The training runs are shown in Fig. 2 below.

From Fig. 2 it is not particularly clear what configuration will perform the best on the test set. However DQN agents with gamma 0.5 and 0.75 scored a bit higher than the 3rd configuration.

After the training had ended we showed agents the testing set and performed evaluation. The highest evasion ratio of 68.64% was achieved by DQN agent with $\gamma = 0.5$ and $lr = 0.001$. Full results are shown in Table 3.

In the second setup, we essentially switched the targeted classifiers feature space and agent's observation space. Here we used the LightGBM classifier's feature space as PE file representation for the agent and as targeted classifier we chose MalConv model. This setting is marked as *M-2*. In this experiment, we validated that DQN is better for our problem and we use it for all future experiments. Complete results are shown in Table 4.

We trained the three top agents up to 15,000 episodes, the training progress is shown in Fig. 3. The highest peak in terms of mean episode reward was recorded by configuration with $\gamma = 0.75$ and $lr = 0.01$. Compared to Fig. 2 the mean episode reward varied dramatically between training iterations and was significantly lower.

After we trained the agents, we evaluated their performance on the test set. The results are shown in Table 5. The results are significantly worse than in our M-

Table 4 Training results of the search for hyperparameters for M-2 setup. Table is sorted by episode reward in descending order

Agent	Gamma	lr	Episode reward	Episode length
DQN	0.999	0.01	37.43	9.39
DQN	0.5	0.01	35.62	9.46
DQN	0.75	0.01	35.59	9.46
DQN	0.75	0.001	35.27	9.56
DQN	0.75	0.0001	33.84	9.63
DQN	0.5	0.0001	33.29	9.58
DQN	0.999	0.0001	32.07	9.48
DQN	0.999	0.001	31.95	9.59
DQN	0.5	0.001	31.77	9.66
PPO	0.5	0.0001	21.19	10.1
PPO	0.5	0.001	20.14	10.18
PPO	0.999	0.0001	19.64	10.17
PPO	0.75	0.0001	19.63	10.16
PPO	0.75	0.001	17.6	10.29
PPO	0.999	0.001	16.84	10.32
PPO	0.75	0.01	15.57	10.35
PPO	0.5	0.01	14.1	10.39
PPO	0.999	0.01	11.2	10.53

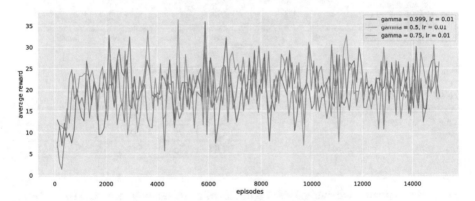

Fig. 3 Training runs of the three most promising agents from the M-2 setup

Table 5 Test results of the 3 most promising agents from the M-2 setup

Agent	Gamma	lr	Evasion rate [%]
DQN	0.999	0.01	12.56
DQN	0.5	0.01	11.94
DQN	0.75	0.01	**13.32**

1 setting and that might indicate that the MalConv classifier is more resilient to adversarial attacks than LightGBM.

In the third setting, we simulated the approach of `gym-malware` framework authors using EMBER feature space for agent's observation and EMBER classifier

Table 6 Training results of the search for hyperparameters for the M-3 setup. Table is sorted by episode reward in descending order

Agent	Gamma	lr	Episode reward	Episode length
DQN	0.75	0.0001	115.41	5.66
DQN	0.75	0.001	114.38	5.53
DQN	0.75	0.01	113.13	5.67
DQN	0.5	0.01	112.62	5.61
DQN	0.999	0.01	110.96	5.85
DQN	0.999	0.001	108.65	5.91
DQN	0.999	0.0001	108.36	5.95
DQN	0.5	0.001	105.6	6.17
DQN	0.5	0.0001	103.01	6.17

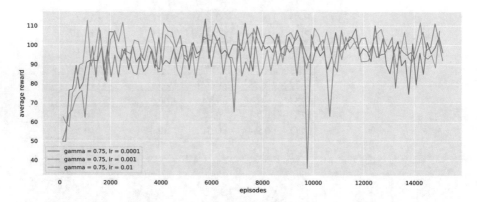

Fig. 4 Training runs of the three most promising agents from the M-3 setup

Table 7 Test results of the three most promising agents from the M-3 setup

Agent	Gamma	lr	Evasion rate [%]
DQN	0.75	0.0001	58.50
DQN	0.75	0.001	**60.37**
DQN	0.75	0.01	57.25

as a targeted model. We marked this approach *M-3*. Interestingly, we did not achieve a higher average reward during training than with our first setup. The results of the search for optimal hyperparameters can be found in Table 6.

We again took the three best performing agents from the ranking and trained them up to 15,000 training episodes. The training process is pictured in Fig. 4. We did not find big differences between configurations during training runs, although the agent with $\gamma = 0.75$ and $lr = 0.001$ recorded the biggest drops in average reward during training.

We evaluate these three agents on the test set and the result are in Table 7 below. At the beginning of Sect. 3 we argued that having an agent with observation space equal to the feature space of the target classifier might give the agent an unfair advantage. However, in our evaluation, we actually did score a lower evasion rate than with our M-1 setup.

Table 8 Overall results of all three setups from the first experiment against seven real-world AVs [%]

	AV-1	AV-2	AV-3	AV-4	AV-5	AV-6	AV-7
M-1	2.76	1.8	5.54	1.47	4.87	1.47	8.06
M-2	**6.12**	**31.69**	**26.52**	**13.03**	**16.94**	**47.09**	**17.67**
M-3	3.74	14.68	3.01	1.13	4.61	31.79	2.03

At the end of the evaluation, we wanted to verify real-world performance. Therefore, we took the best agents from all three setups, modified all samples from the test set, excluding the misclassified samples, and then showed them to commercially available AVs. We used DQN agents with $\gamma = 0.5$ and $lr = 0.001$ from M-1 setup, $\gamma = 0.75$ and $lr = 0.01$ from M-2 setup together with $\gamma = 0.75$ and $lr = 0.001$ from M-3 setting. The results we achieved are shown in Table 8.

From the Table 8 we can see that the overall best setting is the DQN agent ($\gamma = 0.75, lr = 0.01$) from M-2 setup. This is an unforeseen result since when testing against the original target classifier (MalConv), this configuration achieved in most cases lower evasion rate (Table 5) than against real-world AVs.

4.2 Adversarial Benign Examples

In the second experiment, we implemented an exploratory integrity attack with interjected classes, i.e., we modified benign files to mislead the target classifier into falsely predicting malware. This is an unusual setup since most researches focus on the opposite scenario. Even though it is less popular, we think that a scheme where one company develops both the AV and other software could potentially modify their software to increase the false positive of rival AV developers.

In this experiment, we defined two environments with LightGBM and MalConv classifier as a targeted model. This time we used only the 1024 bytes feature space for the first setup and did not experiment with other than DQN agents. We labelled the first and second settings as *B-1* and *B-2*, respectively.

Table 9 shows results of DQN agents from B-1 setting. The difference between the results from the first experiment and these is striking, with the latter performing significantly worse. On the other hand, results from B-2 setup (Table 10) where we attack the MalConv model are looking better than from the first experiment.

As in experiment one, we took the first three configurations from both setups and trained them up to 15,000 episodes. The first setup with LightGBM model as target classifier is shown in Fig. 5. All agents struggled to increase the false positive rate of the classifier with a mean reward not exceeding 20.0.

The second setup performed a lot better in modifying benign files against the MalConv classifier with two agents configurations that exceeded the mean reward of 70.0. The training runs are shown in Fig. 6.

The test results from Table 11 verified the statement mentioned above about the B-1 setup. The evasion rate of 3.45% is the lowest recorded evasion rate across the

Table 9 Training results of the search for hyperparameters for the B-1 setup. Table is sorted by episode reward in descending order

Agent	Gamma	lr	episode_reward_mean	episode_len_mean
DQN	0.5	0.01	13.76	10.34
DQN	0.999	0.01	13.7	10.35
DQN	0.75	0.01	12.28	10.4
DQN	0.999	0.001	12.0	10.43
DQN	0.5	0.001	10.0	10.54
DQN	0.999	0.0001	8.9	10.56
DQN	0.5	0.0001	7.5	10.61
DQN	0.75	0.001	7.4	10.62
DQN	0.75	0.0001	7.2	10.64

Table 10 Training results of the search for hyperparameters for the B-2 setup. Table is sorted by episode reward in descending order

Agent	Gamma	lr	episode_reward_mean	episode_len_mean
DQN	0.75	0.01	73.28	7.55
DQN	0.75	0.0001	68.36	7.79
DQN	0.5	0.01	67.84	7.74
DQN	0.999	0.0001	67.73	7.84
DQN	0.5	0.0001	65.0	8.07
DQN	0.75	0.001	64.85	7.77
DQN	0.5	0.001	64.27	7.87
DQN	0.999	0.01	62.66	8.0
DQN	0.999	0.001	59.28	8.24

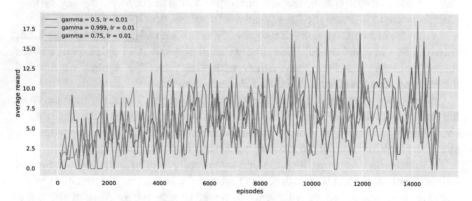

Fig. 5 Training runs of the three most promising agents from the B-1 setup

experiments. This reveals that creating benign files classified as malware against the LightGBM classifier is more complicated than the reversed scenario. However, this is not the case with the MalConv classifier. The results from Table 12 show that

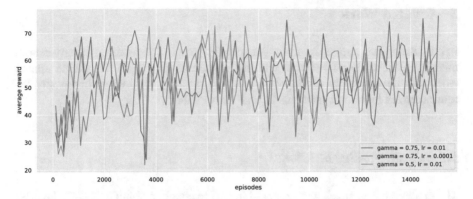

Fig. 6 Training runs of the three most promising agents from the B-2 setup

Table 11 Test results of the three most promising agents from the B-1 setup

Agent	Gamma	lr	Evasion rate [%]
DQN	0.5	0.01	2.19
DQN	0.999	0.01	2.19
DQN	0.75	0.01	**3.45**

Table 12 Test results of the three most promising agents from the B-2 setup

Agent	Gamma	lr	Evasion rate [%]
DQN	0.75	0.01	**36.62**
DQN	0.75	0.0001	26.76
DQN	0.5	0.01	29.58

Table 13 Overall results of two setups from the second experiment against seven real-world AVs [%]

	AV-1	AV-2	AV-3	AV-4	AV-5	AV-6	AV-7
B-1	6.25	**0.33**	0.0	**0.34**	0.0	0.0	0.0
B-2	**14.29**	0.0	0.0	0.0	0.0	0.0	0.0

we achieved higher evasion rates with benign files than with malware in the first experiment. These results might also be partially caused by the differing sizes of benign and malware datasets in our experiments.

In the end, we tested one agent from both setups, which achieved the highest evasion rate against seven AVs programs. The results can be found in Table 13. We can see that for most AVs, our RL agent struggled to mislead antivirus into classifying benign files as malware. Together with our test results, the results against real-world AVs hint that performing exploratory integrity attacks with interjected classes in the domain of malware detection is much harder than traditional scenario.

5 Related Work

In this section, we summarize up-to-date publications which focus on the adversarial machine learning in conjunction with malware detection topic. We divide related work into three sections based on the author's strategies: gradient-based attacks, reinforcement learning based attacks and other methods.

5.1 Gradient-Based Attacks

The gradient-based attack was proposed in [21] to attack the MalConv [25] classifier which utilizes raw bytes of binaries. Their attack modified in average less than 1% of padding bytes at the end of the file and achieved a 60% evasion rate.

Grosse et al. in [16] performed a white-box attack on their neural network classifier. In their attack, they computed necessary perturbation using a gradient of their network and then changed the corresponding features. They successfully mislead their deep neural network in more than 63% of cases.

Another research that is based on gradient-based attacks is presented in [22]. The authors generated small chunks of bytes called payloads which they injected either into unused parts of sections or at the end of the file to ensure the functionality after injection. Their white-box attack targeted against MalConv and achieved an almost perfect evasion rate of 99%.

More attacks on MalConv were carried out in [10] by Luca Demetrio et al. They used a technique called integrated gradients to explain what parts of binaries contribute to prediction. They uncovered that MalConv learns weak features from the DOS header, and perturbing only a few bytes is enough to obscure detection in 52 of 60 malware samples.

Yang et al. in [35] treated binaries as greyscale images. Firstly the authors marked key parts of binaries by "00", then they processed them by a convolutional neural network (CNN). On CNN, a fast gradient sign method was applied to find perturbations within marked sections. The perturbations were then converted to "closest" dead-code instructions (wait, nop, ...) or API calls. Their approach recorded a decrease of over 60% in the accuracy of deep learning detectors and an evasion rate of over 30% on VirusTotal.

5.2 Reinforcement Learning-Based Attacks

The authors of [13] proposed a deep reinforcement model called RLAttackNet to attack their deep neural classifier. They achieved an evasion rate of 19.13% and used adversary samples to retrain their malware classifier to increase its area under the receiver operating characteristic curve from 0.989 to 0.996.

Anderson et al. presented a reinforcement learning framework called gym-malware [2]. They targeted gradient boosted decision tree (GBDT) trained on 100,000 executables. They experimented with both score-based and black-box attacks, with the latter being more successful. After closer inspection of their implementation, we have found that the authors did not perform a complete black-box attack because they used identical feature space both for their RL agent and targeted classifier.

Another reinforcement learning approach is presented in [28]. The authors present stateless RL model, which means that the order of actions applied by RL agent does not matter. They also try to remove unnecessary actions and thus interpret the targeted model. Their work shows promising results with an evasion rate of 74%–97% on ML detectors (LigthGBM Ember [1] and MalConv) and 32%–48% on commercial AVs. They also study the transferability of attacks between targeted models and found out that among ML detectors, it's over 80%. However, between ML and commercial AVs, only up to 7%.

The author of [26] used Android permissions as feature space. This is a key point because modifying permissions doesn't cause any malfunction of a given application. Using a reinforcement learning agent, they achieved an average evasion rate of 44.28% in a white-box scenario against 8 ML classifiers and 53.20% in a grey-box strategy. They managed to reduce this evasion ratio by 15.22%–29.44% by retraining with adversarial samples.

5.3 Other Methods

In [8], the authors used the feature space of their classifier to tailor the attack to evade detection. The feature space was in the form of application programming interface (API) calls of input sample. Greedy search was used to find a set of API calls to add or remove. Later they retrained the classifier with adversarial samples and introduced security regularization to improve their detection ratio further. However, the authors did not propose an algorithm to convert an adversarial set of API calls back to the real-world executable.

Demetrio et al. [11] performed a black-box attack at MalConv and GBDT by injecting small chunks of benign codes into malware binaries either at the end of the file or inside newly created sections. The authors tackled this as an optimization problem to maximize the evasion rate while minimizing the size of injected code. They achieved an evasion ratio of more than 90% on the MalConv classifier and 60%–80% on GBDT. Their attack also bypassed at average 12 out of 70 AVs on VirusTotal [33].

Hu and Tan in [17] proposed a generative adversarial network (MalGAN) to create evasive samples. The authors achieved an almost perfect evasion ratio on random forest, logistic regression and other ML classifiers. They also showed that their model MalGAN could be quickly retrained to bypass new detectors. Needless to say, they worked only with API calls as a representation of executable binary and did not mention how to translate the results back to binaries.

Several authors focused on data poisoning, e.g., Chen et al. poisoned Android training dataset in [9] which led to misclassification of around 70% samples.

In [12], authors trained generative sequence-to-sequence recurrent neural network language model on benign binaries to generate benign bytes, which are later appended at the end of malware executable to bypass detection. This black-box approach led to an evasion rate of more than 72% on three different ML classifiers.

Unique concept so-called grey-box attack is showed in [4] where the attacker has knowledge about feature space of targeted classifier but does not have access to its predictions. They trained their model utilizing the same feature space as the targeted model to substitute it. Using Monte Carlo search, they found a set of operations (limited to size 5) to evade their substitute model in 56% of cases. The authors found out that simple changes such as certificate signature change were enough in 71% of successful mispredictions. Same operations were tried against the targeted classifier but achieved an evasion ratio of less than 9%.

Another paper was published by Fleshman et al. [14] where they presented multiple adversary attacks. First, the authors tried up to 10 random changes. Second, using a binary search algorithm to find critical regions for malware classifications and alter their contents, and last injecting malicious code at the end of otherwise benign binaries. The results varied depending on the modifications performed. The random changes and byte occlusion proved ineffective against ML-based models n-gram and MalConv, whereas the accuracy of four tested AVs suffered. The injection of malicious code successfully bypassed most classifiers, with the lowest evasion rate of 77% recorded by the 4th AV.

6 Conclusion

We successfully implemented a reinforcement learning approach to adversarial machine learning on the space of PE binaries. We tested numerous agent and environment configurations which we evaluated in two separate experiments. Firstly we focused on generating malware adversarial samples that would evade detection, i.e., exploratory integrity attack. In the first experiment, we achieved an evasion rate of 68.64% by the DQN agent ($\gamma = 0.5, lr = 0.001$) with observation states consisting of 1024 raw bytes from the beginning of PE binary and LightGBM model as target classifier. We compared this result with environment setup mimicking gym-malware setup with which we recorded an evasion rate of 60.37%, thus exceeding this setting. Further, we reached an evasion rate of only 13.32% with setup consisting of the DQN agent ($\gamma = 0.75, lr = 0.01$), 2381 features extracted from the binary as the agent's observation space and MalConv classifier as a targeted model. However, when testing the same models against real-world AVs, we accomplished the best result with the last mentioned setup, scoring as high as 47.09% evasion rate against AV-6 and consistently outperforming the other setups.

In the second experiment, we applied the opposite scheme where we created benign adversarial samples, thus increasing the false positive rate of the target model. We recorded an evasion rate of 36.62% with DQN agent ($\gamma = 0.75, lr = 0.01$) with observation space made by 2381 features extracted from the PE binary and MalConv as target model. The lower result was achieved when targeting the LightGBM classifier, only 3.45%. These relatively low results were confirmed when testing against real-world AVs. The highest evasion rate of 14.29% was accomplished by the agent targeting the MalConv classifier. This result was registered against AV-1 with other AVs unaffected by the modifications. The agent targeting the LightGBM model managed to mislead additionally AV-2 and AV-4 but in less than 0.5% of cases. These results indicate that creating false positive samples is far more demanding than the typical approach of creating false negative adversarial samples. This is almost certainly caused by the design of antivirus programs which typically focus on maintaining good accuracy while minimizing false positive rate [20].

6.1 Future Work

In the future, we would like to explore reinforcement learning in more depth, trying different agent implementations or broadening hyperparameter search space. We aim to implement a better protocol for testing the executability of binaries since we have found that our testing method is sensitive, e.g., to installed library version on our virtual machine or folder path of the exe. More work should be devoted to designing modifications of PE files and studying which are responsible for evasion. To improve our results against real-world AVs, we would like to target specific antivirus in the training process directly. An exciting area of adversarial machine learning that we did not cover is retraining the classifier with adversarial samples to increase its resistance against such attacks. Further in the future, we wish to publish the source code for this work as open-source.

Acknowledgments This work was supported by the Student Summer Research Program 2021 of FIT CTU in Prague and by the OP VVV MEYS funded project CZ.02.1.01/0.0/0.0/16 019/0000765 "Research Center for Informatics" and by the Grant Agency of the CTU in Prague, grant No. SGS21/142/OHK3/2T/18 funded by the MEYS of the Czech Republic.

References

1. H. S. Anderson and P. Roth. EMBER: An Open Dataset for Training Static PE Malware Machine Learning Models. *ArXiv e-prints*, April 2018.
2. Hyrum S Anderson, Anant Kharkar, Bobby Filar, David Evans, and Phil Roth. Learning to evade static pe machine learning malware models via reinforcement learning. *arXiv preprint arXiv:1801.08917*, January 2018.

3. Ömer Aslan Aslan and Refik Samet. A comprehensive review on malware detection approaches. *IEEE Access*, 8:6249–6271, 2020.
4. John Boutsikas, Maksim E. Eren, Charles Varga, Edward Raff, Cynthia Matuszek, and Charles Nicholas. Evading malware classifiers via monte carlo mutant feature discovery, 2021.
5. Greg Brockman, Vicki Cheung, Ludwig Pettersson, Jonas Schneider, John Schulman, Jie Tang, and Wojciech Zaremba. Openai gym. *CoRR*, abs/1606.01540, 2016.
6. E Carrera. Pefile, 2017.
7. Bedang Sen Chandni Vaya. Pesidious, malware mutation using deep reinforcement learning and gans. https://github.com/CyberForce/Pesidious#malware-mutation-using-deep-reinforcement-learning-and-gans, 2020.
8. Lingwei Chen, Yanfang Ye, and Thirimachos Bourlai. Adversarial machine learning in malware detection: Arms race between evasion attack and defense. In *2017 European Intelligence and Security Informatics Conference (EISIC)*, pages 99–106. IEEE, 2017.
9. Sen Chen, Minhui Xue, Lingling Fan, Shuang Hao, Lihua Xu, Haojin Zhu, and Bo Li. Automated poisoning attacks and defenses in malware detection systems: An adversarial machine learning approach. *computers & security*, 73:326–344, 2018.
10. Luca Demetrio, Battista Biggio, Giovanni Lagorio, Fabio Roli, and Alessandro Armando. Explaining vulnerabilities of deep learning to adversarial malware binaries, 2019.
11. Luca Demetrio, Battista Biggio, Giovanni Lagorio, Fabio Roli, and Alessandro Armando. Functionality-preserving black-box optimization of adversarial windows malware. *IEEE Transactions on Information Forensics and Security*, 16:3469–3478, 2021.
12. Mohammadreza Ebrahimi, Ning Zhang, James Hu, Muhammad Taqi Raza, and Hsinchun Chen. Binary black-box evasion attacks against deep learning-based static malware detectors with adversarial byte-level language model, 2020.
13. Yong Fang, Yuetian Zeng, Beibei Li, Liang Liu, and Lei Zhang. Deepdetectnet vs rlattacknet: An adversarial method to improve deep learning-based static malware detection model. *Plos one*, 15(4):e0231626, 2020.
14. William Fleshman, Edward Raff, Richard Zak, Mark McLean, and Charles Nicholas. Static malware detection amp; subterfuge: Quantifying the robustness of machine learning and current anti-virus. In *2018 13th International Conference on Malicious and Unwanted Software (MALWARE)*, pages 1–10, 2018.
15. Leilani H. Gilpin, David Bau, Ben Z. Yuan, Ayesha Bajwa, Michael A. Specter, and Lalana Kagal. Explaining explanations: An approach to evaluating interpretability of machine learning. *CoRR*, abs/1806.00069, 2018.
16. Kathrin Grosse, Nicolas Papernot, Praveen Manoharan, Michael Backes, and Patrick McDaniel. Adversarial examples for malware detection. In *European symposium on research in computer security*, pages 62–79. Springer, 2017.
17. Weiwei Hu and Ying Tan. Generating adversarial malware examples for black-box attacks based on gan, 2017.
18. Ling Huang, Anthony D. Joseph, Blaine Nelson, Benjamin I.P. Rubinstein, and J. D. Tygar. *Adversarial Machine Learning*. AISec '11. Association for Computing Machinery, New York, NY, USA, 2011.
19. Microsoft Karl Bridge. Pe format - win32 apps. "https://docs.microsoft.com/en-us/windows/win32/debug/pe-format", 8 2019.
20. Eugene Kaspersky. Doing the homework. https://eugene.kaspersky.com/2012/06/20/fighting-false-positives/, 2012.
21. Bojan Kolosnjaji, Ambra Demontis, Battista Biggio, Davide Maiorca, Giorgio Giacinto, Claudia Eckert, and Fabio Roli. Adversarial malware binaries: Evading deep learning for malware detection in executables. In *2018 26th European signal processing conference (EUSIPCO)*, pages 533–537. IEEE, 2018.
22. Felix Kreuk, Assi Barak, Shir Aviv-Reuven, Moran Baruch, Benny Pinkas, and Joseph Keshet. Deceiving end-to-end deep learning malware detectors using adversarial examples, 2019.

23. Richard Liaw, Eric Liang, Robert Nishihara, Philipp Moritz, Joseph E Gonzalez, and Ion Stoica. Tune: A research platform for distributed model selection and training. *arXiv preprint arXiv:1807.05118*, 2018.
24. Volodymyr Mnih, Koray Kavukcuoglu, David Silver, Alex Graves, Ioannis Antonoglou, Daan Wierstra, and Martin A. Riedmiller. Playing atari with deep reinforcement learning. *CoRR*, abs/1312.5602, 2013.
25. Edward Raff, Jon Barker, Jared Sylvester, Robert Brandon, Bryan Catanzaro, and Charles Nicholas. Malware detection by eating a whole exe, 2017.
26. Hemant Rathore, Sanjay K Sahay, Piyush Nikam, and Mohit Sewak. Robust android malware detection system against adversarial attacks using q-learning. *Information Systems Frontiers*, pages 1–16, 2020.
27. John Schulman, Filip Wolski, Prafulla Dhariwal, Alec Radford, and Oleg Klimov. Proximal policy optimization algorithms. *CoRR*, abs/1707.06347, 2017.
28. Wei Song, Xuezixiang Li, Sadia Afroz, Deepali Garg, Dmitry Kuznetsov, and Heng Yin. Mab-malware: A reinforcement learning framework for attacking static malware classifiers, 2021.
29. Richard S Sutton and Andrew G Barto. *Reinforcement learning: An introduction*. MIT press, 2018.
30. Romain Thomas. Lief - library to instrument executable formats. https://lief.quarkslab.com/, April 2017.
31. Daniele Ucci, Leonardo Aniello, and Roberto Baldoni. Survey of machine learning techniques for malware analysis. *Computers & Security*, 81:123–147, 2019.
32. Virusshare dataset. https://virusshare.com/.
33. Virustotal. https://www.virustotal.com/.
34. Christopher John Cornish Hellaby Watkins. Learning from delayed rewards, 1989.
35. Chun Yang, Jinghui Xu, Shuangshuang Liang, Yanna Wu, Yu Wen, Boyang Zhang, and Dan Meng. Deepmal: maliciousness-preserving adversarial instruction learning against static malware detection. *Cybersecurity*, 4(1):1–14, 2021.
36. Yanfang Ye, Dingding Wang, Tao Li, and Dongyi Ye. Imds: Intelligent malware detection system. In *Proceedings of the 13th ACM SIGKDD international conference on Knowledge discovery and data mining*, pages 1043–1047, 2007.

Auxiliary-Classifier GAN for Malware Analysis

Rakesh Nagaraju and Mark Stamp

Abstract Generative adversarial networks (GAN) are a class of powerful machine learning techniques, where both a generative and discriminative model are trained simultaneously. GANs have been used, for example, to successfully generate "deep fake" images. A recent trend in malware research consists of treating executables as images and employing image-based analysis techniques. In this research, we generate fake malware images using auxiliary classifier GANs (AC-GAN), and we consider the effectiveness of various techniques for classifying the resulting images. Our results indicate that the resulting multiclass classification problem is challenging, yet we can obtain strong results when restricting the problem to distinguishing between real and fake samples. While the AC-GAN generated images often appear to be very similar to real malware images, we conclude that from a deep learning perspective, the AC-GAN generated samples do not rise to the level of deep fake malware images.

1 Introduction

Malware is malicious software that is intentionally designed to do harm. The potential dangers of malware include access to private data, which in turn can lead to confidential or financial data theft, identity theft, ransomware, and other problems. Those affected by malware attacks can range from large corporations and government organizations to a typical individual computer user. According to McAfee Labs, "419 malware threats were encountered per minute in the second quarter of 2020, an increase of almost 12% over the previous quarter" [34]. Malware plays a major role in computer crime and information warfare, and hence malware research plays a prominent—if not dominant—role in the field of cybersecurity.

R. Nagaraju · M. Stamp (✉)
San Jose State University, San Jose, CA, USA
e-mail: rakesh.nagaraju@sjsu.edu; mark.stamp@sjsu.edu

© The Author(s), under exclusive license to Springer Nature Switzerland AG 2022   27
M. Stamp et al. (eds.), *Artificial Intelligence for Cybersecurity*, Advances in
Information Security 54, https://doi.org/10.1007/978-3-030-97087-1_2

A recent trend in malware research consists of treating executables as images, which opens the door to the use of image-based analysis techniques. For example, a malware detector that uses image features known as "gist descriptors" is considered in [54]. Other image-based approaches that have been used with success in the malware domain include convolution neural networks (CNN) and extreme learning machines (ELM); see [24] and [55], respectively.

A generative adversarial network (GAN) is a powerful machine learning concept where both a generative and discriminative networks are trained simultaneously [54]. GANs have previously been studied in the context of malware images. For example, in [31] a transfer learning-based GAN method is used to classify previously unknown malware—so-called zero-day malware. In this approach, GANs are used to generate fake malware images that serve to augment the training data, thereby reducing the required number of training samples.

In this research, we focus on generating realistic fake malware images using GANs, and we consider classification of the resulting fake and real images. Specifically, we use auxiliary classifier GAN (AC-GAN), which enables us to work with multiclass data. We first convert malware executables from a large and diverse malware datasets into images. We train AC-GAN models on these images, which enables us to generate fake malware images corresponding to each family. To determine the quality of these fake samples, we train various models, including CNNs and ELMs, to distinguish between the real and fake samples. The performance of these models provide an indication of the quality of our fake malware images—the worse the models perform, the better, in some sense, are our fake malware images. We also consider the quality of the discriminative models trained using AC-GANs. In all cases, we experiment with various combinations of real and fake malware images.

The remainder of this paper is organized as follows. Section 2 covers relevant related work. In Sect. 3, we outline the methodologies used in this project. Section 4 provides details on the datasets and our specific implementation. Our experimental results appear in Sect. 5, while in Sect. 6, we conclude the paper and provide a brief discussion of possible avenues for future work.

2   Related Work

In this section, we selectively survey some of the previous work related to malware classification using machine learning techniques. The limitations and advantages of various approaches are considered.

Most malware detectors are based on some form of pattern matching. An inherent weakness of such techniques is that a malware writer can evade detection by altering the underlying pattern. Even statistical and machine learning-based malware detectors can be susceptible to a wide variety of code obfuscation techniques [54]. Hence, the challenge is to find an efficient approach that provides strong results along with robustness, even under such attack scenarios.

In [27] deep learning techniques are considered for malware classification. The results from two different experiments show that deep learning techniques achieves better accuracy than standard malware detectors. However, these models are costly, particularly in terms of training.

A semi-supervised malware detection approach is proposed [43]. Here, the authors use a technique that they refer to as "learning with local and global consistency" to reduce dependency on labeled data. In [11], another popular deep learning model, Word2Vec, is used for malware representation. Paired with a gradient search algorithm, this method achieves an accuracy of about 94%. However, for both this model, the training time is high.

In [31], the authors show that the generative aspect of GANs can be used to improve malware classification. The article [21] proposes a GAN-based model, denoted as MalGAN, that generates fake malware, which the authors claim are undetectable by state-of-the-art techniques. In [25], MalGAN is extended to "improved MalGAN," which additionally learns benign features. These approaches were trained on a variety of features, including opcodes. Experiments in [26] show that a deep convolution GAN can enable training with limited data, while in [31], deep learning GAN models are used to produce images that appear to be malware samples visualized as images [22].

In [37], a conditional GAN is used to produce results comparable to previous research, while additionally providing more control over the image generation. One problem in this case, is that the discriminator model cannot be used to classify the sample labels, as the labels are passed as a parameter to the model.

In [21, 25], malware detection models are trained on a variety of features, including opcodes. Specifically, in [21], detectors based on neural networks are generated by considering malware features such as opcodes. It should be noted that the extraction and processing of opcodes is a relatively costly process.

A recent trend in malware research consists of treating executables as images, which opens the door to the use of image-based analysis techniques. In [35], the authors develop a procedure to convert executable binary files into grayscale images. In [13], the authors determine the parts of an executable (.text, .data, etc.) based on image structure. As mentioned above, a malware detector that relies on image features known as gist descriptors is described in [54], where experiments show that using malware images results in a relatively robust detection technique.

Deep learning techniques including recurrent neural networks (RNN) and convolutional neural networks (CNN) are applied to malware images in [46]. Good accuracies are observed for these approaches, which further supports the use of images for malware analysis. Other image-based malware research involves CNNs and extreme learning machines (ELM); see [24] and [55], respectively.

The literature to date clearly shows that deep learning models applied to malware images can yield strong results. In this vein, we build on previous GAN-based malware research.

3 Methodology

The goal of this research is to create realistic-looking fake malware images, and then analyze these images using various learning techniques. We achieve this using GANs, in particular, AC-GANs. The real malware images are fed through AC-GAN which, as part of its training, learns to generate fake malware images (generator) as well as to discriminate between real and fake (discriminator). Once, we generate these fake malware images, we analyze their quality by various means.

3.1 Data

We use two distinct datasets in this research. First, the MalImg dataset contains more than 9000 malware images belonging to 25 distinct families [35]. The MalImg dataset has been widely studied in image-based malware research. We have also constructed a new malware image dataset that we refer to as MalExe. The MalExe dataset is derived from more than 24,000 executables belonging to 18 families—we obtained the executables from [15].

The malware families in the MalImg and MalExe datasets are listed in Tables 1 and 2, respectively. Since the MalExe files are executable binaries, we convert them

Table 1 Details of MalExe dataset

Family	Type	Description
Alureon	Trojan	Provides access to confidential data [5]
BHO	Trojan	Performs malicious activities [8]
CeeInject	VirTool	Obfuscated code performs any actions [12]
Cycbot	Backdoor	Provides control of a system to a server [14]
DelfInject	VirTool	Provides access to sensitive information [16]
FakeRean	Rogue	Raises false vulnerabilities [19]
Hotbar	Adware	Displays ads on browsers [20]
Lolyda.BF	Password Stealer	Monitors and sends user's network activity [28]
Obfuscator	VirTool	Obfuscated code, hard to detect [36]
OnLineGames	Password Stealer	Acquires login information of online games [38]
Rbot	Backdoor	Provides control of a system [40]
Renos	Trojan Downloader	Raises false warnings [42]
Startpage	Trojan	Change browser homepage/other malicious actions [45]
Vobfus	Worm	Download malware and spreads it through USB [50]
Vundo	Trojan Downloader	Downloads malware using pop-up ads [51]
Winwebsec	Rogue	Raises false vulnerabilities [53]
Zbot	Password Stealer	Steals personal information through spam emails [57]
Zeroaccess	Trojan Horse	Downloads malware on host machines [58]

Table 2 Details of MalImg dataset

Family	Type	Description
Adialer.C	Dialer	Perform malicious activities [1].
Agent.FYI	Backdoor	Exploits DNS server service [2].
Allaple.A	Worm	Performs DoS attacks [3].
Allaple.L	Worm	Worm that spreads itself [4].
Alureon.gen!J	Trojan	Modifies DNS settings [6].
Autorun.K	Worm:AutoIT	Worm that spreads itself [7].
C2LOP.gen!g	Trojan	Changes browser settings [9].
C2LOP.P	Trojan	Modifies bookmarks, popup adds [10].
Dialplatform.B	Dialer	Automatically dials high premium numbers [17].
Dontovo.A	Trojan downloader	Download and execute arbitrary files [18].
Fakerean	Rogue	Pretends to scan, but steals data [19].
Instantaccess	Dialer	Drops trojan to system [23].
Lolyda.AA1	PWS	Steals sensitive information [29].
Lolyda.AA2	PWS	Steals sensitive information [29].
Lolyda.AA3	PWS	Steals sensitive information [29].
Lolyda.AT	PWS	Steals sensitive information [30].
Malex.gen!J	Trojan	Allows hacker to perform desired actions [33].
Obfuscator.AD	Trojan Downloader	Allows hacker to perform desired actions [36].
Rbot!gen	Backdoor	Allows hacker to perform desired actions [41].
Skintrim.N	Trojan	Allows hacker to perform desired actions [44].
Swizzor.gen!E	Trojan downloader	Downloads and installs unwanted software [47].
Swizzor.gen!I	Trojan downloader	Downloads and installs unwanted software [48].
VB.AT	Worm	Spreads automatically across machines [49].
Wintrim.BX	Trojan downloader	Download and install other software [52].
Yuner.A	Worm	Spreads automatically across machines [56].

to images using a similar approach as in [24, 35]. We discuss this conversion process in more detail below.

Figure 1 shows samples of images from the Adialer.C family of the MalImg dataset and images of the Obfuscator family from the MalExe dataset. For these samples we observe a strong similarity of images within the same family, and obvious differences in images between different families. This is typical, and indicates that image-based analysis should be useful in the malware field.

3.2 AC-GAN

A generative adversarial network (GAN) is a type of neural network that—among many other uses—can generate so-called "deep fake" images [37]. A GAN includes a generator model and a discriminative model that compete with each other in a

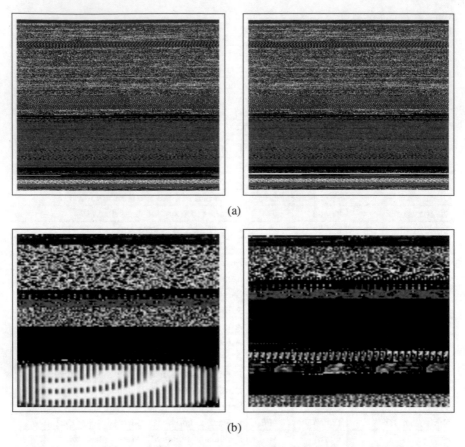

(a)

(b)

Fig. 1 Images from the MalImg and MalExe datasets. (a) Examples of `Adialer.C` from MalImg. (b) Examples of `Obfuscator` from MalExe

min-max game. Intuitively, this competition will should make both models stronger than if each was trained separately, using only the available training data. The GAN generator generates fake training samples, with the goal of defeating the GAN discriminative model, while the discriminative model tries to distinguish real training samples from fake.

However, a standard GAN is not designed to work with multiclass data. Since we have multiclass data, we use auxiliary-classifier GAN (AC-GAN), which is an enhanced type of GAN that includes a class label in the generative model. Additionally, the discriminator predicts both the class label and the validity (i.e., real or fake) of a given sample. A schematic representation of AC-GAN is given in Fig. 2.

For the research in this paper, the key aspect of AC-GAN is that it enables us to have control of the class of any image that we generate. We will also make use of

Fig. 2 Schematic
representation of AC-GAN

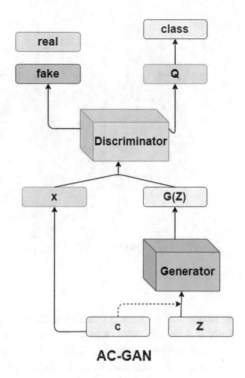

AC-GAN

AC-GAN discriminative models, as they will serve as a baseline for comparison to other deep learning techniques—specifically, CNNs and ELMs.

3.3 Evaluation Plan

Once, we have trained and tested our AC-GAN model, we need to evaluate the quality of the fake images. To do this, we compare the AC-GAN classifier to CNN and ELM models trained on real and fake samples. The remainder of this section is devoted to a brief introduction to CNNs and ELMs.

3.3.1 CNN

A convolutional neural network (CNN) is loosely based on the way that a human perceives an image. We first recognize edges, the general shape, texture, and so on, eventually building up to the point where we can identify a complex object.

A CNN is a feed-forward neural network that includes convolution layers in which convolutions (i.e., filters) are applied to produce higher level feature maps. CNNs typically also include pooling layers that primarily serve to reduce the

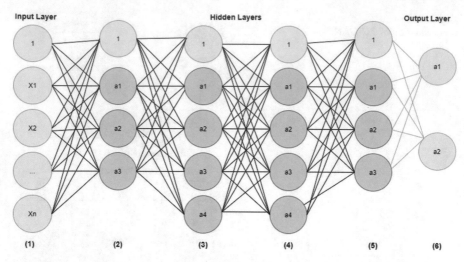

Fig. 3 A generic CNN

dimensionality of the problem via downsampling. CNNs also typically have a final fully-connected layer, where all inputs from previous layers are mapped to all possible outputs. A generic CNN architecture is given in Fig. 3.

For our experiments, we will use the specific CNN architecture and hyperparameters specified in [24]. The CNN experiments performed in our research involve malware images, and the specific architecture that we adopt was optimized for precisely this problem.

3.3.2   ELM

A so-called extreme learning machine (ELM) is a feedforward deep learning architecture that does not require any back-propagation. The weights and biases in the hidden layers of an ELM are assigned at random, and only the output weights are determined via training. Due to this simple structure, an ELM can be trained using a straightforward equation solving technique—specifically, the Moore-Penrose generalized inverse. Thus, ELMs are extremely efficient to train. A schematic representation of a generic ELM can be seen in Fig. 4.

For our experiments, we will use ELM models with parameters as specified in [24]. As with the CNN experiments mentioned above, the experiments performed in our research involve malware images, and the specific ELM architecture that we use was optimized for this specific problem.

To evaluate the quality of our AC-GAN generated images, we first divide the real and fake images into training and testing sets. Then we train a CNN (respectively, ELM) on the training dataset. Once, the CNN (respectively, ELM) has been trained, we predict class labels and determine the accuracy of the predictions. The worse the

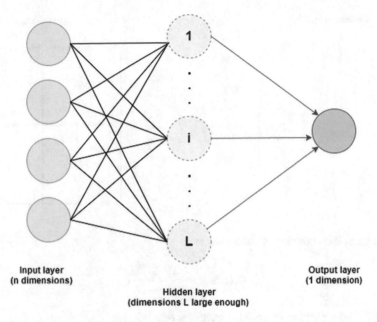

Fig. 4 Schematic representation of ELM

classification accuracy of the CNN (respectively, ELM), the better are our AC-GAN generated fake images. We also want to compare the accuracy of the CNN and ELM models to the AC-GAN discriminator. Note that we consider each real family and each fake family as a separate class, in effect doubling the number of classes from the original dataset.

3.4 Accuracy

Throughout this paper, we use accuracy as the metric to quantify the success of the various experiments considered. Accuracy is simply the ratio of the number of correct classifications versus the total number of classifications.

For a binary classification problem, the confusion matrix is of the form in Fig. 5, where

$$TP = \text{true positives}$$

$$FP = \text{false positives}$$

$$TN = \text{true negatives}$$

$$FN = \text{false negatives}$$

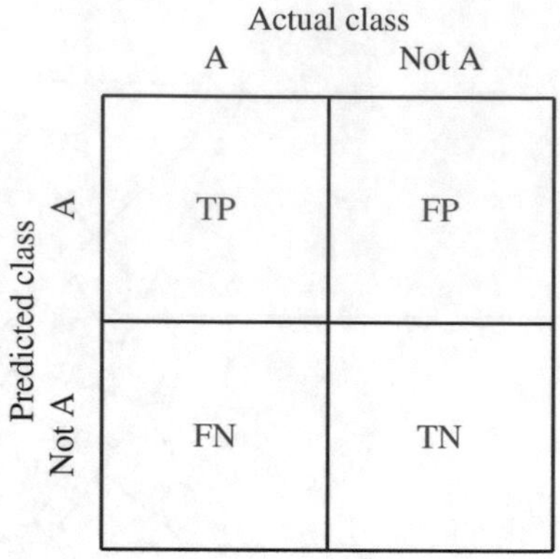

Fig. 5 Confusion matrix

In this case, the accuracy is computed as

$$\text{accuracy} = \frac{\text{TP} + \text{TN}}{P + N}$$

where P is the number of positive samples, that is,

$$P = \text{TP} + \text{FN}$$

and N is the number of negative samples, that is,

$$N = \text{TN} + \text{FP}.$$

This calculation of accuracy easily generalizes to the multiclass case.

4 Implementation

In this section, we present details on the implementation of the techniques discussed in Sect. 3. All of our learning techniques have been implemented in Python using PyTorch and Keras, with the experiments run on Google Colab Pro under a local Windows OS. The precise specifications are given in Table 3.

In the remainder of this section we provide details on the pre-processing applied to the datasets used in our experiments, we outline our AC-GAN training process, and we discuss the training and testing of our CNN and ELM evaluation models. Then in Sect. 5 we present out experimental results.

Table 3 Environment specifications

Specification	Description
Local machine	Windows OS
	Intel(R) Core(TM) i7-9750H CPU @ 2.60 GHz
	16.0 GB RAM
	NVIDIA GeForce RTX 2060 14 GB GPU
Google Colab Pro	24 hours available runtime
	25 GB memory
	T4 and P100 GPUs
Software	PyTorch
	Keras
	Numpy
	Scipy
	PIL

4.1 Dataset Analysis and Conversion

As mentioned above, In this research, we experiment with two distinct datasets, which we discuss in the next section. In both cases, we use the `ImageDataGenerator` and `Dataloader` modules from Keras (in PyTorch) to extract images and labels from the data. Additionally we use the `transforms` functions to compose our pre-processing requirement.

4.1.1 Datasets

The first dataset we consider is the well-known MalImg dataset, which was originally described in [32]. This dataset has become a standard for comparison in image-based malware research. The MalImg dataset contains 9339 grayscale images belonging to 25 classes, where all samples are in the form of images, not executable files.

We refer to our second malware image dataset as MalExe, and it is of our own creation. This dataset contains 24,558 malware images belonging to 18 classes. These samples are in the form of `exe` files.

Since the MalExe samples are executable binary files, we must converting them to images. We perform this transform as follows. iWe also construct images by specify a desired size of each (square) images as $n \times n$. We then read the first n^2 bytes from a malware binary, and these bytes are viewed as $n \times n$ images of type png. For example, if we specify 64×64 images, each image is based on the first 4096 bytes of the corresponding `exe` file. In this conversion process, we only convert samples that contain a sufficient number of bytes. In Table 4, we see the image counts obtained for the MalExe dataset for various image sizes considered. Note that for 512×512

Table 4 MalExe dataset counts

Specified image size	Count	Families
Standard	24,652	18
32 × 32	24,557	18
64 × 64	24,371	18
128 × 128	23,369	18
512 × 512	9963	17

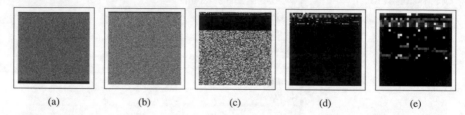

(a)   (b)   (c)   (d)   (e)

Fig. 6 Image conversions of an Alureon sample. (**a**) Real. (**b**) 512 × 512. (**c**) 128 × 128. (**d**) 64 × 64. (**e**) 32 × 32

image, we only have 9963 samples from 17 classes—the family Zeroaccess has no samples with at least $512^2 = 2^{18}$ bytes.

Figure 6 illustrate images of various sizes for one specific sample from the Alureon family. We see that that these different image construction techniques can provide distinct views of the same data.

In Fig. 7 we give bar graphs showing the distribution of samples for the MalImg and MalExe datasets. We note that the MalImg dataset is highly imbalanced, with the majority of the images belong to Allaple.A, Allaple.L, and Yuner.A. To deal with this imbalance, we shuffle the data during training and use balanced accuracy while testing.

Next, we want to scale the pixel values to the range $[-1, 1]$ in order to match the output of the generator model. This is achieve by simply calculating the mean pixel value of an entire image and then subtracting this mean from each pixel and normalizing, which gives us a floating point value in the closed interval from -1 to $+1$ in place of each pixel value.

4.2 AC-GAN Implementation

In this section, we provide additional detail on our implement of AC-GAN. Recall that our model is generated using Python, PyTorch, and Keras modules. Also, recall that an AC-GAN includes both a generator and a discriminator.

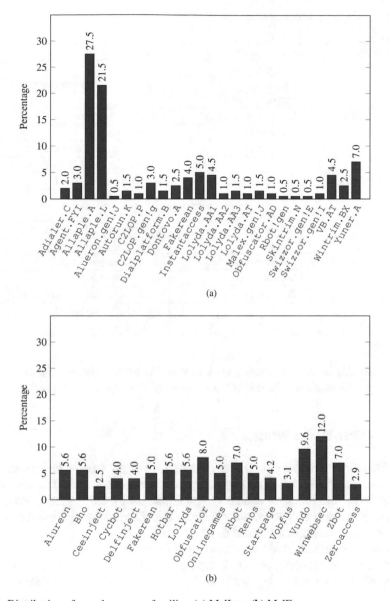

Fig. 7 Distribution of samples among families. (**a**) MalImg. (**b**) MalExe

4.2.1 AC-GAN Generator

Our AC-GAN generator produces a single channel grayscale image by plotting random points on a latent space—the latent space simply consists of noise drawn from a Gaussian distribution with $\mu = 0$ and $\sigma = 1$. Additionally, the model includes the class label as a parameter. The generator is composed as a sequential

Table 5 AC-GAN generator construction parameters

Layer	Functions	Parameters
Embedding	`Embedding()`	classLabels × 100
	`Sequential()`	
	`Linear()`	in-features: 100; out-features: 131,
	`Sequential()`	
1st convolutional	`BatchNormal2d()`	in: 128; momentum: 0.1
	`Upsample()`	Scale factor: 2.0
	`Conv2d()`	in: 128; out: 128; kernel: (3,3);
		stride: (1,1); padding: (1,1)
2nd convolutional	`BatchNormal2d()`	in: 128; momentum: 0.1
	`LeakyReLU()`	negativeslope: 0.2
	`Upsample()`	Scale factor: 2.0
	`Conv2d()`	in: 128; out: 64; kernel: (3,3);
		stride: (1,1); padding: (1,1)
3rd convolutional	`BatchNormal2d()`	in: 64; momentum: 0.1
	`LeakyReLU()`	negativeslope: 0.2
	`Conv2d()`	in: 64; outchannels: 1; kernel: (3,3);
		stride: (1,1); padding: (1,1)
Output	`Tanh()`	Scale factor: 2.0

model. To this sequential model, we add a series of deconvolutional layers. The specific parameters used for the AC-GAN generator are given in Table 5.

4.2.2 AC-GAN Discriminator

The discriminator model discriminates between the original and fake images, while predicting the class label. The generator and discriminator both deal with cross-entropy loss—the generator attempts to minimize binary cross-entropy loss, while the discriminator tries to maximize this loss. The discriminator parameters used in our experiments are given in Table 6.

Once we have initialized the generator and discriminator models, the models are then trained. This training process is typical of any AC-GAN, and hence we omit the details here. After training, we plot loss graphs to verify training stability.

4.3 Evaluation Models

To evaluate our AC-GAN generator results, we train CNN and ELM models on the real and fake images. The better (in some sense) our AC-GAN generated fake images, the worse the CNN and ELM models should perform.

Table 6 AC-GAN discriminator construction parameters

Layer	Functions	Parameters
Input	Sequential()	
1st deconvolutional	Conv2d()	in: 1; out: 16; kernel: (3,3);
		stride: (2,2); padding: (1,1)
2nd deconvolutional	LeakyReLU()	negativeslope: 0.2
	Dropout2d()	rate: 0.25
	Conv2d()	in: 16; out: 32; kernel: (3,3);
		stride: (2,2); padding: (1,1)
3rd deconvolutional	LeakyReLU()	negativeslope: 0.2
	Dropout2d()	rate: 0.25
	BatchNormal2d()	in: 32; momentum: 0.1
	Conv2d()	in: 32; out: 64; kernel: (3,3);
		stride: (2,2); padding: (1,1)
4th deconvolutional	LeakyReLU()	negativeslope: 0.2
	Dropout2d()	rate: 0.25
	BatchNormal2d()	in: 64; momentum: 0.1
	Conv2d()	in: 64; out: 128; kernel: (3,3);
		stride: (2,2); padding: (1,1)
	LeakyReLU()	negativeslope: 0.2
	Dropout2d()	rate: 0.25
	BatchNormal2d()	in: 128; momentum: 0.1
Adversarial	Sequential()	
	Linear()	in-features: 8192; out-features: 1
	Sigmoid()	
Auxiliary	Sequential()	
	Linear()	in-features: 8192; out-features: 18
	Sigmoid()	

4.3.1 CNN Implementation

CNN models include a fully-connected layer, a convolution layer (or layers), and a pooling layer (or layers). The parameters used in our specific implementation are given in Table 7. The parameters that awe use in our CNN models are as specified in [55]. Note that due to the imbalance in the MalImg dataset, we use balanced accuracy.

4.3.2 ELM Implementation

Any ELM includes an initial input layer, a final output layer, and in between these two layers, there is a hidden layer. The hidden layer weights are assigned at random,

Table 7 CNN construction parameters

Layer	Functions	Parameters
1st convolutional	`Sequential()`	
	`Conv2d()`	Filters: 30; in = image-size; out = 840;
		Kernel: (3,3); activation: relu
1st pooling	`MaxPooling2D()`	Size: (2,2)
2nd convolutional	`Conv2d()`	filters = 15; in = 840; out = 4065;
		Kernel: (3,3); activation: relu
2nd pooling	`MaxPooling2D()`	Size: (2,2)
	`Dropout()`	Rate: 0.25
	`Flatten()`	
	`Dense()`	Units: 128; out: 376,448; activation: relu
	`Dropout()`	Rate: 0.5
Other	`Dense()`	Units: 50; out: 6450; activation: relu
	`Dense()`	Units: num-of-classes; activation: softmax
—	Loss	Categorical cross entropy
—	Optimizer	`Adam`

with only the output layer weights determined via training. For an ELM, the only parameter is the number of hidden units, and we use the value specified in [24], namely 5000.

5 Experimental Results

Here, we first consider the use of AC-GAN to generate fake malware images of various sizes. As part of these experiments, we also consider the discriminative ability of AC-GAN discriminator model.

As a followup on our AC-GAN experiments, we conduct CNN and ELM experiments in Sect. 5.2. The purpose of these experiments is to determine how well these deep learning techniques can distinguish between real malware images and the AC-GAN generated fake images.

5.1 AC-GAN Experiments

We consider AC-GAN experiments to generate fake malware images of sizes 32 × 32, 64 × 64, and 128 × 128. In each case, we experiment with both the MalImg and MalExe datasets.

5.1.1 AC-GAN with 32 × 32 Images

Our objective here is generate and classify malware images of size 32 × 32. For the MalImg dataset, which is in the form of images, we resize all of the images to 32×32. We train our AC-GAN model for 1000 epochs with the number of batches set to 100. Since there are 9400 MalImg samples in total, we have 94 samples per batch, and hence about 94,000 iterations. Training this model requires about 24 hours on Google Colab Pro.

In contrast, for the MalExe dataset we read the first 1024 bytes from each binary, and treat these bytes as a 32 × 32 image. We train an AC-GAN model on this dataset for 500 epochs with the number of batches set to 50. Since there are 42,266 samples in the MalExe dataset, we have about 492 samples per batch and requires about 246,000 iterations. Training this model also takes about 24 hours on Google Colab Pro.

Figure 8a shows the training loss plots for our AC-GAN generator and discriminator models when training on the MalImg dataset. Figure 8b shows the corresponding loss plots for the MalExe dataset.

From Fig. 8a, we see that both the generator and discriminator stabilizes at around epoch 100 for the MalImg experiment. The generator spikes up occasionally, but has generally stable loss values, while the discriminator loss is more consistent throughout. In contrast, from Fig. 8b we see that the MalExe model remains relatively unstable throughout its 500 iterations.

Our AC-GAN discriminator achieves an accuracy of about 95% in the MalImg experiment. In contrast, on the MalExe dataset, the AC-GAN discriminator only attains an accuracy of about 89%.

Figure 9 shows a comparison of real and AC-GAN generated fake 32 × 32 images for the families C2LOP.P and Allaple.L from the MalImg dataset. Figure 10 shows a comparison between real and fake images for the Alureon and Zeroaccess families from the MalExe data. Visually the real and fake images share some characteristics, with the MalExe fake images being better than the MalImg case. However, the resolution appears to be too low in all cases. Hence, we perform further AC-GAN experiments based on higher resolution images.

5.1.2 AC-GAN with 64 × 64 Images

Our AC-GAN experiments for 64 × 64 images are analogous to those for 32 × 32 images, as discussed in Sect. 5.1.1. Again, the training time for each dataset is about 24 hours. Figure 11a and b gives the training loss plots for the MalImg and MalExe experiments, respectively.

From Fig. 11a, we see that the training loss stabilizes at around epoch 250 for the MalImg case, while the MalExe experiment stabilizes at around epoch 100. In contrast to the 32 × 32 case, the MalExe model becomes reasonably stable after about 125 epochs.

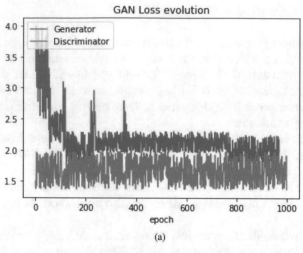

(a)

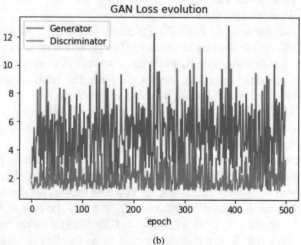

(b)

Fig. 8 Loss plots for 32 × 32 images. (**a**) MalImg. (**b**) MalExe

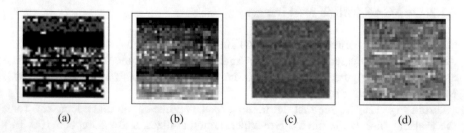

(a)   (b)   (c)   (d)

Fig. 9 Real and fake examples from MalImg (32 × 32). (**a**) C2LOP.P. (**b**) C2LOP.P_fake. (**c**) Allaple.L. (**d**) Allaple.L_fake

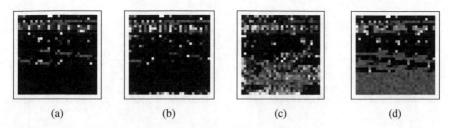

Fig. 10 Real and fake examples from MalExe (32 × 32). (**a**) `Alureon`. (**b**) `Alureon_fake`. (**c**) `Zeroaccess`. (**d**) `Zeroaccess_fake`

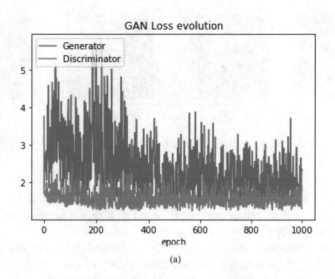

(a)

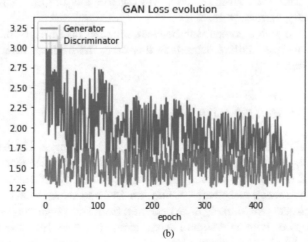

(b)

Fig. 11 Loss plots for 64 × 64 images. (**a**) MalImg. (**b**) MalExe

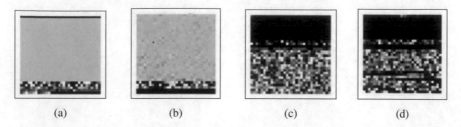

Fig. 12 Real and fake examples from MalImg (64 × 64). (**a**) `Lolyda.AA3`. (**b**) `Lolyda.AA3_fake`. (**c**) `Agent.FYI`. (**d**) `Agent.FYI_fake`

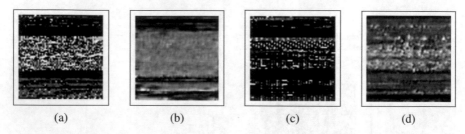

Fig. 13 Real and fake examples from MalExe (64×64). (**a**) `Zbot`. (**b**) `Zbot_fake`. (**c**) `Vobfus`. (**d**) `Vobfus_fake`

The classification accuracy for the MalImg dataset is about 94%, while the AC-GAN achieves a classification accuracy of about 88% on the MalExe dataset. These results are essentially the same as in the 32 × 32 case.

Again, we compare real and AC-GAN generated fake images. Figure 12 shows the comparison between real and fake images of class `Lolyda.AA3` and `Agent.FYI` from the MalImg dataset. We observe that the fake samples in this case are, visually, extremely good.

In Fig. 13, we give a comparison between real and fake images of class `Zbot` and `Vobfus` for the MalExe dataset. In this case, the MalExe fake samples are surprisingly poor.

5.1.3 AC-GAN with 128 × 128 Images

We consider AC-GAN experiments based on 128 × 128 images. These experiments are again analogous to those for the 32 × 32 and 64 × 64 cases discussed above. Figure 14a and b shows the training loss plots for AC-GAN trained on the MalImg and MalExe datasets, respectively. While the MalImg experiments stabilize, the MalExe experiment would likely have benefited from additional iterations.

In this case, we attain a maximum classification accuracy from the AC-GAN of about 92% for MalImg and about 85% for MalExe. Figure 15 shows comparisons

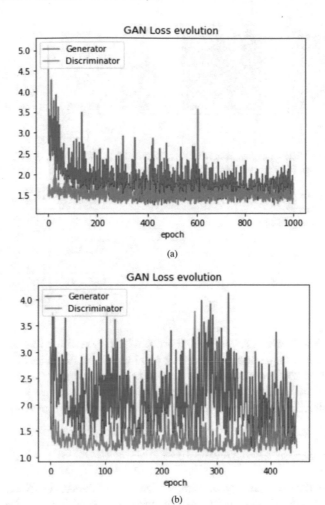

(a)

(b)

Fig. 14 Loss plots for 128×128 images. (**a**) MalImg. (**b**) MalExe

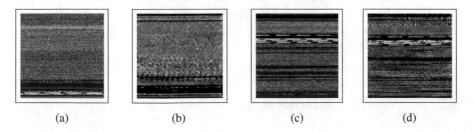

(a)   (b)   (c)   (d)

Fig. 15 Real and fake examples from MalImg (128×128). (**a**) `Yuner.A`. (**b**) `Yuner.A_fake`. (**c**) `VB.AT`. (**d**) `VB.AT_fake`

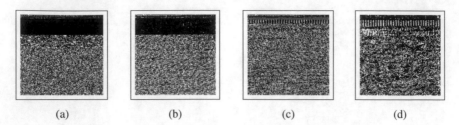

(a)   (b)   (c)   (d)

Fig. 16 Real and fake examples from MalExe (128 × 128). (**a**) Alureon. (**b**) Alureon_fake.
(**c**) Zeroaccess. (**d**) Zeroaccess_fake

Table 8 AC-GAN
discriminator accuracy
(rounded to nearest percent)

Image size	Dataset	Accuracy
32 × 32	MalImg	95%
	MalExe	89%
64 × 64	MalImg	94%
	MalExe	88%
128 × 128	MalImg	92%
	MalExe	85%

of real and fake Yuner.A and VB.AT from MalImg. As in the 64 × 64 case, we
see that the fake images appear to be very good approximations for this dataset.

Figure 16 shows a comparison of real and fake Alureon and Zeroaccess
images from the MalExe data. In contrast to the 32 × 32 and 64 × 64 cases, here
the fake MalExe images are very good approximations to the real images.

5.1.4   Summary of AC-GAN Results

Table 8 gives the discriminative accuracies for each of the AC-GAN experiments in
Sects. 5.1.1–5.1.3. We see that the results are fairly consistent, irrespective of the
size of the images.

With respect to the visual inspection of the fake images in Figs. 9 and 10 (for
the 32 × 32 case), Figs. 12 and 13 (for the 64 × 64 case), and Figs. 15 and 16 (for
the 128 × 128 case), we observed a clear improving trend for larger image sizes.
However, there is a price to be paid for this increased fidelity, as the training time
increases significantly with image size.

5.2   CNN and ELM Experiments

As a first step towards evaluating the quality of the AC-GAN generated images, we
experiment with CNN and ELM. Specifically, we test the ability of these two deep
learning techniques to distinguish between real malware images and our AC-GAN

generated fake images by treating the real data and fake images as distinct classes in multiclass experiments. For example, if we consider 10 classes from the MalImg dataset, then for our CNN and ELM experiments, we will have 20 classes consisting of the 10 original families plus another 10 classes consisting of fake samples from each of the original 10 families. In the following sections, we separately consider experiments for 32×32, 64×64, and 128×128 image sizes.

5.2.1 CNN and ELM for 32 × 32 Images

Here, we consider 32×32 real and fake images and perform experiments for the MalImg and MalExe datasets. For MalExe, we consider all 18 classes and therefore, including classes for the fake images, we have a total of 36 classes. Our dataset consists of 100 samples for each class, and hence we have 3600 images. We train our CNN for 3000 epochs and we generate an ELM with 5000 hidden units. The CNN test accuracy is only about 51%, in spite of a training accuracy of 100%, which is a sign of overfitting. The ELM performs slightly worse, achieving an accuracy of 48%.

Figures 19 and 20 give the confusion matrices for our CNN and ELM experiments on the MalExe dataset. In both cases, we observe that most of the fakes are largely misclassified, but this is not the case for all families. For example, in the CNN experiments, the fake Vundo samples are classified correctly with 100% accuracy, whereas the real Vundo samples are only classified correctly 33% of the time.

For MalImg, we consider all 25 real classes, which gives 50 classes and a total of 5000 images. Again, our CNN is trained for 3000 epochs and we construct an ELM with 5000 hidden units. For the MalImg dataset, our CNN again has a very high training accuracy, but achieves a test accuracy of only about 56%, while our ELM achieves an accuracy of about 37%. The confusion matrices for these experiments are in Figs. 21 and 22. Again, we see that the fakes are misclassified at a much higher rate than the real samples.

5.2.2 CNN and ELM for 64 × 64 Images

In this section, we consider similar experiments as in the previous section, but based on 64×64 images. In this case, we consider 10 of the MalImg families and the corresponding fake samples, for a total of 20 classes for each dataset. We again consider 100 images from each class, and we use 70% of the samples for training and reserve the remaining 30% for testing.

We train a CNN for 3000 epochs with a batch size of 500 while for the ELM we use 50,000 hidden units. For the CNN, we attain 100% training accuracy, but only about 82% test accuracy, which is again a sign of overfitting. For the ELM, we attain an accuracy of 64%. Figures 23 and 24 show the confusion matrices for these experiments. From the confusion matrices, we can see that some images are misclassified as fakes, while some families are consistently classified as other families. For both the CNN and ELM, we see that most images are misclassified, with the exception of specific families. The 64×64 results—in the form of confusion matrices—for the MalExe dataset are in Figs. 25 and 26.

5.2.3 CNN and ELM for 128 × 128 Images

In this MalImg experiment, we consider all families in the dataset. In this case, we train the CNN for 5000 epochs and generate an ELM with 20,000 hidden units. Again, we treat real and fake images as a separate set of classes. We consider all 18 classes in our MalExe experiments.

On the MalExe dataset, we achieve 43% test accuracy with the CNN, and 52% accuracy with out ELM. Figures 27 and 28 show the confusion matrices for our CNN and ELM experiments on the MalExe data. Similar to other experiments on MalExe, we see mostly miscalculation for the CNN. For the ELM, we note that `Rbot` fake, and `Ceeinject` fake are particularly poor results. The results of these 128×128 experiments again indicate that AC-GAN produces strong fake images.

For the 128×128 MalImg experiments, we consider all classes, we train the CNN for 3000 epochs, and we generate an ELM with 20,000 hidden units. The results for these MalImg experiments are given in Figs. 29 and 30. The CNN achieves only 43% test accuracy, while ELM performs better, but still only attains an accuracy of 52%.

5.2.4 Discussion of CNN and ELM Experiments

In Fig. 17 we compare the test accuracies of our CNN and ELM experiments to our AC-GAN classifier. Here, we observe that the AC-GAN models are able to produce much higher classification rates in all cases. This shows that while the AC-GAN generator is able to produce images that are difficult for other deep learning techniques to distinguish, the AC-GAN discriminator is not so easily defeated by these fake images. These results suggest that AC-GAN is not only a source for generating fake malware images, but it is also a powerful model for discriminating between families—both real and fake (Fig. 18).

Finally, we consider the narrower problem of distinguishing real samples from fake samples. In Figs. 31, 32, and 33, we have "condensed" the confusion matrices of Figs. 19, 20, 21, 22, 23, 24, 25, 26, 27, 28, 29, and 30 to better highlight the ability of our CNN and ELM models to distinguish real from fake. Each of these condensed confusion matrices includes the eight (exhaustive) cases listed in Table 9.

If we are only concerned with the ability of our models to distinguish between real and fake samples, then any real sample that is classified as real—either the

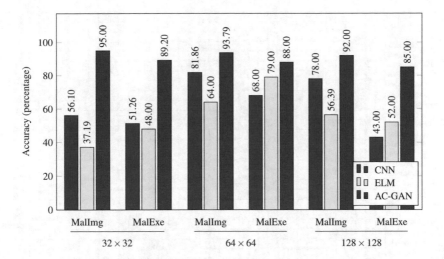

Fig. 17 Test accuracy for all experiments

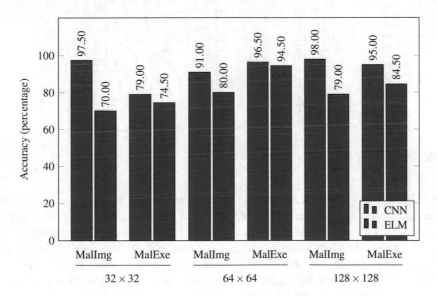

Fig. 18 Distinguishing between real and fake

correct real family or a different real family—is considered a correct classification. Similarly, any fake sample that is classified as any class of fake is considered a correct classification. The results in Fig. 18 are easily obtained from the condensed confusion matrices in Figs. 31, 32, and 33. From this perspective, we see that our CNN models always outperform the corresponding ELM model, and in most cases,

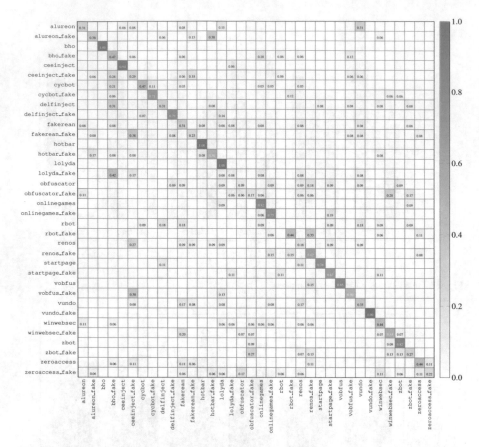

Fig. 19 CNN confusion matrix (MalExe 32 × 32)

the CNN models perform remarkably well. These results indicate that in spite of the relatively low accuracies obtained in the multiclass case, most of the errors are within the real and fake categories, and not between real and fake samples. In particular, for the CNN models, real and fake samples from a specific family are rarely confused with each other. This provides strong evidence that the real and fake categories are substantially different from each other. Perhaps surprisingly, these results strongly suggest that AC-GAN generated fake malware images do not satisfy the requirements of "deep fakes," at least not from the perspective of evaluation by deep learning techniques.

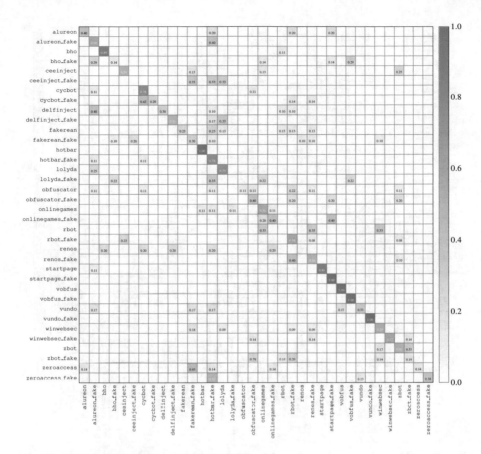

Fig. 20 ELM confusion matrix (MalExe 32 × 32)

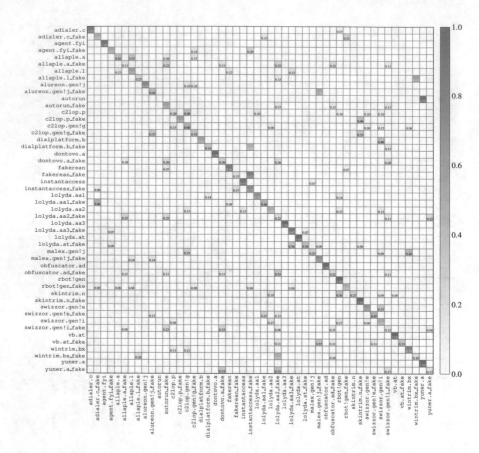

Fig. 21 CNN confusion matrix (MalImg 32 × 32)

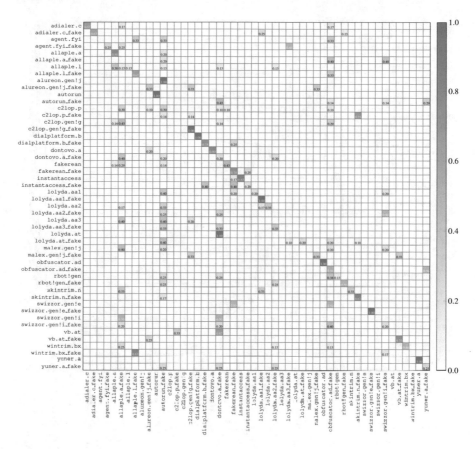

Fig. 22 ELM confusion matrix (MalImg 32 × 32)

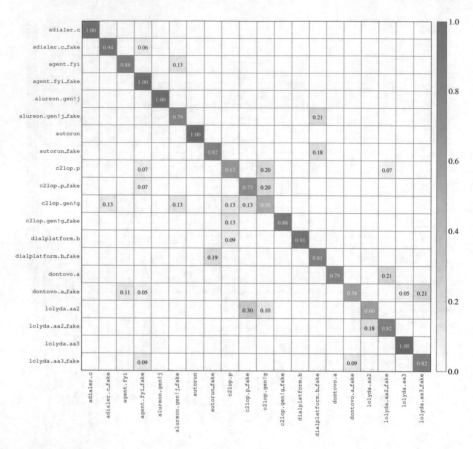

Fig. 23 CNN confusion matrix (MalImg 64 × 64)

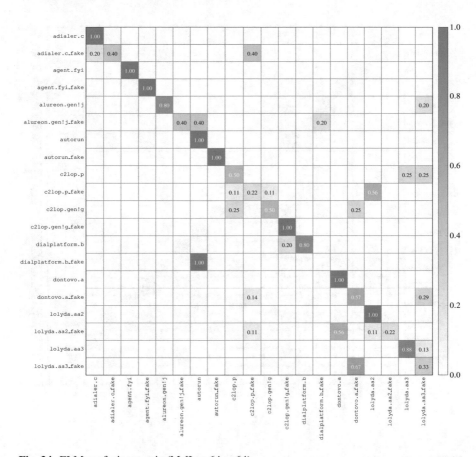

Fig. 24 ELM confusion matrix (MalImg 64 × 64)

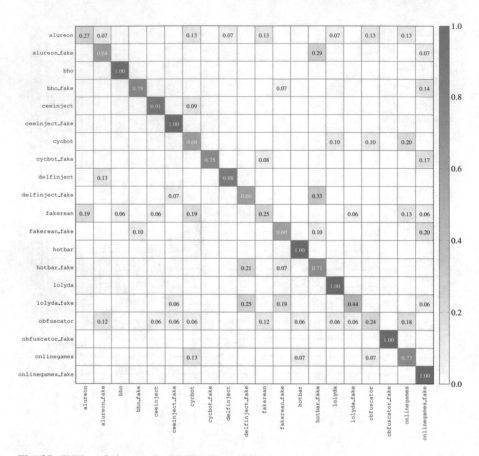

Fig. 25 CNN confusion matrix (MalExe 64 × 64)

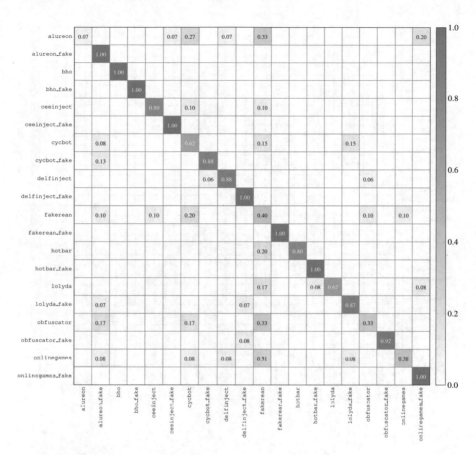

Fig. 26 ELM confusion matrix (MalExe 64 × 64)

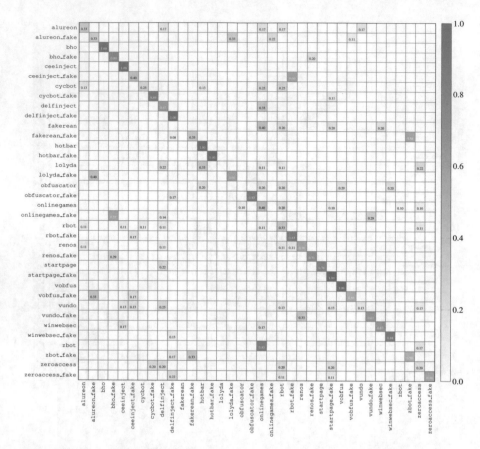

Fig. 27   CNN confusion matrix (MalExe 128 × 128)

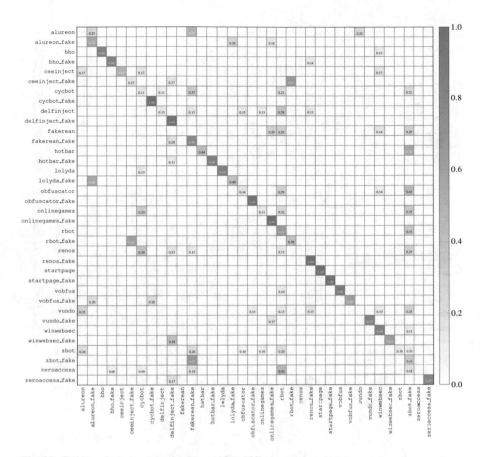

Fig. 28 ELM confusion matrix (MalExe 128 × 128)

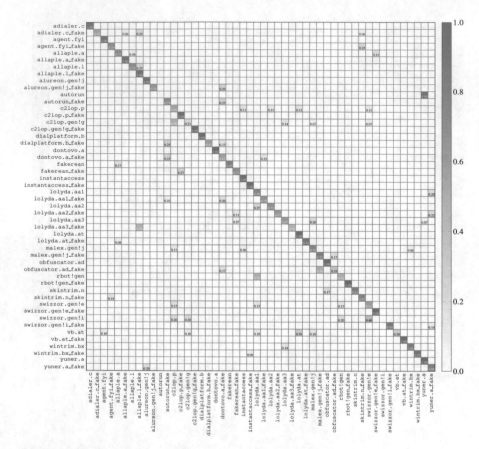

Fig. 29 CNN confusion matrix (MalImg 128 × 128)

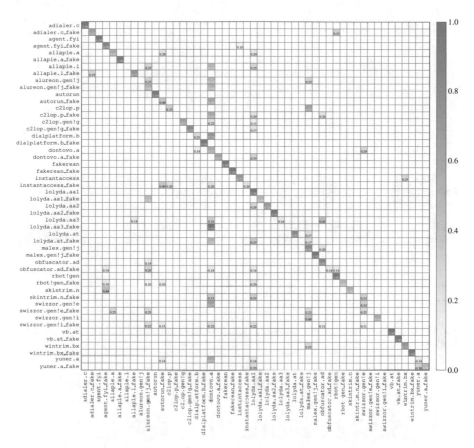

Fig. 30 ELM confusion matrix (MalImg 128 × 128)

Table 9 Condensed confusion matrix cases

Actual class	Classification	Description
real	real-same	Real sample classified correctly
	fake-same	Real sample classified as fake of the same family
	real-other	Real sample classified as a different real family
	fake-other	Real sample classified as a different fake family
fake	real-same	Fake sample classified as real of the same family
	fake-same	Fake sample classified correctly
	real-other	Fake sample classified as a different real family
	fake-other	Fake sample classified as a different fake family

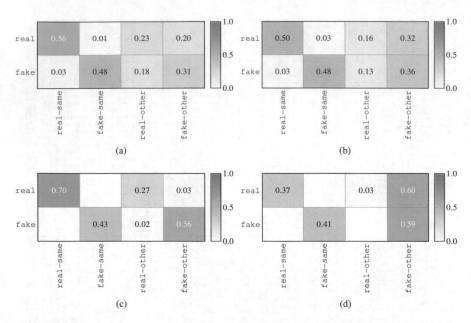

Fig. 31 Condensed confusion matrices (32 × 32). (**a**) CNN MalExe. (**b**) ELM MalExe. (**c**) CNN MalImg. (**d**) ELM MalImg

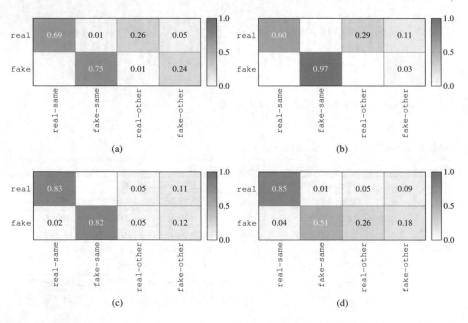

Fig. 32 Condensed confusion matrices (64 × 64). (**a**) CNN MalExe. (**b**) ELM MalExe. (**c**) CNN MalImg. (**d**) ELM MalImg

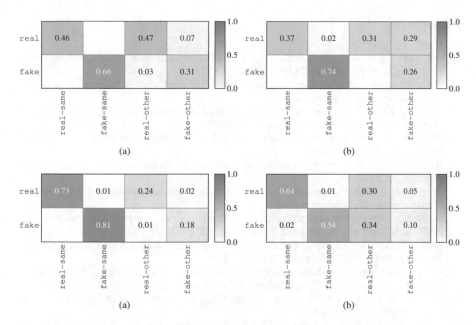

Fig. 33 Condensed confusion matrices (128 × 128). (**a**) CNN MalExe. (**b**) ELM MalExe. (**c**) CNN MalImg. (**d**) ELM MalImg

6 Conclusion and Future Work

In this research we considered AC-GAN in the context of malware research. We experimented with a standard malware image dataset (MalImg) and a larger and more balanced malware image dataset of our own construction (MalExe). We evaluated the images generated by our AC-GAN using CNN and ELM models.

We were not able to reliably classify our AC-GAN generated fake malware images from genuine malware images using either CNNs or ELMs, but the AC-GAN discriminator provided good accuracy. However, we also found that CNNs can distinguish between real and AC-GAN generated fake samples with surprisingly high accuracy.

For future work, more experiments aimed at classifying real and fake malware images would be useful. Additional state-of-the-art deep learning models, such as ResNet152 and VGG-19, could be considered [39]. In addition, the quest for true "deep fake" malware images that cannot be reliably distinguished from real malware images appears to be a challenging problem.

In addition, it would be interesting to explore adversarial attacks on image-based malware detectors. For example, tt would be interesting to quantify the effectiveness of such attacks. That is, assuming that an attacker is able to corrupt the training data, what is the minimum percentage of the data that must be modified to achieve a desired level of degradation in the resulting model?

References

1. Adialer.c. https://www.microsoft.com/en-us/wdsi/threats/malware-encyclopedia-description? Name=Trojan:Win32/Adialer.C&threatId=-2147460766.
2. Agent.fyi. https://www.microsoft.com/en-us/wdsi/threats/malware-encyclopedia-description? Name=Exploit:Win32/Siveras.A.
3. Allaple.a. https://www.microsoft.com/en-us/wdsi/threats/malware-encyclopedia-description? Name=worm:win32/allaple.a&ThreatID=2147574777.
4. Allaple.l. https://www.microsoft.com/en-us/wdsi/threats/malware-encyclopedia-description? Name=Worm:Win32/Allaple.L.
5. Alureon. https://www.microsoft.com/en-us/wdsi/threats/malware-encyclopedia-description? Name=Win32/Alureon.
6. Alureon.gen!j. https://www.microsoft.com/en-us/wdsi/threats/malware-encyclopedia-descript ?Name=Trojan:Win32/Alureon.gen!J.
7. Autorun.k. https://www.microsoft.com/en-us/wdsi/threats/malware-encyclopedia-description ion?Name=Worm:Win32/Autorun.K&threatId=-2147369124.
8. Bho. https://www.microsoft.com/en-us/wdsi/threats/malware-encyclopedia-description?Na me=Trojan:Win32/BHO&threatId=-2147364778.
9. C2lop.gen!g. https://www.microsoft.com/en-us/wdsi/threats/malware-encyclopedia-descripti on?Name=Trojan:Win32/C2Lop.gen!G&threatId=139219.
10. C2lop.p. https://www.microsoft.com/en-us/wdsi/threats/malware-encyclopedia-description? Name=Trojan:Win32/C2Lop.P.
11. Bugra Cakir and Erdogan Dogdu. Malware classification using deep learning methods. In *Proceedings of the ACMSE 2018 Conference*, pages 1–5, 2018.
12. Ceeinject. https://www.microsoft.com/en-us/wdsi/threats/malware-encyclopedia-description? Name=VirTool%3AWin32%2FCeeInject.
13. Gregory Conti, Sergey Bratus, Anna Shubina, Andrew Lichtenberg, Roy Ragsdale, Robert Perez-Alemany, Benjamin Sangster, and Matthew Supan. A visual study of primitive binary fragment types. https://www.semanticscholar.org/paper/A-Visual-Study-of-Primitive-Binary-Fragment-Types-Conti-Bratus/b406e34d0c203deadfb028f14607bfe88e5763ac, 2010.
14. Cycbot. https://www.microsoft.com/en-us/wdsi/threats/malware-encyclopedia-description? Name=Win32/Cycbot&threatId=.
15. Dennis Dang, Fabio Di Troia, and Mark Stamp. Malware classification using long short-term memory models. https://arxiv.org/abs/2103.02746, 2021.
16. Delfinject. https://www.microsoft.com/en-us/wdsi/threats/malware-encyclopedia-description? Name=PWS:Win32/DelfInject&threatId=-2147241365.
17. Diaplatform.b. https://www.microsoft.com/en-us/wdsi/threats/malware-encyclopedia-descrip tion?Name=Dialer:Win32/DialPlatform.B.
18. Dontovo.a. https://www.microsoft.com/en-us/wdsi/threats/malware-encyclopedia-description ?Name=TrojanDownloader:Win32/Dontovo.A&threatId=-2147342037.
19. Fakerean. https://www.microsoft.com/en-us/wdsi/threats/malware-encyclopedia-description? Name=Win32/FakeRean.
20. Hotbar. https://www.microsoft.com/en-us/wdsi/threats/malware-encyclopedia-description? Name=Adware:Win32/Hotbar&threatId=6204.
21. Weiwei Hu and Ying Tan. Generating adversarial malware examples for black-box attacks based on GAN. https://arxiv.org/abs/1702.05983, 2017.
22. Nathan Inkawhich. PyTorch DCGAN tutorial. https://pytorch.org/tutorials/beginner/dcgan_ faces_tutorial.html.
23. Instantaccess. https://www.microsoft.com/en-us/wdsi/threats/malware-encyclopedia-descrip tion?name=dialer:win32/instantaccess.
24. Mugdha Jain. Image-based malware classification with convolutional neural networks and extreme learning machines. https://scholarworks.sjsu.edu/etd_projects/900/, 2019.

25. Masataka Kawai, Kaoru Ota, and Mianxing Dong. Improved MalGAN: Avoiding malware detector by leaning cleanware features. In *2019 International Conference on Artificial Intelligence in Information and Communication*, ICAIIC, pages 040–045, 2019.
26. Jin-Young Kim, Seok-Jun Bu, and Sung-Bae Cho. Malware detection using deep transferred generative adversarial networks. In *International Conference on Neural Information Processing*, pages 556–564, 2017.
27. David Kornish, Justin Geary, Victor Sansing, Soundararajan Ezekiel, Larry Pearlstein, and Laurent Njilla. Malware classification using deep convolutional neural networks. In *2018 IEEE Applied Imagery Pattern Recognition Workshop*, AIPR, pages 1–6, 2018.
28. Lolyda. https://www.microsoft.com/en-us/wdsi/threats/malware-encyclopedia-description? Name=PWS%3AWin32%2FLolyda.BF.
29. Lolyda.aa1. https://www.microsoft.com/en-us/wdsi/threats/malware-encyclopedia-descrip tion?Name=PWS:Win32/Lolyda.AA&threatId=-2147345828.
30. Lolyda.at. https://www.microsoft.com/en-us/wdsi/threats/malware-encyclopedia-description? Name=PWS:Win32/Lolyda.AT&ThreatID=2147627867.
31. Yan Lu and Jiang Li. Generative adversarial network for improving deep learning based malware classification. In *2019 Winter Simulation Conference*, WSC, pages 584–593, 2019.
32. Adam Lutz, Victor F. Sansing III, Waleed E. Farag, and Soundararajan Ezekiel. Malware classification using fusion of neural networks. In Misty Blowers, Russell D. Hall, and Venkateswara R. Dasari, editors, *Disruptive Technologies in Information Sciences II*, pages 165–170. SPIE, 2019.
33. Malex.gen!j. https://www.microsoft.com/en-us/wdsi/threats/malware-encyclopedia-descript ion?Name=Trojan:Win32/Malex.gen!J.
34. McAfee 2020 2nd quarter report. https://www.mcafee.com/enterprise/en-us/lp/threats-reports/ apr-2021.html.
35. Lakshmanan Nataraj, Sreejith Karthikeyan, Gregoire Jacob, and Bangalore S Manjunath. Malware images: Visualization and automatic classification. In *Proceedings of the 8th International Symposium on Visualization for Cyber Security*, pages 1–7, 2011.
36. Obfuscator. https://www.microsoft.com/en-us/wdsi/threats/malware-encyclopedia description ?Name=Win32/Obfuscator&threatId=.
37. Augustus Odena, Christopher Olah, and Jonathon Shlens. Conditional image synthesis with auxiliary classifier GANs. https://arxiv.org/abs/1610.09585, 2017.
38. Onlinegames. https://www.microsoft.com/en-us/wdsi/threats/malware-encyclopedia-descrip tion?Name=PWS%3AWin32%2FOnLineGames, journal=Onlinegames.
39. Pratikkumar Prajapati and Mark Stamp. An empirical analysis of image-based learning techniques for malware classification. In *Malware Analysis Using Artificial Intelligence and Deep Learning*, pages 411–435. Springer, 2020.
40. Rbot. https://www.microsoft.com/en-us/wdsi/threats/malware-encyclopedia-description?Na me=Win32/Rbot&threatId=.
41. Rbot!gen. https://www.microsoft.com/en-us/wdsi/threats/malware-encyclopedia-description? Name=Backdoor:Win32/Rbot.gen.
42. Renos. https://www.microsoft.com/en-us/wdsi/threats/malware-encyclopedia-description? Name=TrojanDownloader:Win32/Renos&threatId=16054.
43. Igor Santos, Javier Nieves, and Pablo G Bringas. Semi-supervised learning for unknown malware detection. In *International Symposium on Distributed Computing and Artificial Intelligence*, pages 415–422, 2011.
44. Skintrim.n. https://www.microsoft.com/en-us/wdsi/threats/malware-encyclopedia-description ?Name=Trojan:Win32/Skintrim.N.
45. Startpage. https://www.microsoft.com/en-us/wdsi/threats/malware-encyclopedia-description? Name=Trojan:Win32/Startpage&threatId=15435.
46. Guosong Sun and Quan Qian. Deep learning and visualization for identifying malware families. *IEEE Transactions on Dependable and Secure Computing*, 18(1):283–295, 2021.
47. Swizzor.gen!e. https://www.microsoft.com/en-us/wdsi/threats/malware-encyclopedia-descrip tion?Name=TrojanDownloader%253aWin32%252fSwizzor.gen!E&navV3Index=3, key=Swizzor.gen!E.

48. Swizzor.gen!i. https://www.microsoft.com/en-us/wdsi/threats/malware-encyclopedia-descrip
    tion?Name=TrojanDownloader:Win32/Swizzor.gen!I.
49. Vb.at. https://www.microsoft.com/en-us/wdsi/threats/malware-encyclopedia-description?Na
    me=Worm:Win32/VB.AT.
50. Vobfus. https://www.microsoft.com/en-us/wdsi/threats/malware-encyclopedia-description?
    Name=Win32/Vobfus&threatId=.
51. Vundo. https://www.microsoft.com/en-us/wdsi/threats/malware-encyclopedia-description?
    Name=Win32/Vundo&threatId=.
52. Wintrim.bx. https://www.microsoft.com/en-us/wdsi/threats/malware-encyclopedia-descrip
    tion?Name=TrojanDownloader:Win32/Wintrim.BX.
53. Winwebsec. https://www.microsoft.com/en-us/wdsi/threats/malware-encyclopedia-descrip
    tion?Name=Win32/Winwebsec.
54. Sravani Yajamanam, Vikash Raja Samuel Selvin, Fabio Di Troia, and Mark Stamp. Deep
    learning versus gist descriptors for image-based malware classification. In Paolo Mori, Steven
    Furnell, and Olivier Camp, editors, *Proceedings of the 4th International Conference on
    Information Systems Security and Privacy*, ICISSP 2018, pages 553–561, 2018.
55. Songqing Yue. Imbalanced malware images classification: A CNN based approach. https://
    arxiv.org/abs/1708.08042, 2017.
56. Yuner.a. https://www.microsoft.com/en-us/wdsi/threats/malware-encyclopedia-description?
    Name=Worm:Win32/Yuner.A&ThreatID=2147600986.
57. Zbot. https://www.microsoft.com/en-us/wdsi/threats/malware-encyclopedia-description?
    Name=PWS:Win32/Zbot&threatId=-2147368817.
58. Zeroaccess. https://www.symantec.com/security-center/writeup/2011-071314-0410-99.

Assessing the Robustness of an Image-Based Malware Classifier with Smali Level Perturbations Techniques

Giacomo Iadarola, Fabio Martinelli, Antonella Santone, and Francesco Mercaldo

Abstract Signature-based approaches adopted by current antimalware have well-known problems. Although they can provide relatively fast and reliable detection of previously known threats, they are not able to catch new malware and also generalize their knowledge to different variants of the same known malware. Deep learning approaches have been adopted to address this problem, and one of the most promising attempts is based on the representation of malware as images. In order to understand whether these approaches can be effectively adopted in a real-world situation, we trained an image-based malware detector and evaluate its resilience when morphed samples are considered. The experiments were conducted on 16384 real-world Android Malware, and the experimental analysis demonstrates that standard image-based malware classifiers are vulnerable to simple perturbations attacks.

1 Introduction

Malware analysis and detection are one of the biggest security threats on the internet today, and one of the biggest and most active topics in cybersecurity. In the past few years, the malware industry has covered the top spots among the most used cyber-attack methodologies [23], and the volume of new malware detected has increased

G. Iadarola · F. Martinelli
Institute of Informatics and Telematics, National Research Council of Italy (CNR), Pisa, Italy
e-mail: giacomo.iadarola@iit.cnr.it; fabio.martinelli@iit.cnr.it

A. Santone
Department of Medicine and Health Sciences, University of Molise, Campobasso, Italy
e-mail: antonella.santone@unimol.it

F. Mercaldo (✉)
Department of Medicine and Health Sciences, University of Molise, Campobasso, Italy

Institute of Informatics and Telematics, National Research Council of Italy (CNR), Campobasso, Italy
e-mail: francesco.mercaldo@unimol.it; francesco.mercaldo@iit.cnr.it

69
M. Stamp et al. (eds.), *Artificial Intelligence for Cybersecurity*, Advances in Information Security 54, https://doi.org/10.1007/978-3-030-97087-1_3

by 233% compared to five years ago. Most of the commercial antimalware employ a preventive analysis mechanism called *Signature-based Detection*, which identifies a threat by looking for a specific pattern or sequences of bytes (called *signatures*) from a database of well-known threats. Many databases are built and continuously updated with the list of signatures of all known families. Although this approach is relatively fast and boasts a low number of false positives [2], it cannot recognize programs that have never been scanned before. This fragility poses a serious threat for the final users and resources protected by antimalware because the malware market changes continuously and the malware database cannot be updated at the same pace. The detection phase takes time, thus, there is always going to be a time window in which the new malware is not included in the well-known threats yet, and it is able to spread across the internet without being recognized as malicious. Indeed, most of the time, the cyber-criminals just need to modify old malware to overcome antimalware detection, heading to rise a huge number of *variants* for each malware.

In order to overcome some of the classical approaches limitations, Artificial Intelligence (AI) has been adopted in malware analysis and has provided increasingly accurate methods for malware detection. In this evolving environment, ensuring the reliability of AI models is becoming fundamental to increase trust in this technology, and extend the adoption of AI models to sensitive and relevant tasks (such as handling our private data, protecting critical infrastructure, driving vehicles, etc.). Despite their usefulness, some AI models provide output decisions without clear explanations of the inference phase and are emerging several criticisms related to the adoption of these techniques as a black box. Moreover, the overall reliability of these models is sometimes debatable. The dataset used to train them is limited, and it may not represent correctly the real-world scenario. Fallacies in the training phase may lead to a domino effect and cause damages when the model is deployed in real scenarios.

The Explainable AI is the research topic that studies how to design and implement AI applications that can be understood by human experts, hence, applications that are more robust and easy to debug. One more approach to design secure AI applications regards studying attacks to such models in order to detect security vulnerabilities. In this work, we adopt this approach and design an attack to a standard image-based malware classifier, in order to assess its robustness on malware source code perturbations. Exploiting Deep Learning (DL) models, such as the Convolutional Neural Network (CNN), for image classification in malware detection is not a new topic, and many relevant papers propose to convert malware binaries to images [7, 10, 25]. Following these approaches, the malware represented as an image can be analyzed by state-of-the-art image classifiers, which has shown to achieve excellent results in standard image classification tasks. The conversion of executables into images, as a means of identifying specific patterns capable of characterizing the most common classes of Malware, is an increasingly widespread and in-depth approach. Nevertheless, DL models suffer from instability problems [1]: small perturbations to the input can easily mislead the DL model.

In this work, we developed a CNN model to detect malware represented as images, achieving more than 90% of accuracy in the test. The model architecture is similar to others proposed in the literature (see Sect. 2.4). In order to assess the robustness of the classification, we perform standard perturbations techniques to modify the input. Despite their simplicity, the perturbations are sufficient for misleading the CNN model. This chapter is based on a couple of recent publications [12, 13], and provide insightful information on the background of the experiments and complete explanations of the methodology to perform the attack.

The chapter proceeds as follows: next section reports background knowledge on the covered topics, such as the CNN, Static Malware Analysis, the Image-based malware classifiers (with related works), and a short overview on the Dalvik EXecutable (DEX); the experiments and perturbations are performed on Android Malware, thus the DEX is the most technical topic on which our methodology is based on. Then, the Methodology is introduced in Sect. 3, followed by the Experiments in Sect. 4, which reports also information on the methodology implementation. Finally, Sect. 5 reports a short discussion on the results, limitations and future works.

2 Background and Related Works

In the following, some relevant Static Malware Analysis and the CNNs are briefly introduced, but we refer to the literature for further information on these interesting topics. The last subsection reports some relevant Image-based malware classification papers related to our work.

2.1 Static Malware Analysis

The *Malware*, literally *Malicious Software*, are programs whose goal is to interfere with the normal functioning of the system, without having the user's consent and to damage it for the benefit of the attacker. These are generally grouped into classes, or *families*, based on their common (malicious) behaviour. Countless families have been found, with different aims and methods of attack, and the number is constantly growing.

We can split the malware analysis techniques into two groups: static and dynamic analysis. Static analysis techniques base their operation on the sole evaluation of the program content and, in no case, its execution. For instance, one typical approach involves the use of the source code by disassembling the binary file, and then the build of an execution flow, such as a *Control Flow Graph* [3]. The information collected can be used individually or in combination with other extracted features to increase the degree of accuracy in the classification. Although static analysis is faster than dynamic analysis, it is highly vulnerable to code obfuscation techniques. Over time, more complex and sophisticated malware has been developed which, through the use of cryptographic techniques, are able to decrypt its own content at runtime. In this way, their syntax is always different, but the semantics are unchanged with

respect to the original program. We refer to the literature for more information on
the subject [5, 27].

2.2 Convolutional Neural Network

The CNN models are one of the Deep Neural Networks models that have demonstrated high accuracy in the field of image classification and recognition. Its main
feature is given by the use of specific layers, called *convolutional*, which have the
task of applying the mathematical convolution operator to the input image, intending
to collect the relevant information in a structure called *feature*.

The first convolutional network was developed by Yann LeCun and is called
LeNet5 [18]. Although more complex and more effective versions have been
developed over time, their basic structure has remained very similar to that modelled
by LeCun. Each of them carries out three fundamental operations: convolution,
subsampling and classification.

2.2.1 Convolution

In the field of image processing, the convolution operation consists of the combination of each image pixel with the neighbouring ones, on the basis of the weights
stored inside a finite matrix, which is called *kernel* or *filter*. The operation takes the
following form:

$$g(a, b) = w * f(a, b) = \sum_{da=-A}^{A} \sum_{db=-B}^{B} w(da, db) f(a + da, b + db) \tag{1}$$

where $g(a, b)$ is the pixel with coordinates (a, b) of the processed image, $f(a, b)$
is the corresponding pixel of the starting image, and $w \in R^{A \times B}$ is the kernel. For
instance, see Fig. 1.

The application of graphic filter results in "sliding" the kernel on the pixels of the
input image. The convolution operation within a CNN is managed through a specific
layer, called *convolutional layer*, which contains an activation function with the aim

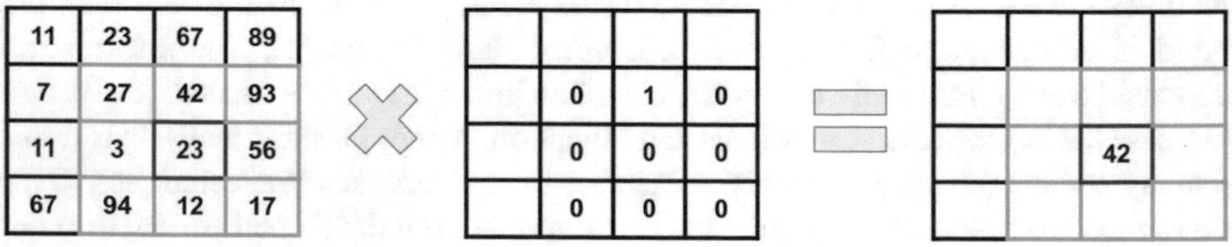

Fig. 1 Kernel application on a image pixels matrix

Fig. 2 Example of max
pooling

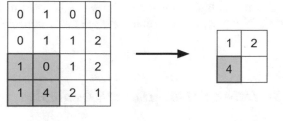

Feature Map   Pooled Feature Map

of making the calculation non-linear; the convolution itself is a linear operation. Although there are numerous functions available for this purpose, the most used is the Rectified Linear Unit [21] (ReLU). This function extracts the positive part of its argument and can be expressed as follows:

$$relu(t) = t^+ = max(0, t) \quad t \in R \tag{2}$$

2.2.2 Subsampling

Sub-sampling, also called *pooling*, is an operation that reduces the size of a two-dimensional matrix, while preserving the most relevant information. To do this, the original matrix is split into sub-matrices of fixed size, and only one element is extracted from them. This can be chosen by applying different strategies, among which, the most used variant is called *max pooling* and extracts the highest value.

The neural layer that performs this operation is called *max pooling layer* and always follows the convolution layer. For a better understanding of how the max pooling operation works, see Fig. 2.

In addition to reducing the size of the feature maps, subsampling offers the advantage of making the classification invariant to small transformations, translations and distortions of the starting image. Consequently, a CNN is able to classify objects regardless of the position they occupy within a graphical context [22].

2.2.3 Classification

The classification is carried out by a series of layers, called *dense layers*, which are formed by a variable number of perceptrons, each connected with all those of the next layer, forming a dense network of connections. The last dense layer is often associated with a specific activation function, called *normalized exponential* or *softmax*; it expresses the belonging of the input image to one of the starting classes in probabilistic terms. During the training process, these hidden layers modify the weights of their neurons, as happens in a canonical DNN, through the application of the backpropagation algorithm, learning to classify the input image.

The neural layers illustrated above can be present in a variable number within the CNN model, while maintaining the order convolution, subsampling, and then classification.

2.3 Dalvik VM and Dalvik EXecutable

Android is an operating system, widely adopted in mobile environments for smartphones and IoT devices. It is based on a Linux kernel integrated by various C/C++ libraries. Its applications are developed in the Java programming language, but they do not run on the standard *Java Virtual Machine* (JVM). Android has its own version of a virtual machine, which is called *Dalvik Virtual Machine* (DVM). This virtual machine is optimized for mobile devices resources, that are limited compared to common laptops and computers.

The DVM bytecode is not produced directly from the compilation of the Java code, but it comes from a translation process on the JVM bytecode. The executable file in Dalvik format is called *Dalvik EXecutable* or *DEX*, and it is produced starting from a *.class* file.

The DEX format is designed to optimize memory usage and, for this reason, adopts an approach based on sharing data, avoiding its replication. Its main optimization mechanism is based on the use of a constant pool, no longer private and usable only by the belonging class, but shared and referenced globally. By doing so, it is possible to subdivide the strings within the constant pool based on the type and eliminate any redundancy, keeping only one occurrence for each of them. Literal constants are known to be responsible for 61% of the weight of a class, while only the 33% depends on its body [8] methods. Figure 3 shows the compilation process of the DEX format and the *.class* format in comparison, while Fig. 4 shows the main differences between the *.class* and *.dex* structure. For further information, we refer to [9].

Starting from the version of Android 4.4 (KitKat), Dalvik was replaced by a new VM, called *ART* (Android Run Time). ART uses a mechanism called *Ahead-Of-Time Compilation* (AOTC), which compiles the entire application in ELF format, increases the initial wait time in favour of faster execution time. ART maintains backward compatibility with DVM by using the same DEX format for bytecode encoding.

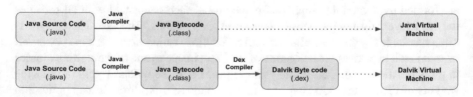

Fig. 3 Compilation process, comparison between JVM and DVM

.class	.dex
Header	Header
Heterogeneous Constant Pool	String Constant Pool
	Type/Class Constant Pool
	Field Constant Pool
	Method Constant Pool
Class	Class Definition
Field	Field List
Method	Method List
Attribute	Code Header & Local Variable

Fig. 4 Structure differences between .dex and .class format

Despite the fact that the DEX contains the executable of an application, it is only a part of the complete APK file. A complete malware analysis should take into account all the files included in the APK file; the payload could be stored in other APK files, or the application may cooperate with external resources to download malicious code at runtime. Nevertheless, the scope of this work is to demonstrate how simple perturbations can drop the efficiency of an image-based malware classifier that focus their analysis on the code (i.e. the DEX). Thus, we restrict our analysis only to the DEX file.

2.4   Image-Based Malware Classification

The first work that proposed an image-based classifier is the one by Nataraj et al. in [20]. The conversion process from malware to images is straightforward: the binary code of the executable content is grouped in 8-bit vectors of *unsigned integer* and shaped into a two-dimensional array. Each vector represents an integer value in the range [0–255], thus, it can be cast to a grayscale pixel. In the Nataraj work, common extraction features techniques were applied to the malware images. The approach was tested on a dataset of 9458 malware belonging to 25 different families, achieving an accuracy of 98%.

Similarly, the paper in [6] convert the DEX files to grayscale images and use GIST descriptor to generate feature vectors. The classification is performed with K-Nearest Neighbor, Random Forest and Decision Tree.

The convolutional part of the Deep Learning model computes the extraction of relevant features from the input. Therefore, many papers propose the adoption of DL models, mainly CNN, to automatize the feature extractions and improve the overall accuracy.

For instance, the work by Zhang et al. [29] utilises a Temporal Convolutional Network model to classify malware represented as images. In detail, it compounds the image with information coming from the DEX file and the AndroidManifest.xml, and they achieved 95% accuracy on a dataset of 5826 malware samples.

On the same topic, many other works were proposed in the literature [7, 10, 14, 25, 26] which use information coming from the DEX files and the APK files to produce images. Then, these images are either used to extract feature vectors or use directly as a plain image. Finally, CNN models are applied to distinguish between benign and malicious applications. Despite their differences, all these papers have in common the use of the DEX file.

The fragility of image-based malware detection techniques is a well-known problem, because the adopted models, usually, are not robust to input perturbations. For instance, many works show that these models are vulnerable to attacks based on *Adversarial Examples* [16, 17, 24]. An Adversarial sample is generated by iteratively changing a correctly classified image by a first DL model. Another DL model chooses the changes to apply, and use the classification from the discriminator (the DL model under attack) to guide its choices and understand if the changes were effective. The process succeeds when its classification differs from that of the original image.

In this work, we perform perturbations that do not change (nor corrupt) the functionalities (malicious behaviour) of the malware itself. In our context, we can not randomly change the input image because it is strictly connected to the malware bytecode: random changes will lead to corrupting the malware functionalities.

3 Methodology

This work aims to verify the robustness of an image-based malware classifier by modifying its input. Moreover, the perturbations injected to decrease the accuracy are modifying the malware code itself, and thus they have to preserve the malware functionalities, otherwise, the malware will not run anymore. In short, if the process was reversed and code was generated from the image, the code has to be syntactically and semantically correct.

Some research papers perform this kind of attack by using Adversarial Examples and applying random filters on the input image. However, they fail to preserve the executability of the modified program [19], losing practical utility. Therefore, the fundamental prerogative of this work is to preserve the aforementioned condition: the methodology will generate fully executable programs and carry the same semantics as the starting ones.

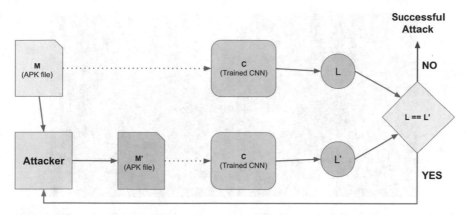

Fig. 5 Attack methodology overview

To do so, we construct a simple CNN model (the most-adopted model for this malware classification task in the literature) to show the feasibility of the attack. Nevertheless, the same approach could be applied to any image-based malware classifier, because we apply the perturbations on the input sample, and interact with the classifier as a black-box.

Formally, given a malware sample M, this is converted into an image and given as input to an image-based classifier C, which classifies the sample into a malware family and assigns the relative L label. The attacking module takes M and produces an executable M' by introducing perturbations. Then, the same model C classifies M' with L' label. If $L \neq L'$, then the attack succeed, otherwise, the attacker keeps introducing perturbations on the M input. The attacker, in each iteration, tries to add a small number of perturbations, in order to achieve the smallest amount of perturbations needed to mislead the sample classification. Figure 5 illustrates the above steps, highlighting the relationships between the parties involved.

This work is focused on the Android environment. We collect a dataset of *APK* files and split the samples into several malware families (e.g. group of malware that exhibit the same malicious behaviour). The malware belonging to the same malware family shares part of the code (usually, the payload), thus they have a pattern in common that can be used to classify them. It is worth noting that a label/class of the dataset is reserved for "Trusted" samples, that is applications considered "benign" and that do not exhibit any of the malicious behaviour symptomatic of the other malware classes.

For each APK sample in the dataset, the DEX is extracted and converted to an image. For the scope of this work, we restrict the analysis on the DEX file only, supported by similar approaches in the literature [15, 28]. We aim to demonstrate how easy is attacking such approaches rather than covering every possible situation of malware detection.

The conversion process generates images that may look like random noise from the human perspective, but they contain all the malware executable data, including

Fig. 6 Comparison between two malware samples converted to images, they both belong to the same malware family (AirPush)

the payload and the structure shared among the other variants of the same malware family. Figure 6 shows clearly that two malware samples that belong to the same family share similar patterns.

To classify the malware into families, we use a Convolutional Neural Network, widely adopted in the literature for this kind of tasks [14]. The extraction of features is delegated to two pairs of convolutional and max-pooling layers having depths of 64 and 128 respectively. The classification takes place with two dense layers, which count 1024 and 512 nodes, interspersed with layers of *dropout*. Each layer of the model uses *relu* as an activation function.

3.1   Untargeted Misclassification

The main goal of the methodology is to generate input samples that mislead the model and reduce the accuracy performance. There are different types of incorrectness, and in this work we focus on "untargeted misclassification". The original output label and the modified samples output label has to differ, but there is no restriction on the resulting label, it can be any one of the available except the original one.

The core of the methodology is the perturbations, which have to modify the input sample but preserve the executability of the starting file. To do this, we resort to perturbation techniques of the DEX code, in order to change or add bytes inside the DEX and, consequently, inside the resulting image.

Due to the inherent difficulties of modifying directly the DEX file and its low-level operations, we decided to use a disassembler to obtain higher-level code, which can be modified and subsequently recompiled into an executable.

To do so, we chose to use the pair of tools *smali* and *baksmali* [4]. These are a compiler and decompiler respectively, which use an assembly-like language to describe programs in ASCII format and produce executable files in DEX format.

The experimental analysis shows that a complete step of decompilation and recompilation of DEX file via baksmali and *smali*, without any modification, causes already an untargeted misclassification situation. The DEX compiler and *smali* one organize the bytecode sections differently, thus, the images converted from the DEX looks different. However, given a DEX obtained through *smali* compilation, the subsequent decompilation/recompilation processes do not involve further changes. Therefore, we pre-process the dataset by decompiling and recompiling all the samples. By doing this, the subsequent perturbations of the code through *smali* is made entirely valid and the misclassification is not compromised by decompilation edits.

The CNN model is trained on the dataset and then used to classify samples, both original malware and perturbed ones. The attack aims to edit the *smali* code and insert the minimum perturbation sufficient to cause the trained classifier to incorrectly predict the malware family of the sample under analysis.

For preliminary results, we implement two perturbations techniques, called NopsBombing and StringBombing. The first one perturbates the code by identifying the *smali* instructions after which it is possible to insert others without changing the logic, and add an arbitrary number of NOP (No Operation) (i.e; an instruction that does not involve any operation). The other technique, the StringBombing, identifies the const-string instructions, and add just before a new const-string with the same destination and random content. In this way, the original operation is executed just after the newly added one and overwrites the content assigned by perturbation. Listing .1 shows the result of the execution of the StringBombing perturbation, in which the instruction inserted by the attack is the one setting the string "*a87bca5*" on line 8, just before the original statement on line 9, setting the string "*Hello world*".

Listing .1 Smali file perturbed with StringBombing, adding line 8 to the original *smali* file

```
1   . method  public  static  main ([ Ljava / lang / String ;) V
2       . registers  2
3
4       . line  5
5       sget−object  p0 ,  Ljava / lang / System ;
6          −>out : Ljava / io / PrintStream ;
7
8       const−string  v0 ,  "a87bca5 "
9       const−string  v0 ,  "Hello  world"
10
11      invoke−virtual  {p0 ,  v0 } ,  Ljava / io / PrintStream ;
12         −>println ( Ljava / lang / String ;) V
13
14      . line  6
15      return−void
16  . end  method
```

4   Implementation and Experiments

This section reports details on the implementation of the methodology and the experimental results to test the attack.

We collected APK samples from a subset of the AMD dataset [14]. The resulting dataset consists of 16384 samples of real-world malware, split into 10 classes, namely *Airpush* (3487 samples), *BankBot, Dowgin* (3384 samples), *DroidKungFu, FakeInst* (2167 samples), Fusob (1275 samples), *Kuguo* (1199), *Mecor* (1820 samples), *Youmi* (1301). One more class of the dataset was reserved for Trusted samples (559 samples). We collect the trusted samples from the Android official store and tested them with VirusTotal. The CNN model was trained on 80% of the dataset, and tested on the 20% samples left. It achieved 92.82% accuracy in test, with a loss of 0.84.

We implement the methodology in a tool, called *DexWave*, which aims to introduce in the DEX files the minimum amount of edits sufficient to produce untargeted misclassifications to a trained model and a given input. The tool is written in Python, it integrates both a tool for producing images starting from executable files and a second tool to verify the achievement of the attack objective, using a *white-box* approach. The tool is open-source and it is available for research purposes on GitHub [11].

The execution begins with the classification of the original DEX file so that the respective label is available. Perturbation techniques are managed in a modular way and can be easily extended by adding new perturbations.

All the perturbations are loaded and applied sequentially, one at a time. Each of them applies just a few edits, and at the end of each of them, the perturbed *smali* code is compiled into a new DEX file, and then converted into an image, and analyzed by the classifier. If the new classification differs from the original, the execution ends successfully. Otherwise, the subsequent perturbations are applied until they are exhausted. By doing so, the number of perturbation techniques applied is the minimum necessary to cause untargeted misclassification.

For example, we report the *DexWave* execution on a malware sample belonging to the *Dowgin* family. Figure 7 shows the image of the original malware in comparison with the perturbed one. In this example, the *DexWave* tool execution ended successfully with the NopsBombing perturbation, the very first perturbation technique; it was not necessary to apply further perturbations, since the Nops adding was enough to mislead the classification. In detail, the malware sample was decompiled into *smali* code, the perturbation was applied on the *smali* code, recompiled and converted again to an image. After that, the sample file was classified as *AirPush*, thus it was classified wrongly even if the semantics of the malware was preserved.

The tool was tested on 30 samples, taken from the test set of the dataset, and they were obfuscated with *DexWave*. For almost half of them (14 out of 30), only the NopsBombing perturbation was enough to mislead the classification; other 12 samples were still correctly classified after the first perturbation, but they were

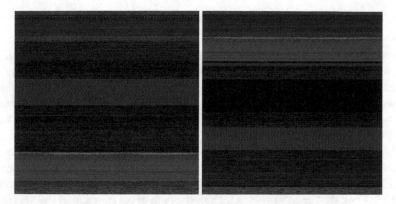

Fig. 7 Comparison between an original malware sample (on the left) and the same malware after the execution of the *DexWave* tool (on the right)

Table 1 Number of perturbations applied per each subset of the 30 samples tested

# Test Samples	# NopsBombing	# StringBombing
14	1	0
12	1	1
3	2	1
1	2	2

wrongly classified after applying also the StringBombing perturbation. Only 4 samples required more than 2 iterations of the tool (i.e. applying again the two perturbations available), but all of the samples ended up being wrongly classified. Table 1 shows the number of perturbations applied on the 30 samples. To avoid an infinite loop, *DexWave* was set to stop after 8 iterations (i.e. applying both two perturbations 4 times each). Nonetheless, it never ended on this termination condition, because it reached 4 iterations at maximum. Despite the simplicity of the obfuscation techniques taken into account, this result confirms the effectiveness of methodology: all obfuscated applications were not correctly classified by the deep learning model, thus, we were able to drop the accuracy to 0%.

Furthermore, we scanned both DEX files (original and perturbed ones) with the online platform *VirusTotal*, which allows scanning of files with multiple AVs. This experiment was useful to test the robustness of commercial AVs to the perturbations applied to the malware samples. For instance, the two DEX files of the samples in Fig. 7 produce a significantly different output: the original file was recognized as malicious content by 20 AVs out of 59, while the one produced by DeWave only 9 AVs out of 59. Thus, the perturbation process has increased the ability of the virus to evade controls by about 20%. The VirusTotal detection results are available on the *DexWave* GitHub repository [11].

5 Conclusion and Future Work

This work presented the process of building a CNN for malware classification and evaluated its performance on a dataset of Android executables. The malware executables, converted to images, were split into malware families. An attack method, based on two simple perturbations techniques, was then applied to obtain untargeted misclassification. Also, a tool was implemented with the aim to automate the aforementioned perturbation process.

The dataset was composed of 16384 malware samples, split into 10 families (9 malware families and one class for trusted samples). We achieved promising classification performances in the test, with an accuracy percentage and a loss value of 92.82% and 0.84 respectively, which may confirm the efficacy of the approaches image-based for malware classification tasks. Nevertheless, we presented a method to slightly modify the malware sample in order to be misclassified by the CNN model and demonstrate the weakness of these image-based malware approaches. Finally, it was verified that the perturbations caused the model accuracy to drop to 0% and also a decrease in the number of commercial AVs able to detect the modified file as a threat.

The proposed methodology has two main limitations: it is not able to target the misclassification, and it could fail when there are no sufficiently effective perturbations. In future works, the authors aim to implement more sophisticated obfuscations techniques and to achieve a targeted misclassification, in order to label all the obfuscated malware into the Trusted family.

Acknowledgments This work has been partially supported by MIUR - SecureOpenNets, EU SPARTA, CyberSANE and E-CORRIDOR projects.

References

1. Vegard Antun, Francesco Renna, Clarice Poon, Ben Adcock, and Anders C Hansen. On instabilities of deep learning in image reconstruction and the potential costs of ai. *Proceedings of the National Academy of Sciences*, 117(48):30088–30095, 2020.
2. Zahra Bazrafshan, Hashem Hashemi, Seyed Mehdi Hazrati Fard, and Ali Hamzeh. A survey on heuristic malware detection techniques. In *The 5th Conference on Information and Knowledge Technology*, pages 113–120. IEEE, 2013.
3. Mihai Christodorescu and Somesh Jha. Static analysis of executables to detect malicious patterns. In *In Proceedings of the 12th USENIX Security Symposium*, pages 169–186, 2003.
4. Repository contributors. Smali. https://github.com/JesusFreke/smali. Accessed: Sept-2021.
5. Anusha Damodaran, Fabio Di Troia, Corrado Aaron Visaggio, Thomas H Austin, and Mark Stamp. A comparison of static, dynamic, and hybrid analysis for malware detection. *Journal of Computer Virology and Hacking Techniques*, 13(1):1–12, 2017.
6. Fauzi Mohd Darus, Noor Azurati Ahmad Salleh, and Aswami Fadillah Mohd Ariffin. Android malware detection using machine learning on image patterns. In *2018 Cyber Resilience Conference (CRC)*, pages 1–2. IEEE, 2018.

7. Yuxin Ding, Xiao Zhang, Jieke Hu, and Wenting Xu. Android malware detection method based on bytecode image. *Journal of Ambient Intelligence and Humanized Computing*, pages 1–10, 2020.
8. David Ehringer. The dalvik virtual machine architecture. *Techn. report (March 2010)*, 4(8), 2010.
9. David Ehringer. The dalvik virtual machine architecture. *Techn. report (March 2010)*, 4(8), 2010.
10. Yong Fang, Yangchen Gao, Fan Jing, and Lei Zhang. Android malware familial classification based on dex file section features. *IEEE Access*, 8:10614–10627, 2020.
11. Federico Gerardi and Giacomo Iadarola. Dexwave - image-based malware classification attacking tool. https://github.com/AzraelSec/DexWave. Accessed: Oct-2021.
12. Federico Gerardi, Giacomo Iadarola, Fabio Martinelli, Antonella Santone, and Francesco Mercaldo. Perturbation of image-based malware detection with smali level morphing techniques. In *2021 International Symposium on Parallel and Distributed Processing with Applications (ISPA)*. IEEE, 2021.
13. Giacomo Iadarola, Rosangela Casolare, Fabio Martinelli, Francesco Mercaldo, Christian Peluso, and Antonella Santone. A semi-automated explainability-driven approach for malware analysis through deep learning. In *2021 International Joint Conference on Neural Networks (IJCNN)*, pages 1–8. IEEE, 2021.
14. Giacomo Iadarola, Fabio Martinelli, Francesco Mercaldo, and Antonella Santone. Towards an interpretable deep learning model for mobile malware detection and family identification. *Computers & Security*, 105:102198, 2021.
15. ElMouatez Billah Karbab, Mourad Debbabi, Abdelouahid Derhab, and Djedjiga Mouheb. Maldozer: Automatic framework for android malware detection using deep learning. *Digital Investigation*, 24:S48–S59, 2018.
16. Bojan Kolosnjaji, Ambra Demontis, Battista Biggio, Davide Maiorca, Giorgio Giacinto, Claudia Eckert, and Fabio Roli. Adversarial malware binaries: Evading deep learning for malware detection in executables. In *2018 26th European Signal Processing Conference (EUSIPCO)*, pages 533–537. IEEE, 2018.
17. Felix Kreuk, Assi Barak, Shir Aviv-Reuven, Moran Baruch, Benny Pinkas, and Joseph Keshet. Adversarial examples on discrete sequences for beating whole-binary malware detection. *arXiv preprint arXiv:1802.04528*, pages 490–510, 2018.
18. Yann LeCun, Léon Bottou, Yoshua Bengio, and Patrick Haffner. Gradient-based learning applied to document recognition. *Proceedings of the IEEE*, 86(11):2278–2324, 1998.
19. Xinbo Liu, Jiliang Zhang, Yaping Lin, and He Li. Atmpa: Attacking machine learning-based malware visualization detection methods via adversarial examples. In *2019 IEEE/ACM 27th International Symposium on Quality of Service (IWQoS)*, pages 1–10. IEEE, 2019.
20. Lakshmanan Nataraj, Sreejith Karthikeyan, Gregoire Jacob, and Bangalore S Manjunath. Malware images: visualization and automatic classification. In *Proceedings of the 8th international symposium on visualization for cyber security*, pages 1–7, 2011.
21. Prajit Ramachandran, Barret Zoph, and Quoc V. Le. Searching for activation functions, 2017.
22. Dominik Scherer, Andreas Müller, and Sven Behnke. Evaluation of pooling operations in convolutional architectures for object recognition. In *International conference on artificial neural networks*, pages 92–101. Springer, 2010.
23. G.S. Shahi, E.F. Pang, and P.P.E. Fong. *Technology in a Changing World*. Lulu Enterprises Incorporated, 2009.
24. Octavian Suciu, Scott E Coull, and Jeffrey Johns. Exploring adversarial examples in malware detection. In *2019 IEEE Security and Privacy Workshops (SPW)*, pages 8–14. IEEE, 2019.
25. Danish Vasan, Mamoun Alazab, Sobia Wassan, Hamad Naeem, Babak Safaei, and Qin Zheng. Imcfn: Image-based malware classification using fine-tuned convolutional neural network architecture. *Computer Networks*, 171:107138, 2020.
26. Xusheng Xiao. An image-inspired and cnn-based android malware detection approach. In *2019 34th IEEE/ACM International Conference on Automated Software Engineering (ASE)*, pages 1259–1261. IEEE, 2019.

27. I. You and K. Yim. Malware obfuscation techniques: A brief survey. In *2010 International Conference on Broadband, Wireless Computing, Communication and Applications*, pages 297–300, 2010.
28. Zhenlong Yuan, Yongqiang Lu, Zhaoguo Wang, and Yibo Xue. Droid-sec: deep learning in android malware detection. In *Proceedings of the 2014 ACM conference on SIGCOMM*, pages 371–372, 2014.
29. Wenhui Zhang, Nurbol Luktarhan, Chao Ding, and Bei Lu. Android malware detection using tcn with bytecode image. *Symmetry*, 13(7):1107, 2021.

Detecting Botnets Through Deep Learning and Network Flow Analysis

Ji An Lee and Fabio Di Troia (iD)

Abstract Botnet attacks pose a serious threat to the Internet infrastructure and its users. Botnets are operated through a command and control (C&C) channel which uniquely distinguishes it from other typical malware threats. The C&C server sends commands to the botnets to execute malicious activities using common Internet protocols, such as Hypertext transfer (HTTP), and Internet Relay Chat (IRC). Since these protocols are common, detecting botnet activities has been a challenge. This paper proposes an approach to identify the IP addresses of C&C servers and infected hosts in a network, without prior knowledge of the addresses or the type of the botnet. The approach is based on the observation that there are unique patterns in the communication between C&C server and bots which could be used to distinguish botnets from the background traffic. Regular botnet activities such as orchestrated attacks, heartbeat signals, or periodic distribution of commands are the main causes that produce such patterns. Deep learning techniques are applied on the extracted patterns to classify potential botnet traffics. The results show this pattern-based botnet detection technique is able to achieve high classification accuracy with low false positive rate.

1 Introduction

A botnet is a collection of machines that have been intentionally infected with malware to carry out various scams and cyber-attacks on the Internet without the authorization of the machines' owners. Once infected, these machines are remotely controlled by a botmaster through communication channels using standard networking protocols. At the core of the botnet are Command-and-Control (C&C) servers that act as headquarters for botnet communication [22]. Cybercriminals use C&C servers to distribute new commands to bots as well as receive execution results. Some of the malicious activities carried out by the bots include identity

J. A. Lee · F. Di Troia (✉)
San Jose State University, San Jose, CA, USA
e-mail: fabio.ditroia@sjsu.edu

© The Author(s), under exclusive license to Springer Nature Switzerland AG 2022
M. Stamp et al. (eds.), *Artificial Intelligence for Cybersecurity*, Advances in
Information Security 54, https://doi.org/10.1007/978-3-030-97087-1_4

theft, security breaches, distribution of SPAM emails, fraudulent financial scams, and perpetrated DDoS (distributed denial of service) attacks [22]. Potentially, any computer machine connected to the Internet has the possibility to become a compromised bot, thus, the impact of a botnet is estimated to cause severe damage. Many studies have been conducted to effectively detect botnet activities and protect machines from botnets. Despite these efforts, botnet attacks continue to pose a serious threat to the Internet infrastructure due to its constantly evolving nature [29]. Some of the previously explored botnet detection techniques include honeypot, passive anomaly analysis, and network traffic based classification [7, 17]. Among these three categories, network traffic based botnet classification is of particular interest for our work. By analyzing botnet behavior, some distinctive traits of botnet traffic may be recognized to help identify botnet activities. For instance, botnets are required to connect with the C&C servers to provide status updates and receive new commands. This unique characteristic suggests that botnets need to periodically communicate with C&C servers to be able to function properly. Using this information, the signs of periodic traffic may serve as a strong indicator for botnet activity. Furthermore, even more features can be specified by reviewing botnet behavior and network traffic for the purpose of botnet detection. The goal of this paper is to propose a deep learning model that detects botnet activities in a network by analyzing its packet captures. This paper tries to find answers to the following problems:

1. Given a dataset that consists of botnet, normal, and background traffic, is it possible to train a deep learning model that accurately classifies botnet traffic?

2. In real-life scenarios, botnets generate a significantly lower proportion of network traffic than non-botnet traffic. How should the dataset imbalance issue be addressed?

3. What are the key features of network traffic that is required to train the deep learning model?

The structure of the remaining Sections of this paper is as follows: Sect. 2 covers background information on the topics covered in this paper. Section 3 analyzes the relevant work on the same domain. Section 4 explains the key details about the CTU-13 dataset used in this project. Section 5 describes the methodology followed in this paper, the specific implementation details for feature extraction, and the deep learning model construction and evaluation. Section 6 summarizes the key findings and reports the overall project result.

2 Background

This section discusses the background domain which this paper is based on. It mainly focuses on botnets, autocorrelation analysis, and deep neural networks.

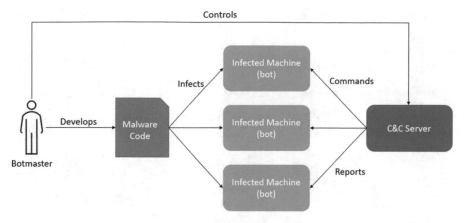

Fig. 1 Overview of the general botnet architecture

2.1 Introduction to Botnets

The term 'botnet' is a compound word from 'robot' and 'network'. It refers to a network of compromised machines that works for a cybercriminal to perform malicious activities over the Internet. Initially, the size of botnets was roughly a few hundreds. However, with the advance of Internet technologies and computing power, the number of bots that comprise a botnet have significantly increased to a few hundreds of thousands [10]. Using this massive network of bots, hackers conduct illegal activities such as personal data theft, server attacks, and distribution of malware to infect more machines [18]. Botnets are controlled by a masterbot through Command and Control (C&C) servers [8]. This control server plays a critical role in distributing commands to the botnets and keeping a list of which botnets are active and inactive. Figure 1 illustrates the architecture of a general botnet system. Botmaster develops a malware program and infects machines through the Internet. The set of infected machines are then operated by a C&C server which is directly controlled by the botmaster.

There are four types of known C&C architectures as shown in Fig. 2. With the direct architecture, botmasters directly infect and control the botnet. However, with the possibility to trace the botmaster from the bots and the limited scaling, it lost popularity within the cybercriminal society. Centralized architecture is identical to the architecture shown in Fig. 2 and was discussed in [16]. Contrary to the direct architecture, the centralized architecture's bots do not lead traces directly to the botmaster. P2P or decentralized architecture evades the single point failure issue by enabling communication between all nodes in the network. A hybrid architecture is an expansion of the P2P network that enables large scaling of the number of bots that a botmaster can operate.

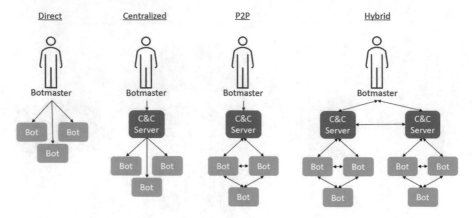

Fig. 2 The four types of known C&C architectures

Fig. 3 Autocorrelation plot
of a periodic signal of lag 7

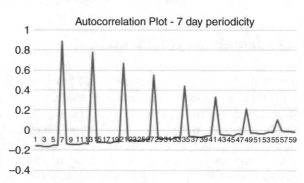

2.2 Autocorrelation Analysis

Given a dataset that consists of observations of a phenomenon at different points
in time, autocorrelation analysis seeks for patterns over the time series. The term
autocorrelation refers to the degree of similarity between a given time and a time-
shifted version of itself. For instance, if it rained heavily every Monday, then the
autocorrelation analysis would find the periodicity of the rain as seven days.

Using the rainy day example, an autocorrelation plot can be constructed. In
Fig. 3, the x-axis is lag and the y-axis is the value of the autocorrelation function.
The plot shows a peak every seven lags: 7, 14, 21, 28, and more. This means that the
original input shows a repeating pattern of seven days. Similar to this example, the
high peaks in the autocorrelation plot are important when using the autocorrelation
tool. An autocorrelation value is considered significant if the value exceeds the
threshold, otherwise known as the confidence interval (CI). The formula to calculate
the CI is shown here

$$CI_{AC_k} = [AC_k - 1.96 \times \frac{AC_{SE,k}}{\sqrt{N}}, AC_k + 1.96 \times \frac{AC_{SE,k}}{\sqrt{N}}]$$

where AC_i is the autocorrelation estimate at lag i and N is the number of time steps in the sample. More details can be found in [11].

2.3 Deep Neural Networks

Deep learning is a subset of machine learning where the structure is constructed theoretically similar to living brains. It is commonly referred to as an artificial neural network (ANN). Two common examples are convolutional neural networks (CNNs) and recurrent neural networks (RNNs). The main difference between the two types of networks is that CNN uses the connectivity patterns between the internal neurons while RNN uses time-series information that is strongly correlated to the order and the neighboring input data. Due to this difference, we use CNN rather than RNN for the deep neural network section of the implementation. There are three types of layers that make up the CNN: convolutional layer, pooling layer, and fully-connected (FC) layer. The convolutional layer is the first layer in the network that is used to extract various features from the input data. This phase consists of mathematical computations of convolution between the input data and a $K * K$ filter. This filter slides through the input array and produces a feature map which provides information about different qualities of the dataset. The pooling layer is generally used to decrease the size of the feature map to increase computational efficiency. In particular, the process of max pooling is the operation of selecting the largest element in the feature map. Lastly, the fully connected layer consists of weights and biases of the neurons and this information is used to connect the neurons between the FC layers. The components of such network are described in Fig. 4.

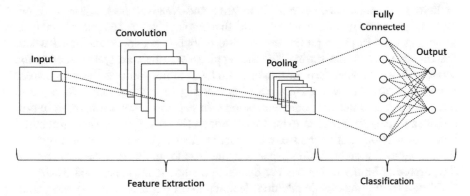

Fig. 4 Components of a convolutional neural network

3   Related Work

In the research field of malware detection, static and dynamic malware analysis have gained popularity over the recent years [21]. While static malware analysis focuses on using signature-based approaches such as file fingerprinting and virus scanning, dynamic malware analysis focuses on analyzing the behavior-based features of the malware samples [17]. Since the focus of static analysis is on the structure of the executable itself, the analysis can be performed without running the actual executable file. Although this is a cost-efficient approach, static analysis becomes vulnerable to malware threats that only reveal themselves during run-time. Dynamic analysis, on the other hand, detects malware by monitoring program activities rather than the program structure itself. Using a behavior-based approach, activities like network communication, API calls, system calls and system resource usages are analyzed. With dynamic analysis, the intention is to understand the working mechanism of a program and use this information to flag any suspicious program behavior. Due to this characteristic, dynamic analysis is resilient and flexible to more sophisticated and obfuscated types of malware. An example of a study that relies on dynamic analysis is [2], where the authors detect sniffing programs within a network. Sniffing is a type of network attack where an attacker tries to seek for vulnerabilities in a network by gathering as much information as possible about the targeted network. Sniffing is conducted by software programs called Sniffers that passively analyzes the incoming and outgoing traffic in a network. Due to this passive behavior, detecting Sniffers has been a challenge. The paper also suggests a measurement-based approach to pinpoint hosts running Sniffers by flooding the network with packets and comparing the round-trip response time among hosts. The findings of this study show that monitoring the behavior of programs can serve as an accurate indicator for finding malicious hosts. As cyber criminals and their malicious programs become more innovative and creative, the mission to detect malware paved the way for a hybrid model that employs static and dynamic analysis in conjunction. In [3], a particular use case of hybrid model is explained where dynamic analysis is used during the training phase and static analysis is used in the scoring phase. Another form of a hybrid model is to use static analysis to inspect network packet data while using machine learning to monitor network traffic to pick up malicious network communications. In [9], the authors utilized a hybrid a malware analysis model to extend the work in [2]. A sniffing detection method that uses network traffic probed with machine learning techniques is proposed. According to the authors, this paper was the first to apply machine learning for the purpose of Sniffer detection. The detection method in the paper used ICMP and HTTP for traffic probing. In addition, features like CPU load and variable period lengths were used for performance evaluation. This extended paper achieved a comparable outcome with the best results obtained in [2]. The work in [12] proposes a botnet detection approach using mining of network flow characteristics. Given a network flow dataset, four features were extracted: the ratio of incoming packets, the ratio of outgoing packets, original packet length, and the ratio of bot-response packets. These features were used by the naïve bayesian classifier to achieve 99% accuracy and 96.9% F-measure performances. However, the authors concluded that the four features were insufficient to accurately represent botnet communication patterns and additional features needed to be identified to further improve the accuracy of the model. In [19], the authors utilized 29 different features of various network protocols and its payload data to detect botnet activity in a network. Using a large number of features, the main focus of this paper was to establish a connection between a host's periodic communication patterns and botnet activity. The author pointed out that since botnets inevitably produce periodic network traffic while communicating with the C&C servers, this characteristic will be a strong indicator of botnet infestation. The datasets used in [19] are network captures that consist of only malware and botnets traffic. Autocorrelation analysis on the features extracted from the dataset was processed and the authors concluded the paper by presenting autocorrelation plots that show signs of periodic behavior of botnet traffic. The work in [19] opens the possibility to utilize the trait of periodic behavior in botnet traffic to detect infected hosts. However, the dataset used in [19] is limited to only botnet traffic where normal and background traffic are absent. The question of whether the same approach will work for real-world network traffic dataset is yet to be answered. Our paper extends the work of [19] by incorporating periodicity as one of the features used to train the deep learning model for botnet traffic classification. In contrast with the previous research, our work uses a network flow dataset that includes a mixture of botnet, normal, and background traffic to prove its efficacy in real-world botnet attack.

4 Dataset

The CTU-13 dataset was collected in 2011 by researchers at CTU University in Czech Republic for the purpose of generating a large capture of botnet traffic mixed with both normal and background traffic captures [6]. The thirteen captures that comprise CTU-13, also referred to as scenarios, are collected using seven different real botnet samples. While the dataset is now aging, it is not less representative of modern botnet attacks. For example, the Virut botnet was identified recently after being considered eradicated for many years [4]. A complete description of the seven botnet samples is provided in Table 1. The distinguishing feature of the CTU-13 dataset is that each packet has been manually examined and labeled as either botnet, normal, or background traffic. From examining the percentage of botnet traffic in all thirteen scenarios, botnet traffic makes up only a small percentage of overall traffic. This imbalance, however, accurately simulates real-world botnet infection and will be used as input without artificial manipulation for the purpose of this research.

For each scenario, the three types of network traffic were captured in a Packet Capture (pcap) file. From processing each pcap file, information such as NetFlows and WebLogs were also obtained. While pcap files carry detailed information,

Table 1 Types of CTU-13 botnet samples and its description

Botnet Type	Scenario Nr.	Description
Neris	1, 2, 9	Neris uses HTTP-based communication with the C&C servers. The infected botnets' main activities include click-fraud and distribution of SPAM emails
Rbot	3, 4, 10, 11	Rbot uses IRC-based communication with the C&C servers. Common with most IRC type malwares, the botnet is controlled by the botmaster through a pre-configured IRC server
Virut	5, 13	Virut uses HTTP-based communication with its C&C servers. Main activities of the infected hosts perform distribution of SPAM emails and unauthorized file downloads
Menti	6	Menti uses IRC-based communication with its C&C servers to scan SMTP servers
Sogou	7	Sogou uses unencrypted HTTP-based communication to connect with the C&C servers. Its malicious activities include downloading binary files and compressing them without authorization
Murlo	8	Murlo uses IRC-based communication with the C&C servers to carry out orders such as downloading executable files and scanning vulnerable local network ports
NSIS.ay	12	NSIS.ay uses P2P-based communication with the C&C

analyzing NetFlow files is the primary interest of our paper as it contains core information about traffic as well as its class labels.

4.1 CTU-13 Dataset Features

In a bi-directional NetFlow dataset, fifteen categories, listed in Table 2, are used to describe a network traffic. The dataset is initially sorted by StartTime in ascending order and, by using software programs that support csv parsing like Excel, each category can be filtered to selectively show rows of particular interest. For instance, to search for traffic generated by botnet activities, filtering the Label category for "Botnet" keyword would bring up all relevant rows.

The CTU-13 dataset has been distributed by Stratosphere Lab through their website and is open to the public for research or educational purposes [13].

5 Proposed Methodology

The goal of this project is to detect botnet activity in a given network by examining the incoming and outgoing network traffic data. As illustrated in Fig. 5, the proposed

Table 2 Explanation of features that comprise the CTU-13 network flow dataset

StartTime	StartTime represents the absolute timestamp in which the row has been recorded and is formatted as hh:mm:ss.
Dur	Dur is duration of the corresponding event in seconds for each row
Proto	There are 15 protocol is categorized in the Proto feature: 'tcp', 'udp', 'rdp', 'rtp', 'pim', 'icmp', 'ipx/spx', 'arp', 'igmp', 'rarp', 'unas', 'udt', 'esp', 'ipv6', 'ipv6-icmp'
SrcAddr	Source IP address in ipv4 format
Sport	Port number at the Source
Dir	Direction of the network flow, represented as '->', '<-', '<->', '<?>', 'who', '<?', '?>'
DstAddr	Destination IP address in ipv4 format
Dport	Port number at the destination
State	This feature describes the transaction state according to the protocol and has 230 unique values
sTos	Source Type of Service (0,1,2,...,192, NaN)
dTos	Destination Type of Service (0,1,2,...,192, NaN)
TotPkts	Total number of packets transmitted
TotBytes	Total number of bytes transmitted
SrcBtytes	Number of bytes transmitted from source to destination
Label	Three unique labels to describe the transaction (normal, background, botnet)

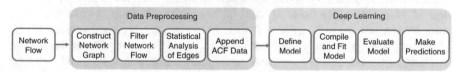

Fig. 5 Overview of the proposed botnet detection mechanism

implementation consists of two essential phases. The first is the data processing phase where the network flow records are rearranged and filtered so that only essential information is left behind for the second deep learning phase. During the first phase, a network graph which consists of nodes and edges is created to show the interconnections between hosts. The final output of the first phase is an array that stores the communication statistics of each edge in the graph, autocorrelation, as well as the label. The last phase is the deep learning stage where a deep learning model is defined, compiled, and fitted to be able to predict botnet activity in a network.

5.1   Data Preprocessing Phase

Infected botnets need to regularly connect with C&C servers to provide status updates and receive new orders. Due to this unique characteristic, the communication log between the botnets and the servers inevitably exhibits periodic patterns that can be used to signal signs of botnet activity. For instance, if there is a host in a local network connecting to an outer network host every n seconds, then this may be evidence that the local network host is an infected botnet that is sending out heartbeat signals to the C&C server. To detect signs of periodicity more efficiently, the original network flow dataset will be filtered to remove excess data.

5.1.1   Filtering Network Flow

Table 3 shows an example of an unaltered CTU-13 network flow record. Initially, there are 15 features that help describe each network transaction between a source and a destination. The explanation for each feature is described in Table 2. Among the 15 features, only 10 are of primary importance, namely, StartTime, Dur, Proto, SrcAddr, DstAddr, State, TotPkts, TotBytes, SrcBytes, and Label. According to previous research on botnets [24], the most frequent protocols used between a botnet and its C&C server are TCP, UDP, HTTP, and ICMP. The communication states that are important for these protocols are CON, URP, and FSPA_FPSA. The state CON indicates Connected in UDP, URP as Urgent Pointer in UDP and FSPA encompassing all flags (FIN, SYN, PUSH, ACK) in TCP. From the original CTU-13 dataset, the rows without Proto as UDP, TCP, HTTP, or ICMP will be filtered out, and, of the remaining rows, only those with connection state CON, URP, or FSPA_FPSA will be kept. This process of removing irrelevant transactions will significantly increase overall compute accuracy and efficiency as well as reduce computational costs.

In this paper, all 13 network flows from the CTU-13 dataset have been used as input, and they all followed the filtering process described in this Section. Table 4 shows an example of the filtered output of the loaded CTU-13 dataset.

5.1.2   Constructing Network Graph

The network graph, also known as network diagram, is useful for understanding the network's physical and logical connection status. It enables viewers to have a visual representation of the network to gain an overall picture of network topology and data flow. By reassembling the result of the previous section's rows of network log into a network graph, the networking nodes and executed transactions will be easily viewable for analysis. A proper network graph would follow a similar topology shown in Fig. 6, where nodes would represent host machines and edges would preserve records of all transactions between two nodes.

Table 3 Example of original CTU-13 network flow records

Protocol	SrcAddr	Dir	DstAddr	State	TotPkts	TotBytes	SrcBytes	Label
udp	147.32.84.165	<->	147.32.80.9	CON	2	203	64	From-Botnet-V46-UDP-DNS
tcp	1.114.187.73	->	147.32.84.229	FSPAC_FSPA	22	1772	951	Background-TCP-Established
udp	147.32.84.13	<->	82.208.56.89	CON	96	8640	4320	To-Normal-V46-UDP-NTP-server

Table 4 Example of a filtered network flow dataset

Proto	SrcAddr	DstAddr	State	TotPkts	TotBytes	SrcBytes	Label
udp	151.51.231.119	147.32.86.44	CON	32,055	8,577,197	8,575,244	Background-UDP-Established
udp	147.32.84.229	66.56.30.27	CON	2	567	95	Background-UDP-Established
udp	147.32.84.229	187.126.122.175	CON	2	567	95	Background-UDP-Established
udp	147.32.84.229	81.234.199.128	CON	2	567	95	Background-UDP-Established
udp	86.52.158.60	147.32.84.229	CON	13	999	422	Background-UDP-Established
⋯	⋯	⋯	⋯	⋯	⋯	⋯	⋯

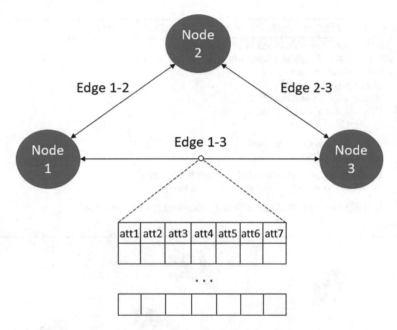

Fig. 6 Basic network graph architecture that consist of nodes and edges

Algorithm 1 describes the pseudo-code for constructing a network graph from a network flow dataset. In essence, the algorithm iterates through an array of network flow records and creates an non-repetitive node in the network graph by looking at the source and destination of the record. Once the nodes exist, the transaction information described in the row is also recorded in the edge between the two nodes on the network graph. The algorithm terminates when all rows of the array have been processed and returns the completed network graph.

Figure 7 is a visualized network graph that clearly shows the connection between each communicating node. Even though this graph was constructed with only a small subset of the filtered CTU-13 dataset to reduce visualization complexity, it still resembles the network graph that would result when the filtered CTU-13 dataset were fully used. The features that are stored in the edges would consist of protocol, duration, total bytes, total packets, state, timestamp, and label.

5.1.3 Statistical Analysis of Edges

The procedures prior to this section has been to remove irrelevant information from the original dataset and rearrange existing information into a graph structure to efficiently analyze data. In this section, a new array is created to store the statistics computed from information recorded in the edges. To remove unsubstantial transactions, only the edges that have more than four rows will be used. Each row

---

Algorithm 1: Pseudo-code for constructing network graph

Input: Network flow data stored in an array (Array)
Output: Network Graph (G) consisting of nodes and edges
1 **for** *each row in Array* **do**
2 | Node1 = row.SourceIP
3 | Node2 = row.DestinationIP
4 | **if** *Node1 not in Graph G* **then**
5 | └ add Node1 to G

6 | **if** *Node2 not in Graph G* **then**
7 | └ add Node2 to G

8 | **if** *edge does not exist between Node1 and Node2* **then**
9 | └ create a new edge between Node1 and Node2

10 └   append row's flow attributes to the edge between Node1 and Node2

11 **return** G

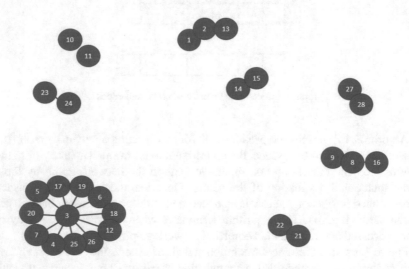

Fig. 7 Example of a network graph plotted from the filtered CTU-13 dataset

of the new array corresponds to a comprehensive summary of transactions between two nodes. Every row consists of twenty six columns: Duration (6), Total Bytes (6), Total Packets (6), Timestamp (6), ACF (1), and Label (1). In particular, the columns that belong to duration, total bytes, total packets, and timestamp are each filled with six statistics, namely, mean, median, standard deviation, minimum, maximum, and range. These numbers are calculated from the information stored at an edge.

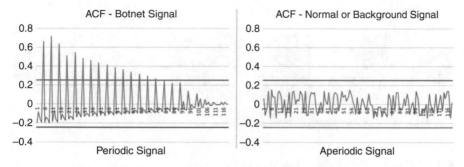

Fig. 8 ACF plot for periodic and aperiodic signals

5.1.4 Autocorrelation Analysis

A key goal of this project is to leverage the periodic communication behavior of the C&C network to detect botnet activities. To achieve this goal, the autocorrelation function (ACF) is used to calculate the periodicity of transactions. To explain with an example, Fig. 8 shows the autocorrelation plot for both periodic and aperiodic signals. In the periodic signal, the autocorrelation value peaked above the upper bound of the confidence interval 15 times. The autocorrelation plot for the aperiodic signal did not see any peaks that passed the confidence interval. In this scenario, the input used for classification would be 15 for periodic signals and 0 for aperiodic signals. A relatively high value of count indicates a strong periodic signal, which would also imply occurrence of cyclic botnet activities.

The final output of the data preprocessing phase should be a single array N which consists of 25 features (X) and a label (y) for each row. To train a deep learning model for classification, this array N will be horizontally split (column-wise) into two arrays: features $N[0:25]$ and labels ($N[25:]$).

5.2 Deep Learning Phase

In machine learning, an input dataset should initially be divided into two categories, that is, training and testing. This separation procedure is important to prevent the model from overfitting while accurately evaluating the model performance [28]. However, before randomly sampling the data into two datasets, a critical character-istic of the original CTU-13 dataset needs to be considered. In fact, the CTU-13 dataset has a highly imbalanced botnet to non-botnet network traffic ratio. As imbalance classification may lead to a biased and misleading deep learning model, using random sampling to divide the dataset is not considered to be an appropriate technique [14].

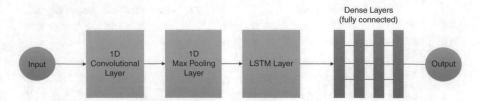

Fig. 9 Overview of deep neural network architecture

5.2.1 Stratified K-Fold Cross Validation

Among various sampling methods, the Stratified K-fold technique [14] performs well with imbalanced datasets. When the sampler divides the dataset into training and testing sets, the percentage of labels that constitutes the original dataset is maintained. For instance, if the botnet to non-botnet label in the original dataset is 1 : 20, both training and testing sets would also keep the same ratio after being split. This ensures botnet traffic to be properly represented while the model is being trained and tested with minimum sampling error or bias.

Both testing and training sets compete to have the maximum number of samples to achieve best learning results. The best validation result would come from having more samples in the test set. However, this inherently triggers a trade-off of having less items in the training set. A solution to this dilemma is to use cross validation technique [15]. In K-fold cross validation, a dataset is partitioned into K sets of equal size and K separate learning experiments are executed. For each learning experiment, a non-repeating partition is selected as a test set while the remaining $K - 1$ partitions are used as train sets. Once all K learning experiments are complete, the performances and test results are averaged. For this project, $K(3)$-fold cross validation, supported by the scikit learn API, was used for a more accurate assessment of the learning model.

5.2.2 Define, Compile, and Fit the Neural Network

The main components of the neural network are the convolution layer, max pooling layer, LSTM layer, and fully-connected layers. In the first convolution layer, a convolution kernel takes a training dataset and extracts hidden features and establishes a relationship between the input dataset and generated features. After the one dimensional convolution layer, max pooling is applied to reduce dimensions in the data to reduce computation overhead. LSTM layer takes the output from the max pooling layer and enables sequential connection among the dataset before feeding it to the dense layers. Finally, the output layer performs classification prediction to label botnet traffic. An overview of the proposed deep network is illustrated in Fig. 9.

5.2.3 Model Evaluation

To evaluate the proposed model, metrics like overall accuracy, precision, recall, and F-measure will be considered in conjunction with four additional performance metrics: true positive (TP), true negative (TN), false positive (FP), and false negative (FN).

In this paper, the botnet traffic will be considered as a positive label and non-botnet traffic will have a negative label. The overall accuracy refers to the number of correctly predicted labels over the total number of samples. Precision is the proportion of true positives over the sum of all positive labels. Recall is the proportion of true positives in the number of all the correctly labeled samples. F1 measure is the weighted mean of precision and recall, with its values ranging from zero to one.

6 Results

The proposed approach was implemented using deep learning models in the Keras Python library with TensorFlow deep learning engine. The implementation involved a combination of $1D$ convolutional network, max pooling, LSTM, and fully-connected layers. A more detailed configuration of these models is provided in Table 6. The order and combination of these layers was selected through multiple rounds of testing various configurations to optimize the performance of the classification result.

With the layers setting in Table 6, the deep neural network repeatedly trained and validated CTU13 input data separated by the type of malware. A total of seven different types of malware, each unique with its own communication pattern, were tested. The performance metrics of each malware type is shown in Table 5. According to the result, the proposed model performed best to accurately classify botnet traffic with the Rbot and Murlo malware type, both achieving over 0.9985 accuracy for the test dataset. While the performance results of different malware types have low variance, the reason for Rbot and Murlo having high detection

Table 5 Classification result of the deep neural network implementation

Malware type	CTU13 scenario #	Accuracy—train	Accuracy—test	Precision	Recall	F1-score
Neris	1, 2, 9	1	0.9976	0.985	0.986	0.9855
Rbot	3, 4, 10, 11	1	0.9985	0.999	0.988	0.9935
Virut	5, 13	1	0.9974	0.982	0.981	0.9815
Menti	6	0.99	0.9747	0.986	0.985	0.9855
Sogou	7	0.98	0.9981	0.982	0.987	0.9845
Murlo	8	0.98	0.9987	0.998	0.981	0.9894
NSIS.ay	12	0.99	0.9905	0.997	0.985	0.9910

Table 6 Classification result of the deep neural network implementation

Layer (type)	Output Shape	Param #
conv1d_1 (Conv1D)	(None, 19,64)	256
max_pooling1d_1 (MaxPooling1)	(None, 9,64)	0
lstm_1 (LSTM)	(None, 32)	12,416
dense_4 (Dense)	(None, 64)	2112
dense_5 (Dense)	(None, 64)	4160
dense_6 (Dense)	(None, 64)	4160
dense_7 (Dense)	(None, 1)	65

Table 7 Comparison of performance metrics with respect to other studies

Research paper	Method	Features	Dataset	Performance metrics
Torres et al. [25]	Recurrent neural network	Size, duration, periodicity	CTU-13	Accuracy: 0.970
				False positive rate: 0.0372
Wang et al. [27]	Social communication detection	Network flow-based	CTU-13	Recall: 0.026
				Precision: 0.80
				Fl-score: 0.088
Chen et al. [20]	Decision tree	Network flow-based	CTU-13	Accuracy: 0.936
				False positive rate: 0.3
Nagarajan [19]	Periodicity in network flow	Periodicity in pcap data	CTU-13	Fl-score: 0.09
Vishwakarma [26]	Data balancing and machine learning techniques	Network flow-based	CTU-13	Accuracy: 0.98
This paper 2021	Pattern-based network flow feature extraction	Statistical network flow-based	CTU-13	Accuracy: 0.9936
				Precision: 0.9898
				Recall:0.9847
				Fl-score: 0.9872

accuracy may be due to availability of a larger input data. The trend in Table 5 shows that the malware type with a larger number of network flows achieved relatively higher performance results. This implies that, with the increase of the number of network flows, the model will perform even better.

The performance results of this paper are compared to those of relevant research papers in Table 7. Using a pattern-based approach and analyzing the features statistically, this paper recorded 0.9936 accuracy, 0.9898 precision, 0.9847 recall, and 0.9872 F1-score. Compared to the results taken from [19, 20, 25, 27], and [26], the proposed approach achieved overall high performance. This comparison result shows that a pattern-based approach enables high detection accuracy while maintaining low false positive rate.

7 Conclusions

In this paper, a novel botnet detection approach is proposed using a pattern-based classification technique. The approach begins by filtering the input network flow to focus on traffic that uses TCP, UDP, and ICMP protocols. Information presented in the filtered network flow is rearranged to enable an intuitive understanding of the network traffic. By leveraging the network graph, features like duration, total bytes exchanged, total packets, timestamp and autocorrelation count were extracted. This approach can be used for all types of botnet architectures and does not require any prior knowledge about the botnet type or C&C server IP address. The proposed approach has been tested with network flow datasets that consist of botnet, normal, and background traffic, to show that detecting botnet traffic in real-life scenarios is possible. A deep neural network was designed to process the statistical features that have been extracted from the network graph. Using the CNN architecture, a classifier for botnet traffic has been created and the statistical features were fed to the model for training and testing. The performance results were compared to the metrics found in relevant research papers to confirm that the proposed approach outperformed those of previous works. The presented method is applicable to various types of botnet families to identify malicious actors in a real-life network environment with high accuracy. For future work, a realistic networked environment can be recreated to simulate real-life implementation. In this case, we expect a considerable increase in background noise, however, we believe that the technique proposed in this paper can still obtain promising results. Other machine learning techniques can be tested relying on our selected features. For instance, Hidden Markov models (HMM) [23], profile hidden Markov models [5], and support vector machines [1] can obtain interesting results in this particular scenario.

References

1. Robert Berwick. An idiots guide to support vector machines (svms). http://web.mit.edu/6.034/wwwbob/svm.pdf, 2003. [Online; accessed August 2021].
2. Krzysztof Cabaj, Marcin Gregorczyk, Wojciech Mazurczyk, Piotr Nowakowski, and Piotr undefinedórawski. Sniffing detection within the network: Revisiting existing and proposing novel approaches. In *Proceedings of the 14th International Conference on Availability, Reliability and Security*, ARES '19, New York, NY, USA, 2019. Association for Computing Machinery.

3. Anusha Damodaran, Fabio Di Troia, Corrado Visaggio, Thomas Austin, and Mark Stamp. A comparison of static, dynamic, and hybrid analysis for malware detection. *J Comput Virol Hack Tech*, 13:1–12, 2017.
4. NHS Digital. Virut botnet. https://digital.nhs.uk/cyber-alerts/2018/cc-2829, 2020.
5. Richard Durbin, Sean Eddy, Anders Krogh, and Graeme Mitchison. Biological sequence analysis: probabilistic models of proteins and nucleic acids, 1998.
6. S. García, M. Grill, J. Stiborek, and A. Zunino. An empirical comparison of botnet detection methods. *Comput. Secur.*, 45:100–123, September 2014.
7. Ibrahim Ghafir and Vaclav Prenosil. Blacklist-based malicious ip traffic detection. In *2015 Global Conference on Communication Technologies (GCCT)*, pages 229–233, 2015.
8. Ibrahim Ghafir, Vaclav Prenosil, Mohammad Hammoudeh, Thar Baker, Sohail Jabbar, Shehzad Khalid, and Sardar Jaf. Botdet: A system for real time botnet command and control traffic detection. *IEEE Access*, 6:38947–38958, 2018.
9. Marcin Gregorczyk, Piotr Żórawski, Piotr Nowakowski, Krzysztof Cabaj, and Wojciech Mazurczyk. Sniffing detection based on network traffic probing and machine learning. *IEEE Access*, 8:149255–149269, 2020.
10. Nabil Hachem, Yosra Ben Mustapha, Gustavo Gonzalez Granadillo, and Herve Debar. Botnets: Lifecycle and taxonomy.
11. Box-Steffensmeier Janet, Freeman John, Hitt Matthew, and Pevehouse Jon. *Time Series Analysis for the Social Sciences*. Cambridge University Press, New York, 2014.
12. G. Kirubavathi and R. Anitha. Botnet detection via mining of traffic flow characteristics. *Computers & Electrical Engineering*, 50:91–101, 2016.
13. Stratosphere Lab. The CTU-13 Dataset. https://www.stratosphereips.org/datasets-ctu13/. [Online; accessed August 2021].
14. Victoria López, Alberto Fernández, and Francisco Herrera. On the importance of the validation technique for classification with imbalanced datasets: Addressing covariate shift when data is skewed. *Inf. Sci.*, 257:1–13, February 2014.
15. M. Lorbach, E.I. Kyriakou, R. Poppe, E.A. van Dam, L.P.J.J. Noldus, and R.C. Veltkamp. Learning to recognize rat social behavior: Novel dataset and cross-dataset application. *Journal of neuroscience methods*, 300:166–172, 2018.
16. Pavan Roy Marupally and Vamsi Paruchuri. Comparative analysis and evaluation of botnet command and control models. In *2010 24th IEEE International Conference on Advanced Information Networking and Applications*, pages 82–89, 2010.
17. H. S. Nair and V. Ewards. A study on botnet detection techniques. *Mathematical Problems in Engineering*, 2, 2012.
18. Emmanuel C. Ogu, Olusegun A. Ojesanmi, Oludele Awodele, and 'Shade Kuyoro. A botnets circumspection: The current threat landscape, and what we know so far. *Information*, 10(11), 2019.
19. Nagarajan Prathiba, Di Troia Fabio, Austin Thomas, and Stamp Mark. Autocorrelation analysis of financial botnet traffic. In *2nd International Workshop on Formal Methods for Security Engineering (ForSE 2018), in conjunction with the 4th International Conference on Information Systems Security and Privacy (ICISSP 2018)*, ICISSP 2018, 2018.
20. Chen Ruidong, Niu Weina, Zhang Xiaosong, Zhuo Zhongliu, and Lv Fengmao. An effective conversation-based botnet detection method. *Mathematical Problems in Engineering*, 2017:166–172, 2017.
21. Rami Sihwail, K. Omar, and K. A. Z. Ariffin. A survey on malware analysis techniques: Static, dynamic, hybrid and memory analysis. *International Journal on Advanced Science, Engineering and Information Technology*, 8:1662–1671, 2018.
22. SéRgio S. C. Silva, Rodrigo M. P. Silva, Raquel C. G. Pinto, and Ronaldo M. Salles. Botnets: A survey. *Comput. Netw.*, 57(2):378–403, February 2013.
23. Mark Stamp. A revealing introduction to hidden markov models. https://www.cs.sjsu.edu/~stamp/RUA/HMM.pdf, 2021. [Online; accessed August 2021].
24. Manoj Rameshchandra Thakur, Divye Raj Khilnani, Kushagra Gupta, Sandeep Jain, Vineet Agarwal, Suneeta Sane, Sugata Sanyal, and Prabhakar S. Dhekne. Detection and prevention of

botnets and malware in an enterprise network. *Int. J. Wire. Mob. Comput.*, 5(2):144–153, May 2012.

25. Pablo Torres, Carlos Catania, Sebastian Garcia, and Carlos Garcia Garino. An analysis of recurrent neural networks for botnet detection behavior. In *2016 IEEE Biennial Congress of Argentina (ARGENCON)*, pages 1–6, 2016.

26. Anand Ravindra Vishwakarma. Network traffic based botnet detection using machine learning, master's project, 2020.

27. Jing Wang and Ioannis Ch. Paschalidis. Botnet detection based on anomaly and community detection. *IEEE Transactions on Control of Network Systems*, 4(2):392–404, 2017.

28. Suleiman Y. Yerima, Mohammed K. Alzaylaee, Annette Shajan, and Vinod P. Deep learning techniques for android botnet detection. *Electronics*, 10(4), 2021.

29. Xing Ying, Shu Hui, Zhao Hao, Li Dannong, and Guo Li. Survey on botnet detection techniques: Classification, methods, and evaluation. *Mathematical Problems in Engineering*, 2021, 2021.

Interpretability of Machine Learning-Based Results of Malware Detection Using a Set of Rules

Jan Dolejš and Martin Jureček

Abstract Machine learning plays an indispensable role in modern malware detection; it provides malware researchers with quick and reliable results. On the other hand, the results can be hard to understand as to why a model classified a given file as malicious or benign. This paper focuses on the interpretability of machine learning models' results using decision lists generated by two rule-based classifiers, I-REP and RIPPER. We use the EMBER dataset, which contains features extracted through static analysis from Portable Executable files, to train various machine learning models. We extract decision lists from the machine learning models' results using our implementation of I-REP and RIPPER. By taking into account accuracies, true positive and false positive rates of the decision lists, we reason whether the generated decision lists make a good representation of the results. To comprehend the interpretability of the machine learning models, we define Human Most Understandable Model and Interpretability Entropy. This allows us to measure and compare the interpretability among the models. The most interpretable machine learning model by RIPPER was Gaussian Naïve Bayes. Results show that RIPPER is relatively successful at interpreting other machine learning models; however, it needs some improvements to increase true positive rate.

1 Introduction

Machine learning (ML) methods have been quite successful in various applications, such as face recognition, weather prediction, image reconstruction, and many more. Security experts use the methods for quick and reliable malware detection [10]. However, many methods, such as deep neural networks, are often considered black boxes as the reasoning behind their decisions may be unclear [13]. Understanding these decisions is essential; we may need to know why a person is considered at

J. Dolejš · M. Jureček (✉)
Faculty of Information Technology, Czech Technical University in Prague, Prague, Czechia
e-mail: dolejj13@fit.cvut.cz; martin.jurecek@fit.cvut.cz

© The Author(s), under exclusive license to Springer Nature Switzerland AG 2022
M. Stamp et al. (eds.), *Artificial Intelligence for Cybersecurity*, Advances in Information Security 54, https://doi.org/10.1007/978-3-030-97087-1_5

high risk of criminal activity [3] or why a benign file was classified as malicious. Understanding the decisions may not only be necessary for data scientists to understand their models better but may be required by some state's law or regulation. The General Data Protection Regulation (GDPR), introduced by the European Parliament in 2016 and has taken effect as law in 2018, makes understanding decision-making based on personal data necessary [14].

However, what does it mean to understand the results of machine learning models? Is it the degree to which we understand the data or the inner workings of the algorithm? As one may expect, the answer to this is not precisely clear. The current literature offers various approaches that may help decide what part of the machine learning or the data mining process should be more understandable. One could gain a better understanding of the machine learning models through interpretability [19], explainability [13], or transparency [24]. Interpretability can help one to understand the decision-making of a model. Explainability, which is often interchangeably used with interpretability, could offer an explanation of why the model made the decision or why the model should make the decision. Transparency mainly relies on the process of understandable data processing or algorithmic deployment. Our work focuses on the interpretability of the results of the machine learning models applied to malware detection. We may also use the term explainability, in which case we consider it equal to interpretability.

In [19], Miller uses the following definition of interpretability: *"the degree to which an observer can understand the cause of a decision"*. In a more machine learning-based context, Doshi-Velez and Kim [9] define interpretability as the *"ability to explain or to present in understandable terms to a human"*. The authors of [5] argue that both definitions could be seen as two different approaches: one that requires a priori interpretable models, and the other that would create explanations to the existing or the future black-box methods.

In our work, we implement two well-known rule-learning algorithms, I-REP and RIPPER and the structures necessary for the representation of a decision list. We discuss possible speed-ups of the RIPPER algorithm, and we incorporate them into our implementation. Some of the speed-ups were necessary, as we use the EMBER dataset, which contains hundreds of thousands of samples.

For a given machine learning model, we try to interpret its results using decision lists generated by the aforementioned rule-learning algorithms. We first discuss the successfulness of the rule-learning process of both algorithms by exploring the success rate of the algorithms (e.g., accuracy, true positive rate) and by taking into account the complexity of the built decision lists. We then discuss the interpretability of the machine learning results using the Interpretability Entropy (see Definition 14). and how much do the predictions of machine learning methods match with the generated decision lists.

Throughout the experiments, we try to understand better the RIPPER algorithm's performance by either changing its pruning metrics or its hyperparameters. We consider whether the order in which the rules were learned is strictly given or whether we can change the positions of the rules.

The rest of the paper is organized as follows: Sect. 2 presents works related to malware analysis. Some of the works make use of rule-learning algorithms (e.g., RIPPER) or approaches that try to either explain or interpret the reasoning behind the decisions of used machine learning methods. Section 3 introduces the theory necessary for rule-learning algorithms. Section 4 describes the specifics of our implementation of the rule-learning algorithms RIPPER and I-REP. Section 5 gives details on the used dataset and its split, transformation, and feature selection. Also, Sect. 5 introduces a metric that can be used to measure the interpretability by the decision lists. It contains the evaluation of the experiments, too.

2 Related Works

In this section, we provide an overview of works related to malware research, rule-learning algorithms, and interpretability.

The authors of [11] combined three different methods for malware detection: hash-based approach, Support Vector Machine-based approach, and rule-based approach. Each of the named methods is intended to be used for malware classes with different distributions. Using static analysis, they extracted n-grams based on the content of a Portable Executable file—n-grams are all substrings of a fixed length n [25]. The paper does not discuss the interpretability of the used model; however, it outlines one of its positive outcomes—reduction of space complexity. In their experiments, they reduced the storage cost from 1.8MB (signature-based approach) to 17.9KB (combined approach).

The work [26] compares the RIPPER algorithm (see Sect. 3.2) with other machine learning algorithms in malware detection. This is done merely on previously unseen samples. The paper does not clarify the number of iterations used for RIPPER. They used static features extracted from the Portable Executable files— used DLLs, DLL function calls, and the number of DLL function calls. The authors of this publication discuss how malware developers could use the information gathered by the classifiers to modify their malware. For example, by changing resource usage.

To explain the reasoning of their model, the authors of [4] created a tool, which not only classifies Android malicious files but also displays the features that contributed the most to the decision. The tool goes by the name DREBIN and uses features extracted through static analysis. These features are then mapped to a vector space, and Support Vector Machines are used for the classification. Features contributing to the classification can be derived from the vector space.

Inspired by deep learning and computer vision, the authors of [17] make use of convolutional neural networks and the Gradient-weighted Class Activation Mapping (Grad-CAM) [29] technique. This technique [29] uses gradient information that serves as an input for the final convolutional layer in the CNN. In [17], an APK file is first converted to a grayscale image representation and then used as an input for the deep learning models. Heatmaps are generated using Grad-CAM to explain the model results. Subsequently, heatmaps are averaged for distinct malware families; the authors refer to this as Cumulative Heatmaps. They can be used for malware analysts to gain more knowledge about the malware (by observing the areas of the code highlighted by the heatmaps), or they can be used to distinguish between better-performing models.

Similar steps towards interpretability in malware detection were taken in [7]. Portable Executable binaries are transformed to grayscale images, and deep transfer learning is employed for the classification task. The authors try to interpret the results of the models as follows. A binary file is first divided into super-pixels, contiguous regions. Then for each region, the coefficients are obtained. The positive coefficient values indicate that a region contributes to the classification decision, and the negative coefficient values indicate that a region does not contribute to the classification decision. The paper, however, does not further explain how malware analysts could use such information.

3 Rule-Based Classification

Several malware detection models based on machine learning techniques, such as neural networks, are considered a black box because it is difficult (for humans) to determine precisely why a given false positive or false negative occurs. Malware researchers prefer interpretable detection systems, such as rule-based methods, since they can be easily understood and better controlled. The goal is to improve the interpretability of the classification models. In this section, we describe the theoretical background for a rule-based system, specifically, decision rules. Some of the definitions provided in this section are later used in the algorithmic description (see Sect. 3.2), and they provide a high-point view of our implementation (see Sect. 4). Decision rules can be expressed as a set of *if-then* rules [20]. For example: *if* a file contains a suspicious function call, *then* mark this file malicious.

Definition 1 (Condition) A condition c is defined as follows,

$$c \equiv x \odot h, \tag{1}$$

where x is a feature, $\odot$ is a relational operator, and h is the value of the feature x.

Usually, the conditions are logically ANDed together, making it necessary for all tests to fire [20].

Definition 2 (Rule & Rule Size) A rule r is defined as follows,

$$r \equiv c_1 \wedge \cdots \wedge c_m, \tag{2}$$

where m is the number of conditions for the rule r. We say that the rule r has a size of m.

However, a conjunction of the conditions is not a necessity, and a single rule may be expressed by a general logical expression [30].

For a given rule, we are interested in its quality, more specifically in its coverage (support) and accuracy (confidence). We say that a rule *covers* a sample if the sample satisfies the rule's conditions [23].

Definition 3 (Rule Coverage) Given a set of samples S, the coverage of a rule r is defined as

$$coverage(r, S) = \{s \mid s \in S, r \ covers \ s\}. \tag{3}$$

The following definition allows to express the coverage of a rule numerically.

Definition 4 (Rule Coverage Size) Given a set of samples S, we define the coverage size of a rule r as

$$coverage\_size(r, S) = \frac{|coverage(r, S)|}{|S|}. \tag{4}$$

Rules are said to be mutually exclusive if no two rules cover the same sample.

Definition 5 (Mutually Exclusive Rules) Given a set of samples S, we say that a rule r_i and a rule r_j are mutually exclusive, if

$$coverage(r_i, S) \cap coverage(r_j, S) = \emptyset, i \neq j, i, j \in \{1, \ldots, n\}, \tag{5}$$

where n is the number of rules for S.

Definition 6 (Exhaustive Rules) Given a set of samples S, we say that rules r_i are exhaustive, if

$$\bigcup_{i=1}^{n} coverage(r_i, S) = S, i \in \{1, \ldots, n\}, \tag{6}$$

where n is the number of rules for S.

However, such rule restrictions are often not required, and we allow the rules to overlap and not cover the whole set. Different problems arise, some rules may contradict each other, or some of the samples may not be covered at all. Two different schemas can be used to solve this: a decision list or a decision set [20].

In a decision list, the rules are ordered as follows:

$$R = [r_1, r_2, \ldots, r_n], \tag{7}$$

where n is the number of rules for a given list. In other words, the rules are kept in the order in which they were added. The same order is later used for classification, too.

The decision set does not require the rules to be ordered; instead, all rules get to vote on classifying a given sample. Unfortunately, once the decision set grows very large, it becomes quite hard to understand. Thus, we will be using a decision list in this work if not stated otherwise.

Definition 7 (Decision List Coverage) Given a set of samples S, the coverage of the decision list R is defined as

$$coverage(R, S) = coverage(r_n, coverage(r_{n-1}, \dots, coverage(r_1, S) \dots)),$$
(8)

where n is the number of rules in R.

3.1 From Trees to Rules

In this section, we will briefly compare another popular machine learning tool—a decision tree. Decision trees are built from nodes, where each node, except the last ones, tests a feature with a given value (see Definition 1). The last node, also called a leaf, represents a decision, for example, classifying samples as benign or malicious [30]. Although the idea behind decision trees is quite simple, they may turn out to be quite complex and hard to interpret [6].

Figure 1 illustrates a simple decision tree. However, its outcome may be a little misleading as it is a simple disjunction, which can be easily described using rules:

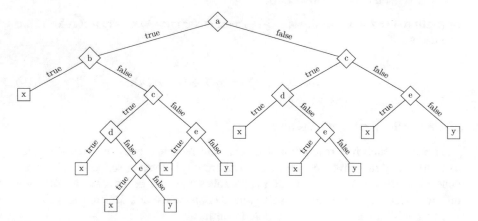

Fig. 1 Decision tree—describing a simple disjunction

$$if\ a \wedge b\ then\ x$$
$$else\ if\ c \wedge d\ then\ x$$
$$else\ if\ e\ then\ x \tag{9}$$
$$else\ y$$

Quinlan [23] designed an algorithm called C4.5rules, which converts a decision tree to a decision list. After its construction, it tries to improve it. Unfortunately, this part of the algorithm is expensive. Cohen [8] showed that the complexity is near $\mathcal{O}(n^3)$, where n is the number of samples.

3.2 Rule-Learning Algorithms

In this section, we briefly discuss one branch of rule-learning algorithms—separate-and-conquer. Unlike the divide-and-conquer technique, separate-and-conquer algorithms first focus on the part of the training set and then try to describe it. In contrast, divide-and-conquer strives to maximize the separation between classes [30].

Incremental Reduced Error Pruning (I-REP) [12] is an algorithm designed by Fürnkranz and Widmer in 1994. It implements two pruning approaches to deal with noisy data: pre-pruning and post-pruning. Pre-pruning ignores some of the training samples in the learning process so that the final decision list would not describe the training set perfectly. Post-pruning corresponds to removing a condition in a given rule. The following metric drives I-REP's pruning,

$$\mathscr{P}_{\text{I-REP}}(p, P, n, N) = \frac{p + (N - n)}{P + N}, \tag{10}$$

where p (n) is the number of positive (negative) samples covered by the current rule from a total number of P (N) positive (negative) samples in the pruning set. The algorithm is described in Algorithm 1.

Thanks to its efficiency, I-REP is well-suited for large training sets. However, in 1995, Cohen showed that I-REP does not learn rules well enough and can be outperformed by previously known algorithms, such as C4.5rules [8]. Cohen has addressed specific issues and explained how I-REP could be improved. With these improvements, Cohen designed a new algorithm called Repeated Incremental Pruning to Produce Error Reduction (RIPPER).

Cohen's team made three modifications—they replaced I-REP's pruning metric, chose a different approach to stop the rule-learning process, and added decision lists optimizations. The following metric replaced I-REP's pruning metric,

$$\mathscr{P}_{\text{RIPPER}}(p, P, n, N) = \frac{p - n}{p + n}. \tag{11}$$

Algorithm 1: I-REP

Input: Positive samples – *Pos*, Negative samples – *Neg*, *SplitRatio*
Output: Decision list *R*
1 $R \leftarrow \{\}$;
2 **while** $Pos \neq \emptyset$ **do**
3 | *SplitExamples*(*SplitRatio*, *Pos*, *PosGrow*, *PosPrune*);
4 | *SplitExamples*(*SplitRatio*, *Neg*, *NegGrow*, *NegPrune*);
5 | $r \leftarrow empty\ rule$;
6 | **while** $NegGrow \neq \emptyset$ **do**
7 | | $r \leftarrow r \wedge FindLiteral(r, PosGrow, NegGrow)$;
8 | | $PosGrow \leftarrow coverage(r, PosGrow)$;
9 | | $NegGrow \leftarrow coverage(r, NegGrow)$;
10 | **end**
11 | $r \leftarrow PruneRule(r, PosPrune, NegPrune)$;
12 | **if** $Accuracy(r) \leq Accuracy(fail)$ **then**
13 | | return *R*;
14 | **end**
15 | **else**
16 | | $Pos \leftarrow Pos \backslash coverage(r, Pos)$;
17 | | $Neg \leftarrow Neg \backslash coverage(r, Neg)$;
18 | | $R \leftarrow R \cup r$;
19 | **end**
20 **end**
21 return *R*;

The following definitions are necessary to understand when RIPPER's rule learning is stopped.

Definition 8 (Rule Description Length) Given the positive real numbers n, k and $p \neq 1$, we define the rule description length as follows,

$$\mathscr{S}(n, k, p) = \frac{1}{2}\left(k \log_2 \frac{1}{p} + (n - k) \log_2 \frac{1}{1 - p} + \log_2 k\right), \quad (12)$$

As described by Cohen [8], this encoding allows two parties (sender and recipient) to work over a set of n elements. The recipient can recognize k elements, and p is known ahead. $\log_2 k$ is the number of bits required to send the number k. The whole metric is scaled by $\frac{1}{2}$ to limit possible redundancy in the features.

Definition 9 (Decision List Exceptions) For a given set of samples S with a positive class P and a negative class N, and for a given decision list R, we define the number of exceptions as follows,

$$\mathscr{E}(R, S) = \log_2 \binom{TP + FP}{FP} + \log_2 \binom{TN + FN}{FN}, \quad (13)$$

where TP (TN) is the number of samples correctly classified as P (N), and FP (FN) is the number of samples incorrectly classified as P (N).

Definition 10 (Total Description Length) For a given set of samples S and a decision list R we define its total description length as follows,

$$\mathscr{T}(R, S) = \sum_{\substack{r \in R \\ in\ order}} \mathscr{S}(n, k_r, \frac{k_r}{n}) + \mathscr{E}(R, S), \tag{14}$$

where n is the total number of possible conditions for S and k_r is rule r's length.

Let minimum description length (MDL) be the current total description length (TDL) of a given decision list. Rule-learning stops if adding a new rule should increase MDL by more than 64 bits. Since Cohen described the RIPPER algorithm mostly with words, we include its pseudocode in Sect. 4 as it may not precisely correspond to the original implementation.

4   Implementation of Rule-Based Classifiers

To efficiently generate decision lists using well-known algorithms mentioned in Sect. 3, we created our implementations of rule-based classifiers (RBCs) in C++. Although some implementations of the algorithms exist, such as `Weka` [15] or `wittgenstein` [21], they are not quick enough to process large amounts of data. We did not want to lose the ability of most of the machine learning tools— quick and easy deployment. Thus, we added `Python` support to our library using `pybind11` [18]. The code has been made publicly available on Github.[1] We further discuss some of the implementation details below.

4.1   Decision List

We implemented basic structures that correspond to the definitions in Sect. 3. Namely, those are the condition (see Definition 1), the rule (see Definition 2), and the decision list, often referred to as the ruleset. At this moment, the available operators for the condition are $\{<=, >=\}$. Both operators are intended to be used for numerical features only.

---

[1] https://github.com/ai-honzik/RuBaC.

4.2   I-REP

As the original paper for I-REP [12] does not cover dealing with numerical features, we used Cohen's [8] suggestions for the algorithm. During the growth phase, the algorithm searches for the best split between the numerical features. Rule growth for both IREP and RIPPER is guided by maximizing FOIL's gain and stops once no negative samples are left in the growing set. Neither of the papers mentioned above tackles the issue of learning the same rule twice. This may happen if the present feature values are the same for both positive and negative growing sets. Our implementation stops the growing phase and proceeds to the next step.

Algorithm 2: I-REP*

Input: Positive samples – Pos, Negative samples – Neg, $Split Ratio$, Largest bit difference
      $- d$, Decision list – R
Output: Decision list R

1 $MDL \leftarrow +\infty$;
2 **while** $Pos \neq \emptyset$ **do**
3     $Split Examples(Split Ratio, Pos, PosGrow, Pos Prune)$;
4     $Split Examples(Split Ratio, Neg, NegGrow, Neg Prune)$;
5     $r \leftarrow GrowRule(PosGrow, NegGrow)$;
6     $r \leftarrow PruneRule(r, Pos Prune, Neg Prune)$;
7     $T DL \leftarrow total\_description\_length(R)$;
8   **if** $T DL < MDL$ **then**
9 | $MDL \leftarrow T DL$
10   **end**
11   **else if** $T DL - MDL > d$ **then**
12 | break
13   **end**
14    $Pos \leftarrow Pos \setminus coverage(r, Pos)$;
15    $Neg \leftarrow Neg \setminus coverage(r, Neg)$;
16    $R \leftarrow R \cup r$;
17 **end**
18 **for** $r_i, i \in \{|R|, \ldots, 1\}$ **do**
19    $MDL \leftarrow total\_description\_length(R)$;
20    $T DL \leftarrow total\_description\_length(R \setminus r_i)$;
21   **if** $T DL < MDL$ **then**
22 | $R \leftarrow R \setminus r_i$;
23   **end**
24 **end**
25 **return** R;

4.3 RIPPER

RIPPER increased the computational complexity with its improvements. The learning process will stop if MDL increases by more than 64 bits. The calculation complexity of TDL mainly lies in the calculation of exception bits. Naively, we could calculate TDL each time; fortunately, we can use memorization to speed up some parts of the calculations.

Rule description lengths can be cached. We only need to calculate the description length of one rule each time throughout the iterations in I-REP*. We can do similar steps for the exception bits. In I-REP*, we only need to remember the remaining samples (samples that were not covered by any rule). We have to do more steps in the optimization phase as rules depend on the previous ones. We need to compare the coverage of the new ruleset with the old ruleset—the new ruleset is the ruleset by which the previous one was replaced, either replacement ruleset or revision ruleset. Let r_n be a new rule, R_n a new decision list, r_o the original rule, R_o the original decision list, and S remaining samples that were not covered by any previous rule. We need to check two cases—increase or decrease of falsely, resp., correctly covered cases.

Algorithm 3: RIPPER

Input: Positive samples – Pos, Negative samples – Neg, $SplitRatio$, Largest bit difference
    – d, Number of iterations – k
Output: Decision list R

1 $R \leftarrow I\text{-}REP * (Pos, Neg, SplitRatio, d, \{\})$;
2 **for** $1 \ldots k$ **do**
3 | $RPos \leftarrow Pos$;   // remaining positive samples
4 | $RNeg \leftarrow Neg$;   // remaining negative samples
5 | **for** $r \in R$ **do**
6 | | $SplitExamples(SplitRatio, RPos, PosGrow, PosPrune)$;
7 | | $SplitExamples(SplitRatio, RNeg, NegGrow, NegPrune)$;
8 | | $rev \leftarrow GrowRule(PosGrow, NegGrow)$;
9 | | $rev \leftarrow PruneRule(rev, PosPrune, NegPrune)$;
10 | | $Revise \leftarrow (R \setminus r) \cup rev$;
11 | | $rep \leftarrow GrowRule(PosGrow, NegGrow, r)$;
12 | | $rep \leftarrow PruneRule(rep, PosPrune, NegPrune)$;
13 | | $Replace \leftarrow (R \setminus r) \cup rep$;
14 | | $DLs \leftarrow \{R, Revise, Replace\}$;   // DLs ...Decision lists
15 | | $R \leftarrow argmin(DLs)$;   // minimise description length
16 | | update r ;   // based on the decision list R
17 | | $RPos \leftarrow RPos \setminus coverage(r, RPos)$;
18 | | $RNeg \leftarrow RNeg \setminus coverage(r, RNeg)$;
19 | **end**
20 | $R \leftarrow I\text{-}REP * (RPos, RNeg, SplitRatio, d, R)$;
21 **end**
22 **return** R;

$$S_n = coverage(R_n, coverage(r_o, S) \setminus coverage(r_n, S)) \tag{15}$$

$$S_o = coverage(R_o, coverage(r_n, S) \setminus coverage(r_o, S)) \tag{16}$$

By calculating the sizes of S_n and S_o we do not have to recalculate TDL. We only need to either increase or decrease the number of falsely, resp., correctly covered cases.

The last part of our speed up lies in using Stirling's approximation of the $n!$. For our case, we can derive the following formula,

$$\log_2(n!) \sim (n + \frac{1}{2}) \log_2(n) - n + \frac{1}{2} \log_2(2\pi). \tag{17}$$

Thus, the exception bits with known input arguments can be calculated in $\mathcal{O}(1)$. We include RIPPER's pseudocode, divided into IREP* (see Algorithm 2) and RIPPER's optimization phase (see Algorithm 3).

5   Experiments

This section discusses the used dataset and data split we used throughout the experiments. We briefly summarize the data preprocessing, too. Furthermore, we discuss the interpretability of ML models by using decision lists. For our experiments, we define what it means for a model to be absolutely or partially interpretable by a decision list and when a model should be considered interpretable. Throughout the experiments, we examine closer the behavior of the RIPPER algorithm.

All of the experiments were run on a single computer platform with two processors (Intel Xeon Gold 6136 CPU @ 3.00GHz), with 755 GB of RAM running the Ubuntu 20.04 LTS operating system.

5.1   Dataset Description

For our experiments, we used the publicly available dataset called EMBER (Elastic Malware Benchmark for Empowering Researchers) [1]. More specifically, we employed the most up-to-date version from 2018. The authors of the dataset dealt with three significant challenges—legal (releasing binaries of monetized software), labeling (potentially requires expert knowledge), and security (releasing malicious binaries is not safe) aspects. Thus, using the static analysis, features were extracted and incorporated into the dataset. This tackles two of the challenges mentioned above; labeling was achieved by using services such as VirusTotal.

The EMBER dataset consists of 1.1M samples, divided into a training set with 900K samples (300K malicious, 300K benign, 300K unlabeled) and a test set with 200k samples (100K malicious, 100k benign). The newer version of the dataset contains the label `avclass` [28] for malicious samples. This label indicates to which malware family a given malicious sample belongs. Throughout the experiments, we ignored the unlabeled samples.

The dataset is stored using the JSON file format, and for each sample, eight groups of raw features are present. General file information—includes information about the virtual size of the file, presence of a debug section, number of symbols, and more. Header information—here, one can find information extracted from the Common Object File Format (COFF) header, for example, file characteristics, machine type, or information from the Optional header. Imported and exported functions—both raw sections include the names of the imported or exported functions, for example `"SHLWAPI.dll":["PathIsUNCW"]`. Section information—contains information about each of the present sections, e.g., their name, size, entropy, and more. The following three sections are independent of the PE file format. Byte and byte-entropy histograms—both sections consist of 256 integer values each, indicating either the number of occurrences or the entropy for each byte. The string section includes information about printable strings, such as their average length, histograms, and more.

5.2   Data Splitting

We merged the train and test sets for our experiments, both predetermined in the EMBER dataset. Furthermore, we created three disjunctive sets as follows: a training set (consists of 40% samples), *first test set* (consists of 40% samples), and *second test set* (consists of 20% samples). The training set will be used to train various machine learning models. *The first test set* will be used to measure the success rate of the models. Moreover, it will be used to generate predictions of the machine learning models, later used to train rule-based classifiers. *The second test set* will be used to measure how well RBCs interpret the ML models, that is, how well are predictions of the models matched with RBCs' predictions.

During the experiments, we used 5-fold cross-validation. First, we partitioned the data into individual folds with the corresponding set sizes. That is, each fold had the corresponding sizes for each of the sets mentioned above—40:40:20. After partitioning, we applied data transformation techniques (e.g., normalization or PCA). The following steps were all done individually for all five folds. We trained each ML model on the training set and evaluated its performance on the first test set (see Table 1) and then on the second test set (see Table 3). RBCs were trained on the predictions of each ML model on the first test set. The performance of RBCs was evaluated on the first test set (see Table 2) and the second test set (see Table 4). Figure 2 describes the data split visually and it also explains the working process used in our experiments.

Table 1 Well-known ML algorithms and their performance on the EMBER dataset using *the first test set*

ML algorithm	RandomForest			PCA		
	ACC	TPR	FPR	ACC	TPR	FPR
DNN	**95.67**	**95.86**	4.52	**96.42**	96.18	3.35
GaussianNaïveBayes	74.67	81.00	31.66	72.45	96.09	51.16
I-REP	84.05	94.61	26.49	78.27	**96.68**	40.12
KNN	95.17	95.32	4.98	95.40	95.16	4.36
RIPPER	88.68	79.06	**1.72**	89.88	82.68	**2.93**
RandomForest	95.56	95.42	4.30	95.38	94.46	3.71
SVM	94.92	94.64	4.80	95.86	95.96	4.23

We highlighted the highest accuracy, the highest true positive rate, and the lowest false positive rate across all the models

5.3 Feature Transformation and Selection

Even though rule-based classifiers can handle both numerical and categorical features, traditional machine learning algorithms and the implementations available from scikit-learn [22], which we used to train our models, require the features to be numerical only. The authors of the EMBER dataset published a code [2] that transforms some of the available raw features into vectorized ones using the hashing trick. We used this code to transform the features and ended up with 2381 new ones.

Before proceeding further, it was necessary to standardize the data. Some machine learning algorithms' behavior may worsen if the data do not appear to be from the normal distribution [27]. We used the class MinMaxScaler from scikit-learn that transforms the features as follows,

$$x_{std} = \frac{x - \min(X)}{\max(X) - \min(X)}, \tag{18}$$

where x is the original value and X is the collection of every value in a given feature.

Consequently, we employed two dimensionality reduction methods—Principal Component Analysis (PCA) and Random Forest (RF), both available in scikit-learn. We picked both techniques as they are simple to use and arguably easy to understand. Even though one cannot simply see the original features with PCA, we can use the correlation matrix to determine which features were used to create the new ones. Unlike PCA, Random Forest keeps original features, thus maintaining higher interpretability. We chose to keep 200 features for PCA as it had less than 4% information loss. We decided to keep the same number of features using RF, too. By choosing this value for RF, we aim to get a different behavior of the models. Note that this may later put RF at a disadvantage as it may use redundant features.

Table 2 Measuring performance of RBCs on ML predictions using the *first test set*

		RBC	ACC	TPR	FPR	DL size	ø r size
PCA	DNN	RIPPER$_0$	91.72	84.55	1.18	287.00	10.03
		RIPPER$_2$	92.89	86.89	1.18	324.40	10.13
		RIPPER$_{2,pr}$	92.61	86.00	0.85	320.80	10.39
		I-REP	79.02	97.02	38.77	54.20	7.73
	GNB	RIPPER$_0$	98.01	97.67	1.03	128.20	10.29
		RIPPER$_2$	**98.49**	98.30	0.98	151.00	10.34
		RIPPER$_{2,pr}$	98.37	98.07	0.79	149.80	10.77
		I-REP	92.24	**99.28**	27.38	**26.60**	**6.84**
	KNN	RIPPER$_0$	91.64	84.48	1.27	300.60	10.11
		RIPPER$_2$	92.39	85.82	1.10	323.20	10.22
		RIPPER$_{2,pr}$	91.83	84.43	0.84	310.60	10.47
		I-REP	77.53	97.37	42.10	45.80	7.70
	RF	RIPPER$_0$	94.29	89.48	1.09	291.80	10.13
		RIPPER$_2$	95.23	91.35	1.04	327.00	10.22
		RIPPER$_{2,pr}$	95.22	91.06	0.79	341.60	10.37
		I-REP	80.00	97.93	37.27	50.60	7.39
	SVM	RIPPER$_0$	91.36	84.11	1.38	312.60	10.21
		RIPPER$_2$	91.84	84.92	1.23	315.00	10.28
		RIPPER$_{2,pr}$	91.49	83.92	0.92	311.40	10.50
		I-REP	78.34	96.39	39.76	47.00	7.79
RF	DNN	RIPPER$_0$	89.51	80.02	0.94	201.60	13.20
		RIPPER$_2$	90.14	81.31	0.96	195.40	13.36
		RIPPER$_{2,pr}$	89.36	79.40	0.61	190.20	13.64
		I-REP	84.48	95.44	26.56	63.40	10.34
	GNB	RIPPER$_0$	96.40	94.50	1.20	201.00	14.96
		RIPPER$_2$	96.48	94.69	1.24	155.60	15.01
		RIPPER$_{2,pr}$	95.90	93.32	0.82	155.80	15.80
		I-REP	90.76	98.38	19.27	41.60	9.39
	KNN	RIPPER$_0$	88.42	77.78	0.88	203.00	13.44
		RIPPER$_2$	88.69	78.28	0.85	183.40	13.59
		RIPPER$_{2,pr}$	88.14	76.97	0.63	186.20	13.87
		I-REP	82.55	95.14	30.11	53.00	9.88
	RF	RIPPER$_0$	92.12	84.88	0.70	178.00	13.30
		RIPPER$_2$	93.32	87.39	0.78	193.80	13.59
		RIPPER$_{2,pr}$	92.92	86.35	**0.56**	191.60	13.94
		I-REP	85.27	96.86	26.24	60.80	10.02
	SVM	RIPPER$_0$	89.78	80.40	0.97	200.60	13.38
		RIPPER$_2$	90.47	81.82	0.99	191.40	13.46
		RIPPER$_{2,pr}$	89.68	79.91	0.68	191.00	13.85
		I-REP	83.76	95.53	27.88	62.20	10.38

We highlighted the highest accuracy, the highest true positive rate, and the lowest false positive rate across all RBCs. Also, we highlighted the smallest decision list size and the smallest mean rule size

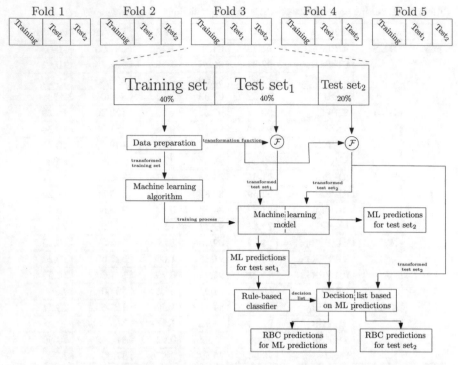

Fig. 2 Visual description of the working process used in this work. Note that the decision list needs to be generated before extracting its predictions for both test sets. Results for machine algorithms for the first test set can be found in Table 1, for the second test set in Table 3. Results for rule-based classifiers and the ML predictions are listed in Table 2, and for the second test set in Table 4

5.4   Evaluation Metrics

To understand how well machine learning algorithms or RBCs perform, we use several different metrics described in this section. We first define terms that are used in the metrics [16].

- True Positive (**TP**)—Correctly predicted malicious samples as malicious
- True Negative (**TN**)—Correctly predicted benign samples as benign
- False Positive (**FP**)—Incorrectly predicted benign samples as malicious
- False Negative (**FN**)—Incorrectly predicted malicious samples as benign

Using the terms above, we can calculate the false positive rate (**FPR**), also referred to as the fall-out rate, the true positive rate (**TPR**), also known as sensitivity and accuracy (**ACC**).

$$FPR \equiv \frac{FP}{FP + TN} \qquad (19)$$

$$TPR \equiv \frac{TP}{TP + FN} \tag{20}$$

$$ACC \equiv \frac{TP + TN}{TP + TN + FP + FN} \tag{21}$$

To better distinguish between individual performances of RBCs, we use additional metrics. We denote the number of rules in a decision list as **DL size** and the mean number of conditions in rules in the decision list as **ø r size**.

5.5 Interpretability of Machine Learning Models

Doshi-Velez's and Kim's [9] definition of interpretability is suitable to our needs the best. Thus, using their definition, we will first define Human Most Understandable Model (HuMUM).

Definition 11 (Human Most Understandable Model) We say that a model is most understandable if it has the ability to explain or to present in understandable terms to a human.

Although this definition is subjective and not rigorous, it will serve more as a naming convention. In our work, we consider decision lists generated by RBCs HuMUM, as they are simple and easily understandable by humans.

We say that a model is absolutely interpretable by Human Most Understandable Model if all of its predictions can be interpreted by HuMUM. That is, HuMUM makes the same predictions as the model would.

Definition 12 (Absolutely Interpretable by HuMUM) We say that the model f is absolutely interpretable by HuMUM g, if the following holds,

$$f(X, y) = y_f,$$
$$g(X, y_f) = y_g, \tag{22}$$
$$y_f = y_g,$$

where X is the training set and y is the label set. Both f and g create new label sets y_f and y_g.

We say that a model is partially interpretable by Human Most Understandable Model, if some of its predictions can be interpreted by HuMUM. That is, HuMUM matches some of the decisions made by the model.

Definition 13 (Partially Interpretable by HuMUM) We say that the model f is partially interpretable by HuMUM g, if the following holds,

$$f(X, y) = y_f,$$

$$g(X, y_f) = y_g, \tag{23}$$

$$y_f \sim y_g,$$

where X is the training set and y is the label set. Both f and g create new label sets y_f and y_g. $y_f \sim y_g$ indicates, that some of the predictions are equal.

To understand when a model is most intepretable, we define Interpretability Entropy.

Definition 14 (Interpretability Entropy) Given the predictions y_f of a model f and the predictions y_g of HuMUM g, we define Interpretability Entropy as follows,

$$\mathscr{H}(T, F) = -\left(\frac{T}{T+F}\right) \log_2 \left(\frac{T}{T+F}\right) - \left(\frac{F}{T+F}\right) \log_2 \left(\frac{F}{T+F}\right), \tag{24}$$

where $T = \sum \delta_{y_f y_g}$ and $\delta_{y_f y_g}$ is Kronecker delta, and $F = |y_f| - T$.

The main goal of HuMUM is to minimize Interpretability Entropy, which can be in the range of $[0, 1]$. Notice that this does not require models to be as close to being absolutely interpretable by HuMUM. If HuMUM would always make a different prediction, we would still have enough information about the model's behavior.

5.6   Measuring Performance of RBCs on ML Predictions

Using the EMBER dataset, we applied five machine learning algorithms—Support Vector Machines with Radial Basis Function kernel (SVM), Random Forest (RF), Gaussian Naïve Bayes (GNB), k-nearest neighbors (KNN), and Deep Neural Network (DNN) with two hidden layers. As mentioned in Sect. 5.2, 5-fold cross-validation was employed to reduce the number of biases. We tried to fine-tune the hyperparameters of algorithms for which it was possible. The averaged results are shown in Table 1. At first glance, neither of the dimensionality reduction approaches seem to have significantly better performance. We included RIPPER and I-REP in the table as we want to compare their out-of-the-box performance to their performances when trained on the predictions of the ML algorithms (see below). RIPPER creates very few false positives; however, it is unable to detect enough malicious samples.

We used our RBCs implementations and trained them on the predictions of the ML algorithms on *the first test set*. That is, for each ML algorithm and all of its five predictions on the first test set, RBCs were used to describe the outcomes of that ML algorithm. The results were then averaged and are shown in Table 2.

I-REP's performance is poor for both accuracy and TPR. Despite this fact, I-REP produces straightforward decision lists in comparison to RIPPER. RIPPER produces immensely few false positives; however, it struggles to find malicious samples. Notice that RIPPER has generally a lower performance when run on models that used RF as dimensionality reduction instead of PCA. This does not apply to I-REP, though.

For the individual cases, the results indicate that the predictions produced by GNB with PCA were reasonably easy for RIPPER to reconstruct. RIPPER, in this case, achieved both high TPR and low FPR, and the total number of conditions is smallest across RIPPER's decision lists. This could mean that GNB is almost absolutely interpretable by RIPPER. GNB in combination with RF has very similar results, although RIPPER needed approximately 1.5 times more conditions than in the case of PCA. For other models, RIPPER generally required more extensive decision lists for those models which used PCA for dimensionality reduction. The RandomForest model was the second most partially interpretable model by RIPPER in both cases of dimensionality reduction. For the rest of the models, RIPPER obtained worse results. Models that were interpreted by RIPPER with RF as dimensionality reduction had generally fewer conditions in total. However, obtained rules were approximately 1.3 times larger. The model that is most probably the least interpretable is KNN with RF dimensionality reduction.

Table 2 presents experiments that aim to verify whether RIPPER's behavior changes based on its learning parameters. $RIPPER_{pr}$ corresponds to RIPPER without pruning. $RIPPER_k$ corresponds to the number of optimization phases. RIPPER does not seem to be improving significantly with its optimization phases. We can see that in some cases, it does increase its FPR in trade for higher TPR. This is not surprising, as its optimization pruning metric is based on accuracy. This could be problematic for imbalanced datasets as the accuracy metric does not take this into account.

The results indicate that I-REP does not generate good rulesets despite their comprehensibility. This is most probably caused by its pruning metric. RIPPER achieved better results, although its TPR is relatively low for some of the models. Results indicate that RIPPER will probably interpret some of the results better than I-REP. Below we give an example of a rule generated by RIPPER for the RandomForest model in combination with RF dimensionality reduction:

```
BHist0[0] <= 0.185925 &&
BHist87[19] <= 0.003879 &&
EntBHist216[92] <= 0.167021 &&
section vsizes hashed1[159] <= 0.000000 &&
section vsizes hashed38[170] <= 0.000000 &&
imported libs hashed206[189] <= 0.400000 &&
EXPORT_TABLE_va[191] <= 0.000001 &&
RESOURCE_TABLE_size[193] <= 0.000000 &&
CERTIFICATE_TABLE_size[195] <= 0.000001.
```

The structure of the rule is as follows:

```
featureName|colNumber?|[colIndex] operator value &&?,
```

where `colNumber` is for features that are either hashed or built from histograms. If the feature is hashed, `colNumber` corresponds to the column of the hash. If the feature is built from byte histograms, `colNumber` corresponds to that byte, e.g., `BHist87` corresponds to the part of byte histogram for byte `0x57`. `colIndex` is used to access different columns (in this case indices can be from $\{0 \ldots 199\}$). `&&` is used as a conjunction. Parts of the rule marked with `?` are optional. Feature names were created according to the EMBER [2] source code.

Figure 3 shows the process of rule learning for both RIPPER and I-REP. Rules with high TP coverage are primarily generated at the beginning of the learning process. IREP's rules that have high FP coverage are obtained at the beginning of the learning process. We can see that RIPPER achieved a few spikes in the case of covered TP samples for DNN, KNN, and SVM. There are smaller spikes for both RF and GNB, too. As a result, RIPPER can find stronger rules in the later phase of the learning process. The reason that those rules are not found earlier is related to its pruning metric—RIPPER needs to cover a certain number of FP samples to start considering better rules. Note that I-REP's decision list sizes are significantly smaller than RIPPER's. This corresponds to the results in Table 5. Different sizes are also given by the fact that both I-REP and RIPPER have distinct stop conditions. I-REP, either used with PCA or RF, has very similar decision list sizes. The same does not apply to RIPPER, as its stop condition allowed it to generate many more rules for PCA than for RF.

5.7   Interpreting ML Results Using RBCs

Using *the second test set*, we measured the performance of the ML algorithms. Results are shown in Table 3. There is no significant drop in performance for none of the algorithms when compared to Table 1.

To find out how well generated decision lists interpret the ML models, we measured also their performance on *the second test set*. This time, we did not use the predictions of the ML algorithms. We used the original class labels for *the second test set*. In Table 4, we replaced the decision list and rule sizes with two columns, TP match, and FP match. We first create predictions of the ML algorithms and extract only TP and FP samples for both columns. We then generate RBC predictions on these samples and calculate how many of the predictions were the same. As seen in both Tables 2 and 4, I-REP's FPR was relatively high. This was not improved by training it on the ML algorithms' predictions. In fact, in some cases, it had FPR higher than when trained on its own. RIPPER's overall accuracy did not change significantly when trained on better models. It did, however, get closer to the results of ML algorithms that it was trained on. The overall results can be misleading, and that is why we included TP and FP matches, and Interpretability Entropy. Our initial proposal that GNB would be the most partially interpretable model holds. RIPPER did match most of its predictions and did achieve the lowest Interpretability Entropy among all other models. This does not apply to I-REP; even though it matches most

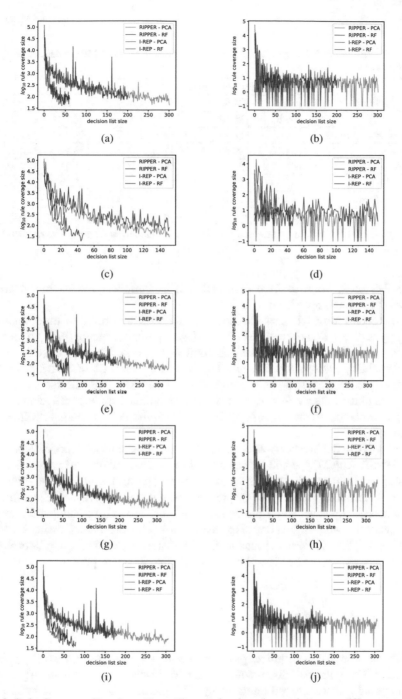

Fig. 3 Rule Coverage Size Over Time. The graphs show how rules cover different samples over time using *the first test set* and one of the five cross-validation folds. The y-axis is log-scaled and represents covered samples; the x-axis represents a decision list size. Value −1 on the y-axis corresponds to no covered samples. (**a**) True positives—DNN. (**b**) False positives—DNN. (**c**) True positives—Gaussian Naïve Bayes. (**d**) False positives—Gaussian Naïve Bayes. (**e**) True positives—KNN. (**f**) False positives—KNN. (**g**) True positives—RandomForest. (**h**) False positives—RandomForest. (**i**) True positives—SVM. (**j**) False positives—SVM

Table 3 Well-known ML algorithms and their performance on the EMBER dataset using *the second test set*

ML algorithm	RandomForest			PCA		
	ACC	TPR	FPR	ACC	TPR	FPR
DNN	**95.62**	**95.82**	4.59	**96.44**	96.20	3.33
GaussianNaïveBayes	74.65	80.97	31.68	72.47	96.11	51.16
I-REP	84.10	94.62	26.41	78.28	**96.70**	40.13
KNN	95.17	95.31	4.96	95.38	95.12	4.36
RIPPER	88.69	79.08	**1.70**	89.89	82.71	**2.93**
RandomForest	95.55	95.40	4.30	95.36	94.45	3.73
SVM	94.92	94.63	4.80	95.86	95.93	4.21

We highlighted the highest accuracy, true positive rate, and the lowest false positive rate across all the models

of GNB's predictions, its high FPR makes it less efficient for machine learning models to be interpreted by it.

The second most interpretable model by RIPPER according to Interpretability Entropy was RandomForest. RIPPER did get similar rates for FPR. It did not match all of the predictions that RandomForest made. This is reflected in the Interpretability Entropy, too. The last three models, namely SVM, DNN, and KNN, and their results seem to be quite hard to interpret by both RIPPER and I-REP. The Interpretability Entropy tells us that models interpreted by RIPPER could retain some information. In the case of I-REP combined with PCA, there is very little one could gain.

The value of the Interpretability Entropy for the RandomForest model interpreted by RIPPER with PCA could be considered an acceptable limit of what we could consider a strong, partially interpretable model. The results in Table 4 show that RBCs have trouble matching the FP predictions of the original model. However, this is not necessarily adversity; if the FP match is close to zero, we could still interpret the results as well as creating explanations of why the model made incorrect predictions. Close to 50%, FP match, resp. TP match, does not give any information whatsoever, as it could be viewed more as guessing than interpreting.

5.8   Pruning and Metrics

I-REP and RIPPER both utilize pruning to handle noisy data. Cohen [8] pointed out that I-REP's incapability to converge towards better solutions is mainly caused by its pruning metric, based on accuracy. The metric is one of the essential parts of I-REP-like algorithms. Naturally, we may ask whether or not we can affect the behavior of the metrics or whether it is more of a trial and error challenge.

The proposed version of RIPPER by Cohen [8] is capable of handling multiclass problems. Using RIPPER, we can reduce the multiclass problem to an alternating

Table 4 Testing how well do RBCs interpret ML algorithms' predictions using *the second test set*

		RBC	ACC	TPR	FPR	TP match	FP match	$\mathcal{H}$
PCA	DNN	$RIPPER_0$	89.35	82.05	3.35	84.76	32.90	0.46
		$RIPPER_2$	90.17	83.98	3.63	86.71	36.17	0.43
		$RIPPER_{2,pr}$	89.95	83.17	3.27	85.93	34.25	0.44
		I-REP	78.68	96.47	39.11	97.14	84.62	0.75
	GNB	$RIPPER_0$	73.24	95.45	48.98	99.07	93.85	0.17
		$RIPPER_2$	73.01	95.66	49.65	99.28	95.02	**0.16**
		$RIPPER_{2,pr}$	73.13	95.59	49.34	99.24	94.60	**0.16**
		I-REP	68.37	**98.71**	61.96	**99.87**	**97.86**	0.40
	KNN	$RIPPER_0$	89.54	82.22	3.15	85.57	25.09	0.47
		$RIPPER_2$	90.03	83.28	3.23	86.65	26.80	0.45
		$RIPPER_{2,pr}$	89.59	82.05	2.87	85.49	24.34	0.46
		I-REP	77.33	96.94	42.29	97.75	83.79	0.77
	RF	$RIPPER_0$	90.12	84.60	4.36	89.04	60.09	0.37
		$RIPPER_2$	90.79	86.18	4.60	90.67	62.73	0.35
		$RIPPER_{2,pr}$	**90.81**	86.02	4.40	90.54	61.67	0.35
		I-REP	78.82	95.91	38.27	97.72	96.30	0.73
	SVM	$RIPPER_0$	89.46	82.30	3.38	84.96	23.75	0.48
		$RIPPER_2$	89.81	82.95	3.32	85.66	23.95	0.47
		$RIPPER_{2,pr}$	89.55	82.02	2.92	84.75	22.26	0.48
		I-REP	78.14	96.32	40.04	96.92	77.62	0.76
RF	DNN	$RIPPER_0$	88.48	79.19	2.22	82.00	19.67	0.51
		$RIPPER_2$	88.97	80.27	2.33	83.11	21.17	0.50
		$RIPPER_{2,pr}$	88.26	78.44	1.91	81.30	18.12	0.52
		I-REP	83.64	94.76	27.47	96.06	73.61	0.63
	GNB	$RIPPER_0$	74.55	78.31	29.20	95.67	88.97	0.26
		$RIPPER_2$	74.48	78.34	29.37	95.72	89.41	0.26
		$RIPPER_{2,pr}$	74.48	77.36	28.40	94.67	87.10	0.28
		I-REP	73.00	86.75	40.73	98.78	96.63	0.44
	KNN	$RIPPER_0$	87.74	77.24	1.76	80.25	13.62	0.54
		$RIPPER_2$	88.01	77.74	1.71	80.78	13.38	0.54
		$RIPPER_{2,pr}$	87.44	76.35	**1.47**	79.40	11.88	0.55
		I-REP	82.35	95.12	30.41	96.19	69.64	0.67
	RF	$RIPPER_0$	89.17	81.86	3.53	85.47	53.49	0.43
		$RIPPER_2$	90.13	84.11	3.86	87.78	56.91	0.39
		$RIPPER_{2,pr}$	89.85	83.21	3.51	86.88	54.01	0.40
		I-REP	83.42	94.88	28.04	96.74	93.07	0.61
	SVM	$RIPPER_0$	88.29	78.78	2.20	82.50	20.62	0.50
		$RIPPER_2$	88.89	80.11	2.33	83.89	21.83	0.48
		$RIPPER_{2,pr}$	88.11	78.20	1.98	81.97	20.01	0.51
		I-REP	83.18	94.70	28.35	96.22	75.50	0.65

We highlighted the highest accuracy, the highest true positive rate, and the lowest false positive rate across all RBCs. Also, we highlighted the highest true positive match, the highest false positive match, and the lowest Interpretability Entropy

two-class problem. Thus, we can understand pruning metrics as two-variable functions. Fortunately, this number is perfect for a better understanding of pruning metrics by graphing them. Figure 4 shows metrics used by I-REP and RIPPER, and other metrics we tried to use throughout the experiments. We simplified I-REP's pruning metric as it can be viewed as a plane for fixed P and N (see Metric 10). Here lies the key reason why I-REP tends to make bad decisions when pruning; points with a different number of malicious and benign samples are often indistinguishable. RIPPER's pruning metric has good characteristics; we could only identify that it does not differentiate between positive samples when no negative samples are present.

We experimented with the metrics in Fig. 4, and the results are shown in Table 5. Below we assign each metric (additionally, we added a function with a saddle point) to its name:

$$\underbrace{\frac{p-n}{\sqrt{p}+\sqrt{n}+1}}_{\text{sqrt}}, \quad \underbrace{\frac{p-n}{p+n+1}+\frac{p}{n+1}}_{\text{impr}}, \quad \underbrace{p^2-n^2}_{\text{saddle}}, \tag{25}$$

where p is the number of positive samples and n is the number of negative samples in the pruning set. Each of the names (sqrt, impr, saddle) is used in Table 5 and indicates what pruning metric was used. For the experiment, we used *the first test set* and RIPPER with k set to zero. Results indicate that used metrics did not achieve significantly better performance. RIPPPER$_{\text{impr}}$'s behavior is comparable to RIPPER$_0$, and for some cases, it reaches better FPR. With the decreasing number of rules, we can see a significant decrease in the performance. With higher TPR and FPR, RIPPER$_{\text{sqrt}}$ achieves similar accuracy rates to RIPPER$_{\text{impr}}$; for most cases with more than 20% decrease in decision list sizes. RIPPER$_{\text{saddle}}$ performs worse than the original pruning metric of I-REP. Surprisingly, it generates more rules than I-REP (see Table 2).

5.9   Does Order of the Rules Matter?

Figure 3 demonstrates a few spikes throughout the learning process of RIPPER. We could potentially shift these spikes to have them occur as soon as possible. As a side effect, we would violate the order in which they were learned. On the other hand, would this change the overall behavior of the model?

Let R be a decision list with rules $r_1, \ldots, r_n$. We want to swap rules r_i and r_j, where $i < j$. Rule r_j now covers at least all samples it covered before the swap. It may also cover new samples then covered by r_i and r_k, where $i < k < j$. This means that the number of TP and FP samples for rule r_j can remain the same or increase. Samples that were covered by r_i before the swap and are not covered by r_j after the swap can still be covered by r_k. Thus, the number of TP and FP samples covered by r_i after swap can remain the same or decrease. The swap does not add any new

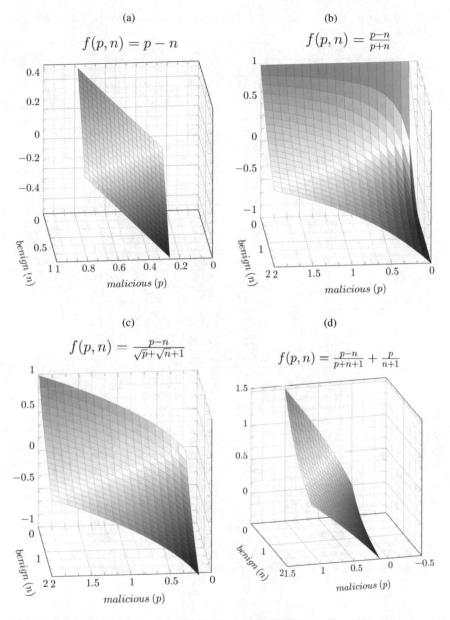

Fig. 4 Understanding pruning metrics as 3D graphs. Pruning metrics should have the following properties: If only malicious files are present, pruning metrics should be at their maximum. For benign files only, they should be at their minimum. Otherwise, they need to compromise, and should take into account a lower number of benign files. (**a**) Simplified IREP's pruning metric. (**b**) RIPPER's pruning metric. (**c**) RIPPER metric with curvature for malicious samples. (**d**) RIPPER metric with more curvature for malicious samples

Table 5 Measuring performance of RBCs with different metrics on ML predictions using *the first test set*

		RBC	ACC	TPR	FPR	DL size	ø r size
PCA	DNN	RIPPER$_{saddle}$	76.69	98.55	44.94	59.00	7.53
		RIPPER$_{sqrt}$	91.34	91.86	9.17	218.60	9.33
		RIPPER$_{sqrt}$	91.67	84.43	1.17	286.40	10.04
	GNB	RIPPER$_{saddle}$	89.57	**99.79**	38.89	**31.40**	**6.69**
		RIPPER$_{sqrt}$	97.09	98.95	8.16	83.40	8.91
		RIPPER$_{sqrt}$	**98.09**	97.78	1.05	133.80	10.26
	KNN	RIPPER$_{saddle}$	75.41	98.84	47.77	58.60	7.56
		RIPPER$_{sqrt}$	84.68	78.64	9.32	171.40	9.40
		RIPPER$_{sqrt}$	92.01	85.33	1.38	317.00	10.10
	RF	RIPPER$_{saddle}$	77.66	99.35	43.21	68.60	7.35
		RIPPER$_{sqrt}$	92.94	94.51	8.57	224.00	9.41
		RIPPER$_{sqrt}$	94.59	90.17	1.15	310.60	10.15
	SVM	RIPPER$_{saddle}$	75.58	98.21	47.10	49.20	7.30
		RIPPER$_{sqrt}$	90.47	90.97	10.02	214.00	9.40
		RIPPER$_{sqrt}$	91.20	83.76	1.34	299.00	10.17
RF	DNN	RIPPER$_{saddle}$	74.96	98.30	48.52	43.60	8.70
		RIPPER$_{sqrt}$	90.01	90.05	10.02	112.40	11.92
		RIPPER$_{sqrt}$	89.25	79.49	0.92	195.80	13.23
	GNB	RIPPER$_{saddle}$	85.60	99.31	31.79	35.20	8.50
		RIPPER$_{sqrt}$	94.90	97.31	8.23	94.60	12.02
		RIPPER$_{sqrt}$	95.83	93.42	1.09	168.00	14.84
	KNN	RIPPER$_{saddle}$	74.55	97.95	48.97	39.60	8.80
		RIPPER$_{sqrt}$	89.15	86.86	8.56	109.20	12.14
		RIPPER$_{sqrt}$	88.24	77.45	0.92	196.00	13.29
	RF	RIPPER$_{saddle}$	75.23	98.98	48.35	48.00	9.07
		RIPPER$_{sqrt}$	91.89	92.54	8.75	117.60	11.80
		RIPPER$_{sqrt}$	92.16	85.01	**0.74**	188.40	13.28
	SVM	RIPPER$_{saddle}$	75.05	98.38	47.99	44.00	9.02
		RIPPER$_{sqrt}$	90.08	88.99	8.86	110.40	11.93
		RIPPER$_{sqrt}$	89.88	80.58	0.94	203.20	13.43

We highlighted the highest accuracy, the highest true positive rate, and the lowest false positive rate across all RBCs. Also, we highlighted the smallest decision list size and the smallest mean rule size

samples that could be covered and only changes the behavior of each individual rule. The overall behavior of R remains the same.

To sort the rules, we would always need to find a rule with a spike that is larger than the previous ones. Unfortunately, we cannot use fast sorting algorithms as we always need to update the number of TP samples of the following rules. The sorting itself would require $\mathcal{O}(n^2)$, where n is the number of rules, and the coverage of each following rule would require $\mathcal{O}(nm)$, where m is the number of samples. This means that the sorting would require $\mathcal{O}(n^3m)$. Therefore, we decided to sort the rules greedily—only once given by their covered TP samples. The results can be seen in Fig. 5—we used the first test set and the predictions of ML algorithms. We can see that this approach smoothened the TP curves for RIPPER. Some rules generated by I-REP had no TP coverage when reordered. This can be seen in the case of the GNB model with RF dimensionality reduction. A similar case can be seen for SVM, again with RF.

Rule ordering could lead to potential speed-ups if used in production, as stronger rules would trigger earlier. Also, it could be used as an additional tool in RIPPER's optimization phase to achieve new properties.

6 Conclusion and Future Work

The interpretability of machine learning methods could be considered one of the leading research goals in the current era. Many works focus on the essence of interpretability itself, whereas other works focus on the domain of specific models. In this paper, we examined the use of rule-learning algorithms to extract decision lists based on the predictions of machine learning models. We used decision lists as they are one of the most understandable models in machine learning.

In our experiments, we used two rule-learning algorithms: I-REP and RIPPER. I-REP had inferior results, and we discussed the reason for this in Sect. 5.8. RIPPER covered most of the predictions well; however, it could not find appropriate rules that would not increase the MDL metric mentioned in Sect. 3.2. Using Doshi-Velez's and Kim's definition of interpretability, we defined Human Most Understandable Model. We defined absolutely and partially interpretable models by HuMUM, together with Interpretability Entropy (see Sect. 5).

We tried to estimate how well do RBCs interpret the results produced by the ML models. We merely did this by taking into account the accuracies, true and false-positive rates of RBCs. This gave us a good idea of what ML models could be interpreted by RBCs better than others. For example, we have correctly assumed that GNB would be more interpretable than KNN by taking into account all of the three metrics. Using this approach is limited since we cannot state how well RBCs interpret the ML models precisely. Thus, we inspected the amounts of matched predictions for the ML models. Using these amounts, we saw where RBCs fail to interpret the ML models. Finally, the Interpretability Entropy allowed us to numerically compare what ML models are more interpretable than others. We conclude that in the case of the EMBER dataset, we could consider the Gaussian Naïve Bayes model almost absolutely interpretable by HuMUM. The random forest model could be viewed as a possible borderline of what we still could consider interpretable. To increase the measure of interpretability for other models, such as deep neural network, we need to improve the performance of the rule-based classifiers.

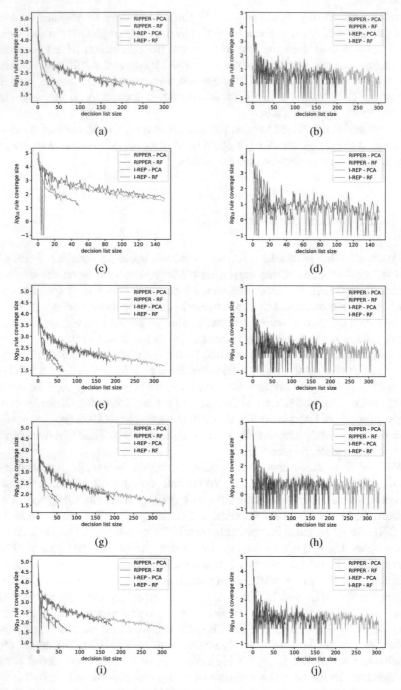

Fig. 5 Changing Order of the Rules. The graphs show how rule ordering affects covered samples over time using *the first test set* and one of the five cross-validation folds. The y-axis is log-scaled and represents covered samples; the x-axis represents a decision list size. Value −1 on the y-axis corresponds to no covered samples. (**a**) True positives—DNN. (**b**) False positives—DNN. (**c**) True positives—Gaussian Naïve Bayes. (**d**) False positives—Gaussian Naïve Bayes. (**e**) True positives—KNN. (**f**) False positives—KNN. (**g**) True positives—RandomForest. (**h**) False positives—RandomForest. (**i**) True positives—SVM. (**j**) False positives—SVM

Throughout the experiments, we tried to inspect the behavior of the RIPPER algorithm. We discussed the importance of the rule order in a decision list and how changing it will not affect the behavior of the whole decision list. We confirmed that the metric plays a significant role in rule learning, and by modifying it, we can either achieve better performance or more comprehensible decision lists.

Although we created our implementations of rule-based classifiers, they are far from being finished. We believe there is still space for speed improvements using memorization. Currently, our implementations run sequentially; we could achieve significant speed-ups by parallelizing certain parts of the implementations, for example, looking for the best condition while growing a rule.

Decision lists generated by rule-learning algorithms could be used as an adversarial tool, too. We could create features given by the conditions and examine when the predictions of an interpreted model differ from the predictions generated by RBCs. This could deepen the understanding of the interpreted model and allow for other methods to be used in its weaker performing parts.

Acknowledgments This work was supported by the Student Summer Research Program 2021 of FIT CTU in Prague and by the OP VVV MEYS funded project CZ.02.1.01/0.0/0.0/16 019/0000765 "Research Center for Informatics" and by the Grant Agency of the CTU in Prague, grant No. SGS21/142/OHK3/2T/18 funded by the MEYS of the Czech Republic.

References

1. Hyrum S. Anderson and Phil Roth. EMBER: An Open Dataset for Training Static PE Malware Machine Learning Models. *ArXiv e-prints*, 2018.
2. Hyrum S. Anderson and Phil Roth. Elastic malware benchmark for empowering researchers, n.d.
3. Julia Angwin, Jeff Larson, Surya Mattu, and Lauren Kirchner. Machine bias, 2016.
4. Daniel Arp, Michael Spreitzenbarth, Malte Hubner, Hugo Gascon, Konrad Rieck, and CERT Siemens. Drebin: Effective and explainable detection of android malware in your pocket. In *Ndss*, volume 14, pages 23–26, 2014.
5. Diogo V. Carvalho, Eduardo M. Pereira, and Jaime S. Cardoso. Machine learning interpretability: A survey on methods and metrics. *Electronics*, 8(8), 2019.
6. Jadzia Cendrowska. Prism: An algorithm for inducing modular rules. *International Journal of Man-Machine Studies*, 27(4):349–370, 1987.
7. Li Chen. Deep transfer learning for static malware classification, 2018.
8. William W. Cohen. Fast effective rule induction. In *Machine learning proceedings 1995*, pages 115–123. Elsevier, 1995.
9. Finale Doshi-Velez and Been Kim. Towards a rigorous science of interpretable machine learning, 2017.
10. Jeff Elder. Is machine learning useful for cybersecurity?, n.d.
11. Zhentan Feng, Shuguang Xiong, Deqiang Cao, Xiaolu Deng, Xin Wang, Yang Yang, Xiaobo Zhou, Yan Huang, and Guangzhu Wu. Hrs: A hybrid framework for malware detection. In *Proceedings of the 2015 ACM International Workshop on International Workshop on Security and Privacy Analytics*, IWSPA '15, page 19–26, New York, NY, USA, 2015. Association for Computing Machinery.
12. Johannes Fürnkranz and Gerhard Widmer. Incremental reduced error pruning. In *Machine Learning Proceedings 1994*, pages 70–77. Elsevier, 1994.

13. Leilani H. Gilpin, David Bau, Ben Z. Yuan, Ayesha Bajwa, Michael Specter, and Lalana Kagal. Explaining explanations: An overview of interpretability of machine learning. In *2018 IEEE 5th International Conference on data science and advanced analytics (DSAA)*, pages 80–89. IEEE, 2018.
14. Bryce Goodman and Seth Flaxman. European union regulations on algorithmic decision-making and a "right to explanation". *AI magazine*, 38(3):50–57, 2017.
15. Mark Hall, Eibe Frank, Geoffrey Holmes, Bernhard Pfahringer, Peter Reutemann, and Ian H. Witten. The weka data mining software: An update. *SIGKDD Explor. Newsl.*, 11(1):10–18, November 2009.
16. Mohammad Hossin and MN Sulaiman. A review on evaluation metrics for data classification evaluations. *International Journal of Data Mining & Knowledge Management Process*, 5(2):1, 2015.
17. Giacomo Iadarola, Fabio Martinelli, Francesco Mercaldo, and Antonella Santone. Towards an interpretable deep learning model for mobile malware detection and family identification. *Computers & Security*, 105:102198, 2021.
18. Wenzel Jakob, Jason Rhinelander, and Dean Moldovan. pybind11 – seamless operability between c++11 and python, 2017. https://github.com/pybind/pybind11.
19. Tim Miller. Explanation in artificial intelligence: Insights from the social sciences, 2018.
20. Christoph Molnar. *Interpretable machine learning*. Lulu. com, 2020.
21. Ilan Moscovitz. wittgenstein, n.d.
22. F. Pedregosa and et al. Scikit-learn: Machine learning in Python. *Journal of Machine Learning Research*, 12:2825–2830, 2011.
23. John Ross Quinlan. *C4.5: Programs for Machine Learning*. Morgan Kaufmann, 1993.
24. Ribana Roscher, Bastian Bohn, Marco F. Duarte, and Jochen Garcke. Explainable machine learning for scientific insights and discoveries. *IEEE Access*, 8:42200–42216, 2020.
25. Igor Santos, Yoseba K. Penya, Jaime Devesa, and Pablo Garcia Bringas. N-grams-based file signatures for malware detection. *ICEIS (2)*, 9:317–320, 2009.
26. Matthew G. Schultz, Eleazar Eskin, F. Zadok, and Salvatore J. Stolfo. Data mining methods for detection of new malicious executables. In *Proceedings 2001 IEEE Symposium on Security and Privacy. S&P 2001*, pages 38–49. IEEE, 2000.
27. scikit-learn Developers. 6.3. preprocessing data, n.d.
28. Silvia Sebastián and Juan Caballero. Avclass2: Massive malware tag extraction from av labels. In *Annual Computer Security Applications Conference*, ACSAC '20, page 42–53, New York, NY, USA, 2020. Association for Computing Machinery.
29. Ramprasaath R Selvaraju, Michael Cogswell, Abhishek Das, Ramakrishna Vedantam, Devi Parikh, and Dhruv Batra. Grad-cam: Visual explanations from deep networks via gradient-based localization. In *Proceedings of the IEEE international conference on computer vision*, pages 618–626, 2017.
30. Ian H. Witten, Eibe Frank, and Mark A. Hall. *Data Mining: Practical Machine Learning Tools and Techniques*. Morgan Kaufmann, 2011.

Mobile Malware Detection Using Consortium Blockchain

George Martin, Dona Spencer, Aditya Hair, Deepa K, Sonia Laudanna, Vinod P, and Corrado Aaron Visaggio

Abstract The purpose of the paper is to explore the problem of detecting malicious codes in malware and a way, based on consortium blockchain, to detect and control the propagation in mobile devices. According to Damballa's Q4 State of Infections report, the antivirus products overlooked 70% of malware signatures within the first hour (Q4 2014 State of Infections Report. Q4 2014 state of infections report. https://www.interwest.com/news/press-releases/1013, accessed August 2021). This is despite the fact that malware detection is carried out via numerous detection techniques such as static analysis, behavioural analysis and sand-boxing. Specially, malware detection in the mobile devices has always been a challenging issue, especially on the efficient and open-source Android platform. Since each company acts as an independent entity and there is a proliferation of antivirus products, the rate of detection and effective identification of the malware is slowed down. In this chapter, we try to establish a relation between the different detection products through better communication, faster updating (via the common ledger) and more efficient and accurate detection of malicious programs. The communication is improved as all the nodes (anti-malware agencies) refer to the same blockchain in the consortium network, hence possessing a common record. Combining the malware signature of all entities into one increases the detection of malware, reduces false positive rates via majority voting and speeds up the spread of signature awareness. The resulting system, as proposed in this paper creates an environment that provides a more precise classification of the application file provided by the user. Therefore, in conclusion, incorporating the blockchain technology, with the anti-malware producers as nodes, improves accuracy, merging the security services provided by the blockchain technology.

G. Martin · D. Spencer · A. Hair · Deepa K · Vinod P
SCMS School of Engineering & Technology, Ernakulam, India
e-mail: george.martin@scmsgroup.org; donaspencer@scmsgroup.org;
aditya.hari@scmsgroup.org; deepak@scmsgroup.org; vinod.p@cusat.ac.in

S. Laudanna · C. A. Visaggio (✉)
University of Sannio, Benevento, Italy
e-mail: slaudanna@unisannio.it; visaggio@unisannio.it

© The Author(s), under exclusive license to Springer Nature Switzerland AG 2022
M. Stamp et al. (eds.), *Artificial Intelligence for Cybersecurity*, Advances in
Information Security 54, https://doi.org/10.1007/978-3-030-97087-1_6

1   Introduction

The volume of malicious programs is growing exponentially with the increasing pervasiveness of software systems, along with their impact and sophistication [17]. Anti-malware products realize the detection of the malicious codes through different techniques that follow different paradigms, as well as: signature-based methodology, heuristics-based, static and dynamic analysis, and machine-learning based classification. Unfortunately, the rate of discovery, detection and the distribution of the necessary information to recognize malicious programs is still low.

In this chapter we propose a solution aimed to improve the collection and classification of signatures from different sources, using different techniques, the fusion of the signatures concerning the same malware and a platform for storing and distributing them based on the blockchain technology. The goal is to build an environment that allows different providers like antivirus vendors, nids, and threat intelligence platforms to exchange malware signatures or indicators of compromise.

Blockchain is a decentralized chain of blocks that contain information, in this case the information about each app, such as its signature and class (malware/benign/unknown). Every member of the blockchain has a copy of the distributed ledger, which makes it hard to tamper the information about malware signatures. Three classes of blockchain can be deployed, namely Public, Private and Consortium. A public blockchain allows any member to join the network, Bitcoin and Ethereum are two popular examples. A private blockchain consists of nodes that are centralized, quite against the very idea of blockchain technology. On the other hand, a consortium blockchain is controlled by a pre-selected set of nodes (pre-approved set of anti-malware agencies). Thus, though blockchain is said to be decentralized, the network we employ in this paper is the consortium blockchain which is not fully decentralized as we only require the approved parties to add the malware signatures onto the block.

The underlying solution consists of two main parts: (i) malware detection along with the production of indicators of compromise and (ii) the managing of a blockchain for fusing, storing and distributing malware signatures. The malware detection is realized through static analysis, by extracting properties from the app under analysis; it is not limited to the source code alone, but also includes the manifest files, .smali files and other resources that form the apk (Android Application Package) [7]. The dataset for the model generation includes 15,000 apk files, which includes 2203 malware apks from the Drebin database [3] and 4039 benign apks of different categories taken from the Android market as the test set. The proposed system is based on a blockchain network in the form of a consortium of members responsible for detecting malware in mobile devices, along with a trusted server that interacts with the client and provides the consortium with the necessary data. It is necessary to deploy a consortium blockchain since the members responsible for detecting malware represent independent anti-malware agencies. The platform should offer these agencies an environment which is immune from malpractice as the system would undoubtedly build a reputation for said agencies depending on how consistently and correctly they classify APK files as malware or goodware. It is absolutely necessary that the system remains immune to malpractice and ambiguity. A centralized system is prone to malpractice such as tainting the results, influencing or unfairly elevating the performance of one or more anti-malware agencies from the others and corrupting records. A consortium blockchain, on the other hand, would be capable of tackling these problems and hence enable these independent agencies to operate effectively while ensuring a healthy competition between each other. Each member in the consortium may make use of a particular feature extracted from the app, like permissions, APIs, Intents and may classify the application at her convenience. The trusted server collects the .apk file from the client and looks up the blockchain ledger which acts as a perpetual data bank holding malicious program details. This trusted server is also responsible for notifying consortium members of the applications it couldn't classify.

The rest of the chapter is structured as follows. Section 2 discusses the motivation and the context through the use cases. Sections 3, 5, 6 provide the necessary background, the architecture of the proposed system and literature, while Sects. 7, 8, 9 and 10 present the dataset, the feature extraction and the training methodology. A discussion of the results follows in Sect. 11. Lastly, Sect. 12 provides concluding remarks along with the directions of the future work.

2 Use Case

A number of android applications exist with malicious components and privacy issues. The traditional method for identifying applications that cause harm, like intruding on the user's privacy, would include the detection of malware based on its features. The unit responsible for analyzing is solely responsible for classifying the application as either malware or goodware without any input or collaboration with other analyzing units.

The Mobile Malware Detection System using Consortium Blockchain in this experiment overcomes such issues. This proposed system includes three separate and independent nodes. Each of which is specialized to perform classification based on a separate feature. The user, after having uploaded an application that doesn't match the records on the ledger, will be notified about the classification performed by the consortium. Each node obtains the file and they proceed to analyze the file individually, performing the extraction of features they are responsible for. After this step, the individual nodes shall prepare the feature vector and feed it to their respective classifiers. The classifier is trained to distinguish a file as malware or goodware from the training data by identifying the unique set of combinations of features differentiating a malware from a regular application. The node responsible for identifying malware based on the permissions set used by the applications will collect the permissions used by the application, such as camera, contacts, location, microphone, sensors, SMS, and storage (small set of permissions that has the potential to cause harm if the developer of the malware intends to), and creates a

Fig. 1 Components of
android application

feature vector. Based on the permissions alone, the classifier tries to understand the true nature of the application and then gives the prediction accordingly. Similarly, the other nodes also perform classification operations using the API calls features and intent features. Using each nodes prediction, the final label for the file is determined. Hence the system takes various agencies' input into account and comes with a result that is determined by more than one component as opposed to a centralized agency. The system provides the environment to simulate a form of classification inclusive of diverse analyzers and better performance despite the lack of any real interaction between them.

3 Android Application Components

This section provides a brief summary of the basic Android application components and how they are used in malware detection.

Figure 1 depicts the basic building blocks of an Android application, known as the components of an Android application. These components are loosely coupled and bound by the manifest file *AndroidManifest.xml*. The manifest file of the application describes the components and their interaction with one another, along with additional information such as the application's metadata, and required permissions.

3.1 Activities

An activity, representing the presentation layer, is a single screen UI (User Interface). For example, the login screen of a messaging application is an activity, the chat visualization on the monitor of the device is another activity. All activities of an app work cohesively but are also independent from each other. Thus, another app, if allowed by the current application may start an activity. For example, a camera app starts an activity to share photos in the messaging app.

3.2 Services

A service is a general-purpose entry point that is used to keep the application running in the background. It has no user interface. The application can be made to run in the background either due to Started service or Bound service. They perform several operations such as data source updating, broadcasting intents and even performing tasks of inactive applications.

3.3 Broadcast Receivers

Broadcast receivers listen for intents such as BOOT_COMPLETED, POWER_CONNECTED, SMS_RECEIVED. They help the application to react to the received intent that matches certain specified criteria. Basically, it allows the app to respond to system-wide broadcast announcements. A system can deliver broadcasts to systems that aren't currently running.

3.4 Content Providers

Content providers basically handle application data and data management issues. Such data may be stored in the file system, database or elsewhere. This component, also known as the data storage, has the responsibility of handling data access beyond the boundary of the application.

4 Role in Malware Detection

Faruki et al. in [6] consider the components that are launched using Intents such as Activities and Services and state how each component may be accessed by other applications. Suarez et al. in [15] carried out experimentation using meta-information and count of the components to determine whether an application was malicious. Xu et al. in [20] state that malware tends to register more broadcast receivers than as seen in benign applications.

Hence, there are many access points that can be maliciously utilized. Since anatomy analysis of malware needs to take a look at the relationships between the features in order to reveal the malicious behavior and identify its patterns, we tried to achieve this by implementing both feature extraction and classification. For example, one application can make use of activity of another application, services can run tasks in the background without the user's knowledge, broadcast receiver can act as a general entry-point and handle communication between Android OS and application, other apps can query and modify data using content provider, and so on. Thus, there is a need to assess certain features such as intents and permissions, to ensure that there are no malicious intentions behind an application.

5   The Blockchain Network

The proposed solution incorporates a form of consortium blockchain architecture. Three types of blockchains exist: public blockchain, private blockchain, and consortium blockchain [23]. The public blockchain is the blockchain built without any restrictions on access. This kind of blockchain does not require a hierarchy, since every node has the authority to perform every action such as reading, writing, and auditing. Private blockchains limit access to an individual or an organization. Such a system, unlike a public blockchain, has an entity responsible for maintaining the network, assigning authorizations to perform different actions. Hence, the management of a private blockchain is centralized, having a trusted party the duty of assigning the rights for performing the tasks, offering the different members of the organization a secure medium to realize the transactions from within. A consortium blockchain removes the constraint of a single entity that centrally controls all the activities, replacing it with a pre-selected group of entities with identical authority performing the various actions across the ledger [11].

The solution makes use of a permissioned blockchain with a consortium architecture. Unlike a permissionless blockchain which does not require permission to join, thereby allowing any individual to be a part of the network, the permissioned blockchain asks for approval from the central authority for a user to participate. The network developed consists of two pre-selected entities: a Trusted Server and the various organizations responsible for classifying files as malware or benign. The trusted server has the authority for running functions to initialize new blocks on the blockchain when the client uploads new apk files and notifies the different organizations in the consortium about the classification of a program. They are also enabled to read the ledger to determine the classification of the program. Once the connected organizations are notified of a new block in the ledger by the trusted server, they would work on figuring out what the classification will be after which they would proceed to run the functions enabled to update the block with their findings. The functions enabled in the network are specific to the participants in the network and can only be implemented by the participant if and only if the participant has the proper authorization. The trusted server and the members of the consortium have different permissions and so this would imply that a permissioned network is necessary.

Though the clients are crucial parts of the system, they are not members of the blockchain. They do not have any authority to perform any action across the blockchain ledger. The necessary actions are carried out by the members of the consortium and the trusted server. In fact, the only interaction the client has with the proposed system is with this trusted server. They are end-users that provide apk files that they suspect may be malicious to this trusted server and the trusted server goes on to inform the consortium to classify the file provided by the client and to write the results onto the ledger. Once the required number of members in the consortium have shared their results on the ledger, the trusted server simply reads this result and provides the user with a response about the classification of the file, i.e. malware or benign. The user does not directly interact with the ledger in any way. The necessary operations are carried out by the trusted server and the members of the consortium. And so, this rules out the need to implement a public blockchain as well. Since the two entities have different permissions to operate on the ledger and any new node can be a part of the system without the necessary approval, the best option was to construct a permissioned blockchain on the top of consortium architecture.

6 Related Works

The malware detection approach adopted by the proposed system is static analysis. Several works based on static analysis have been investigated. Papers [7, 9] discuss about the enhancement or improvement of the methodology. These methods however only help an individual agency improve their detection rates.

This paper utilizes static-based analysis for malware detection to represent the possible models. There are three models incorporated into the network which are based on the respective features as permission, API calls and intents.

Paper [22] provides an evaluation of permission based android malware detection, which is what we employ for one of the nodes of the blockchain network. However, this paper only considers a small dataset. Similarly, paper [25] cites API sequence as the feature of concern, with the similar drawback of considering only 600 benign and 600 malware samples. Paper [2] also deals with API but also states how achieved better results as compared to permission based malware detection techniques. Another paper, [19], considers permissions over inter-component communications as intents are taken as the main feature in the malware detection process.

The studies in [16] and [4] consider a combination of features such as permission and intents, permission and API calls, respectively. They try to bring about a hybrid method for android malware detection by analyzing the required features. Some authors [16] state that the system is improved enough to get better detection of zero-day malware. The other combination of features, as seen in paper [4] shows better results that are based on permission alone.

Paper [21] briefly discusses the different techniques used in android malware detection and states their highlight and limitations. Other papers [14, 18] and [24] discuss about malware detection in Android. But, all the papers stated so far are to enhance a single anti-malware agency and improve a single node's detection rate. On the contrary, we employ blockchain technology to better communicate the signatures and create an environment for improved malware detection with a collection of nodes that can employ the preferred malware detection technique.

Some papers [10, 12] and [8] explore these concerns. In [12], Raje et al. design a full-proof heuristic solution based on the blockchain technology and deep learning to classify Portable Executable (PE) files as malicious or benign and they can achieve nearly 90% accuracy using a two-layer DBN. Meng et al. [10] in their review discusses the applicability of blockchain technology to solve some issues of the intrusion detection systems (IDS). Particularly, they point out that the blockchain technology can be used to improve the performance of an IDS, especially in the aspects of data sharing and trust computation but not all IDS issues can be solved with this technology. On the same topic, Gu et al. in [8] propose a framework, called CB-MMDE, to detect and classify malware on Android-based mobile devices through blockchain technology. Starting from the use of the Drebin dataset [3], they can achieve higher detection accuracy in limited time with lower false-positive and false-negative rates.

7 Methodology

This section describes the working of the system, the different components and how they interact with one another. The diagrammatic representation of the working is shown in Fig. 2.

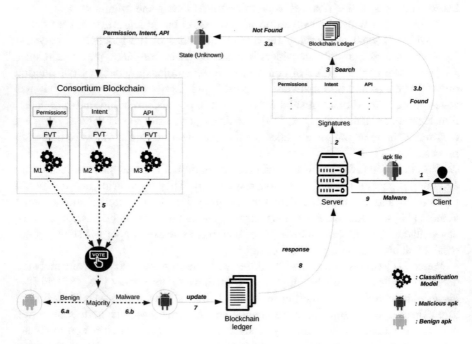

Fig. 2 Working and overall structure of the malware detecting blockchain system

7.1 APK Files

The client that requires a particular apk, file to be classified as malware or benign, provides it to the trusted server. The client establishes an encrypted connection with the server from the browser via an SSL Certificate and uploads the apk file.

7.2 Trusted Server

When the server receives the apk file, decompresses it and proceeds to retrieve the unique permissions, API calls, and intents. The feature vector is then populated with the corresponding values of each feature. These vectors are hashed with the MD5 algorithm and these hashes are used as the signatures to identify malware. The blockchain ledger is checked to see whether the signatures match with a record already present in it. If all the three signatures of the apk file match with the signatures of a particular record in the ledger, the apk file is classified by the "state" of the record. If just two of the signatures match, the vote under each signature is considered. If they match, the state is shared.

No match would imply that the signature of the apk file being considered does not exist on the ledger and so the trusted server becomes unable of classifying the file. In such a situation the server creates a new record of the file on the ledger. The server is the only entity in this system with the privileges to add a new record to the ledger.

7.3 Adding a Record

The server adds a new record to the ledger with information such as MD5 hash of the content of the apk file (this will serve as the identifier of this record), the URL from which this file can be downloaded from the server (this URL will be used by the consortium members of the blockchain) and initializes the state of the file as "UNDETERMINED".

The server makes use of a function defined by the blockchain network to add records onto the ledger and hence the ledger cannot add data to the ledger on its own and is restricted to the features provided by the network. On adding a new record to the ledger, an event is emitted across the network. This event will hold the added record's identifier along with the URL.

7.4   Members of the Consortium

The members connect to the network using certificates containing information, including their private keys, which will be used for their authentication. These members will be continuously listening for events that may occur on the blockchain ledger. When the server adds a new record on the ledger, the event emitted is caught by these listeners. Each event carries the hash of the file content along with the URL from which the file can be downloaded from the server. The members use this URL and download the file. They then compute the MD5 hash on the content themselves and compares the result with the hash they received from the event. If both the hashes match it can be concluded that the apk record created on the blockchain matches the one downloaded by the member. If not, it would imply that somewhere along the way the file has been corrupted. Each member on the consortium uses a particular feature, among permission, API and intent, for classifying the apk file, see Table 1. And so, after ensuring that the file received is authentic, they extract the features they are focusing on from it and create a feature vector. This vector is fed to their respective models and the classification is obtained. The member node then proceeds to cast its vote in the ledger based on the result.

7.5   Blockchain Ledger

The data block of each record holds the necessary details about the analyzed malware. The apkHash, State, URL, as was explained in previous sections, is initialized by the server when the new record is added. Each consortium node identifies the features and classifies the file. Based on the classification, they vote for or against the record of being malware. Their vote, the MD5 hash of the feature vector they have created along with the list of the particular feature used by the application is provided to the record on the ledger under their identity. It should be noted that only the consortium member on the network has the right to update any given record on the blockchain ledger. Figure 3 shows how a single data block would look like.

7.6   Final Response

Once the record has accumulated enough votes, the network determines whether the majority has voted for malware or for benign. The server retrieves this information and passes it to the client.

Table 1 The lists of features provided to the blockchain ledger by each member of the consortium

Feature	Values
Permissions	"android.permission.ACCESS_NETWORK_STATE"
	"android.permission.INTERNET"
	"android.permission.WRITE_EXTERNAL_STORAGE"
	"android.permission.ACCESS_FINE_LOCATION"
	.
	.
	.
	"android.permission.ACCESS_WIFI_STATE"
API	"setOverrideImpression Recording"
	"setOverrideClickHandling"
	"isSize Appropriate"
	"getHeightInPixels"
	.
	.
	.
	"getCustomEventExtrasBundle"
Intents actions	"android.intent.action.BOOT COMPLETED"
	"com.google.firebase.MESSAGING EVENT"
	"com.google.firebase.INSTANCE ID EVENT"
	"android.intent.action.MAIN"
	.
	.
	.
	"com.android.vending.INSTALL_REFERRER"

7.7 Technology Behind Blockchain Network

The mobile malware detection system using consortium blockchain was built making use of Hyperledger composer [5]. Hyperledger composer is an extensive, open development toolset and framework to make developing blockchain applications easier. Hyperledger composer supports the existing Hyperledger Fabric Blockchain infrastructure and runtime, which supports pluggable blockchain consensus protocols to ensure that a transaction is validated according to policy by the designated business network participants.

Hyperledger Composer can be used to quickly model the current business network; it is a private blockchain containing existing assets and the transactions related to them; assets may be tangible or intangible goods, services, or property. However, Hyperledger Composer also supports deploying a business network across multiple organizations allowing the establishment of a consortium. The business network model can be defined in such a manner that transactions can interact with assets. Business networks also include the participants who interact with them,

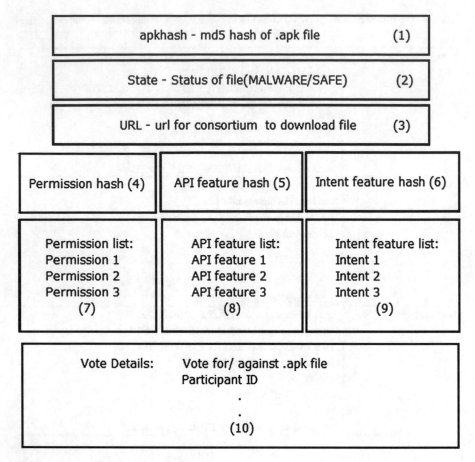

Fig. 3 Content of each block in the ledger

each of which can be associated with a unique identity, across multiple business networks.

Our system considers new blocks added to the ledger as the asset. Hyperledger Composer allows the user to define the assets and the various attributes associated with them. In this case, the asset holds data regarding its state, and the list of features of the apk file. This asset is the one the participants of the network will be working with. Participants are defined in a similar fashion as well with various attributes describing the participants, trusted server and the consortium members.

Hyperledger Composer allows the developers to deploy the functions that can be run by the participants on the ledger. This is considered as the logic of the network. This is the module in which the various operations that can be run by the various participants are defined. This logic module holds functions necessary for writing into the ledger, updating the blocks on the ledger, reading from the ledger. The logic module also allows defining events. Events can be paired with functions so when the function is invoked the event will be passed. In this paper, when the trusted server creates a new block on the ledger, an event is passed to the members of the consortium, notifying them that there is a new block to be considered.

Hyperledger Composer also provides a "permission" module. This module is what can be used to define what functions in the logic module the different participants have the authorization to operate. Functions to initialize a new block, and to read from the ledger are allowed to be invoked only by the trusted server. Whereas updating the predicted state field in a block along with the list of features of the apk file is performed only by the members of the consortium. Hyperledger Composer offers a private, permissioned blockchain. Our system ensures that each entry in the ledger is mapped to the author and is not left anonymous. This ensures that the data provided can be held against the author motivating them to provide credible data and discouraging them from malpractices as it would affect the reputation of the organization.

This permissioned consortium blockchain offers malware analysts a platform for spreading awareness when new malware is trending, by keeping a perpetual memory of all the indicators of compromise collected. The blockchain ensures all the analyses are notified to the members allowing them to take appropriate actions to counter the harm immediately. The interaction of various organizations to identify the class of various cases will prove to be much faster than when organizations work on their own. The mass malware attacks deployed on the internet can be classified and handled much faster, cutting down on the intensity of such attacks. Hyperledger Composer provides all the functionalities to run such a system across various organizations and build a very effective system.

8 Implementation Details

In this section two use cases are presented that illustrate the client's interaction with the ledeger blockchain and describe three algorithms to contemplate the initialization of a block and the voting operation in the corresponding blockchain ledger.

8.1 Scenario 1

Figure 4 depicts the scenario where a user feeds the system with an apk to be tested and the server upon receiving the apk will proceed to check the ledger if a record for this apk has already been created.

Step 1 represents the user requesting the server for classifying the apk. The user uploads the apk onto the server. The server accesses the blockchain network in order to look up the apk on the ledger which is step 2.

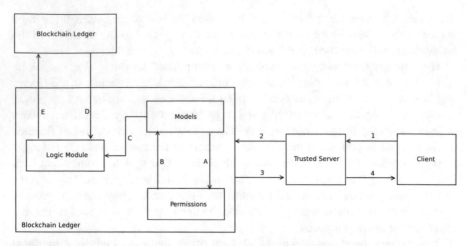

Fig. 4 The signature of the APK file is found on the blockchain ledger

The blockchain network contains three components, namely, Models, Permissions, and Logic Module. The Logic Module defines the various operations that can be performed across the blockchain ledger such as InitializeRecord (initializes a new record on the ledger for an apk that has not been classified as either malware or benign.), MakeVote(executed by each of the members of the consortium to vote for or against the apk being a malware), CheckVote(function necessary to count the votes and write the classification of the apk on the ledger).

The Models of the blockchain ledger defines all the entities included in the system along with their attributes. These entities include the trusted server (with attributes like the server details, id and role for identification and assigning privileges), the members of the consortium (having attributes like id, name, role, description and details), the block on the blockchain ledger with the attributes such as the hash of the apk, the URL for the members of the consortium to download the apk from the trusted server, the list of permissions, api and intents in the apk, the signatures, the votes cast by the members of the consortium, and also the state field which specifies the label of the apk (undefined or malware or benign). The permissions module in the blockchain ledger specifies the privileges each member in the blockchain network posses. The operations in the logic module can only be invoked by a member of the blockchain network if they are authorized in the permission module.

Within the blockchain, after the trusted server interacts with the network as depicted in step 2, the server is identified. Step A determines if the server has permission to read from the ledger. The permissions of the trusted server are confirmed in step B and the operation to read from the ledger is invoked in step C. Step D includes reading the ledger to determine if the apk had already been classified in the blockchain. Once the block corresponding to the apk has been found on the ledger, step E occurs where the details within the block are retrieved from the

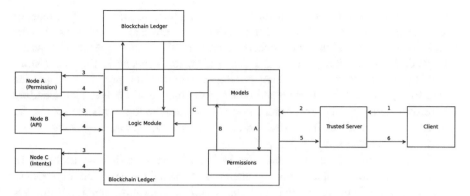

Fig. 5 The signature of the APK file is not found on the blockchain ledger

ledger. The trusted server then receives this information from the network in step 3. The server then proceeds to step 4, informing the user if the apk uploaded is labeled as malware or goodware on the blockchain ledger.

8.2 Scenario 2

Figure 5 considers a scenario where the signature for the apk sent by the user is not found by the trusted server when the server reads the ledger.

Step 1 represents the user uploading the apk to check if it is a malware. The trusted server then proceeds to step 2 to check if the apk had already been classified and is present in the blockchain ledger. Within the blockchain ledger, step A checks the permissions of the trusted server and upon confirming the permissions on step B, the ledger is read in step D by invoking the operation to read from the server into the logic module. Not having found the block for the apk on the ledger, indicating that this apk signature had never been classified as malware by the system before, step E returns empty and an operation to initialize a new block on the ledger is invoked from the logic module. This operation creates a new block and sets the state of the apk in the block as undetermined. At the end of the operation, an event is emitted across the network and is received by the members of the consortium (individually responsible for determining if the apk is malware or not), as shown on step 3. This event carries the URL from which these members can download the apk. Each member upon receiving the event is alerted that a new block has been added to the blockchain ledger. They download the apk from the trusted server and begins to extract the features, that is, the member in charge of the permissions will extract the permissions from the apk, the member responsible for classifying the apk based on the API will extract the API from the apk and the member responsible for classifying the apk based on the intents will extract the intents from the apk. These nodes will determine if the apk is malware based on these extracted features. Step 4 represents each member of the consortium reporting their contributions (the label of the apk, the list of features they extracted to determine the category of the apk) to the blockchain network. Within the blockchain network, the same sequence of steps are followed. After identifying the node interacting with the network, the permissions for this node are determined in step A. Step B confirms the permissions and the operation to update the undetermined block in the blockchain ledger by appending the information provided by the node is invoked from the logic module on step C. The updates are made in the corresponding block in the ledger on step D and step E returns the acknowledging response.

The votes are counted and the apk is determined to either be malware or goodware after enough nodes have updated the blockchain ledger ("enough nodes" suggests that if n is the total number of nodes in the consortium, then if n/2 number of nodes vote the apk a certain label −100 malware or benign, the contributions of the rest of the nodes needn't be provided before informing the user. In this experiment only 3 nodes were part of the consortium and participation of all these nodes is considered). On Step 5, the trusted server receives the most voted label for the apk from the blockchain network. The user is informed if the uploaded apk is malware or not on step 6 by the trusted server.

8.3   Initializing Block for Unknown apk

In Algorithm 1, the operation is employed to create a new block for an apk that has never been classified in the blockchain ledger. This operation will be invoked by the trusted server and requires the parameters, URL and apkHash, to function. The URL is the path defined in the trusted server that can be used by the different members of the consortium to download the apk from the trusted server to further analyze and classify. The apkHash is the MD 5 hash of the apk. The apkHash serves two purposes. First, it is used as the unique key to identify the block in the blockchain ledger. Second, when a member of the consortium blockchain downloads the apk from the server, the member performs an MD 5 hash of the apk on their own. The hash they compute is compared to the apkHash (hash computed at the server). The two must be a match, else it would indicate that the apk has been corrupted. This is very important as it verifies the authenticity of the apk by proving that it has not been tampered with. The operation begins on line 2 by creating a new resource making use of the apkHash as the primary key and this new instance of the block in the blockchain ledger is stored in a variable record. Since this block is new and has not been classified as either malware or benign, on line 3 the state is initialized as undetermined. On line 4 the block is appended with the URL of the apk location on the server. Once these initial attributes of the block is set, on line 5 the new block is added to the chain of blocks on the blockchain ledger. On line 6, the operation proceeds to initialize the arguments of an event. In line 8, the URL is added to the event. Line 9 has the apkHash defined to the event as well. Once the event is defined, it is emitted to the different members of the consortium blockchain on line 10.

Algorithm 1 Initializing block for unknown apk

```
 1: procedure INITIALIZERECORD(url, apkHash)
 2:     record ← newResource(apkHash)
 3:     record.state ← "UNDETERMINED"
 4:     record.url ← url
 5:     record.save()
 6:     event ← newEvent()
 7:     event.type ← NewBlock
 8:     event.apkHash ← apkHash
 9:     event.url ← url
10:     event.emit()
11: end procedure
```

8.4 Updating Block with Vote and Features

In Algorithm 2, the operation is employed to update the block, having a state undetermined, with the vote and data offered to be each of the nodes of the consortium blockchain. After having extracted the features, these nodes classify the apk as either malware or goodware.

Algorithm 2 Updating block with vote and features

```
 1: procedure MAKEVOTE(apkHash, vote, features)
 2:     record ← getResource(apkHash)
 3:     userContribs ← {}
 4:     userContribs.vote ← vote
 5:     userContribs.features ← features
 6:     record[ballots].append(userContribs)
 7:     record.update()
 8: end procedure
```

It is the duty of these nodes to append to the corresponding block the list of features along with the vote for or against the apk being malware. If the apk is classified as malware, the vote cast will be 1. If the apk is classified benign, the vote will be set to 0. This operation is invoked by passing this vote, the features and the apkHash as the parameters. On line 2, the block is retrieved from the blockchain ledger, by making use of the apkHash, which is the unique key that will identify the block and the instance of the block will be stored in the record. The record[ballot] will be a list where an element is one node's contributions. This includes the features along with the vote. Line 4 has the vote being stored in userContrib[vote]. Line 5 has the features stored in userContrib[features]. Line 6, userContrib is appended as an element of the record[ballot] list. This list's length would hence increase with the increase in the number of participating nodes. After the contributions are provided to the record variable, on line 7 record is updated in the corresponding block in the blockchain ledger.

8.5   *Setting the State of the apk After Counting All the Votes*

In Algorithm 3, Once n (The predefined odd number of members in the consortium blockchain that will participate in voting whether the apk is malware or not) number of nodes have updated the block with their votes and feature information, the votes need to be counted. The apkHash is passed as a parameter to identify the block in question.

Algorithm 3 Setting the state of the apk after counting all the votes

```
 1: procedure CHECKVOTE(apkHash)
 2:     n                                    ▷ Total number of members participating in the voting process
 3:     record ← getResource(apkHash)
 4:     voteCount ← 0
 5:     if n == record.ballots.length then
 6:         while n ≠ 0 do
 7:             if record.ballots[n].vote=1 then
 8:                 voteCount ← voteCount + 1
 9:                 n ← n − 1
10:             end if
11:         end while
12:         if voteCount>n/2 then
13:             record.state ← "Malware"
14:         else
15:             record.state ← "SAFE"
16:         end if
17:     end if
18:     record.update()
19: end procedure
```

On line 3, the record variable holds the block instance after retrieving it from the blockchain ledger making use of the apkHash. On line 4, a variable voteCount is initialized to 0. On line 5, a check is performed to see if n number of updates have been made. If so the counting of the votes can begin. Line 6, Looping through the elements in record[ballot], each vote is read. Within the loop, considering each element, if the vote is 1, the voteCount variable is incremented. After checking every vote, the loop ends. On line 10, if voteCount is more than $n/2 + 1$, then that would suggest that a majority of the nodes voted the apk as malware. If voteCount is less than $n/2 + 1$, then the majority voted the apk as benign. If the malware is voted, the state of the record is updated to malware on line 11. Else, the state of the record is updated to safe. Now that the apk is classified, on line 14, the block is updated on the blockchain ledger.

9 Feature Extraction and Model Training

The different nodes that participate in classifying the APK file represent the different malware agencies simulating how these agencies would participate in the classification process as the different members of the consortium blockchain. For the current experiment, each of these nodes are equipped with a machine learning model which requires specific data from the APK, permissions or API features or intent features. This section explains how the models used by these nodes are built and how the data used for training each model was extracted from the APK files. By working together, the system labels the file as malware or goodware, taking the input of all the different nodes that participated by using the models that specialize in either permissions, API features or intent features. The apk can be easily decompressed to obtain the Android Manifest file along with the necessary smali files using the APKTool. This tool is used for reverse engineering third-party, closed, binary Android apps. The Manifest file contains the list of permissions as well as the intents used by the Android application. Furthermore, the apk file don't necessarily be decompressed to obtain the list of permissions as it can be facilitated using the Android Asset Packaging Tool.

This experiment made use of 15,606 apk files for the purposes of training and testing the various machine learning models. A total of 9364 apk files, out of which 3170 files were malware and the remaining 6194 files were benign, for the training purpose. With a total of 3 nodes in the consortium, each node extracted the feature they were assigned. From the list of extracted features, all the unique features were collected. The number of unique APIs collected is extremely large number and so it is necessary to filter the result. Thus Fisher's scoring algorithm, a form of Newton's method used in statistics to solve maximum likelihood equations numerically, was used to score each API feature. The Fisher Score of the ith feature S_i can be calculated as

$$S_i = \frac{\sum n_j (\mu_{ij} - \mu_i)^2}{\sum_j n_j * \rho_{ij}^2} \tag{1}$$

where:

μ_{ij} = mean of the i-th feature in the j-th class
ρ_{ij} = variance of the i-th feature in the j-th class
n_j = number of instances in the j-th class
μ_i = mean of the i-th feature
S_i = score of the i-th feature

The first 5000 API features with the maximum scores were selected. These were then used as the attributes for the Feature Vector Table (FVT). Each row would correspond to a particular file among the 9364 files considered and for each feature that was present in a particular file, the cell corresponding to said file and feature attribute would be marked. The FVT is essentially a sparse matrix indicating the unique features present in each file. Each node will produce an FVT based on the feature they focused on.

Each node made use of five different classifiers-K-Nearest Neighbours, Linear Discriminant Analysis, Logistic Regression, Classification and Regression Tree and Random Forest Classifier. All three nodes would create models for each of the five classifiers making use of the FVTs they have created. When a test set is fed to these models, each of these models will predict the class of the sample independent of each other. The class predicted by the majority of the models is declared as the final prediction.

10 Dataset and Experimentation

The proposed system utilized 15,606 apks: 5373 malicious apks from the Drebin data-set and 10,233 benign apks from 9apps [1]. A web crawler was employed to download applications from 9apps. Various python libraries like Beautifulsoup [13] were used to fetch URLs for downloading the benign apks. The apks from 9apps were scanned with Virus Total and if any of the applications were classified by 5 or more of the antivirus software as malware, they were not included in the benign dataset. The Drebin data-set is considered to be the standard data-set of malicious apks used for malware detection studies as it contains applications from 179 different malware families and was collected through a span of August 2010 still October 2012. Due to the lack of availability of a standard data-set of benign applications we made use of the Android market ensuring that it encompasses various categories of applications.

In this experiment, the performance of the machine learning models used by the different members of the consortium blockchain to classify the different apks are evaluated through the following metrics: accuracy, precision, recall and F1-score. True Positive (TP) is the number of samples that are positive and are predicted to be positive. False Negative (FN) is the number of samples that are positive but are classified as negative. True Negative (TN) is the number of samples that are correctly identified as negative. False Positive (FP) is the number of samples that are negative but are recognized as positive.

$$Accuracy = \frac{TP + TN}{TP + TN + FP + FN} \tag{2}$$

$$Precision = \frac{TP}{TP + FP} \tag{3}$$

$$Recall = \frac{TP}{TP + FN} \tag{4}$$

Table 2 Evaluation of permission-based classifiers

Classifiers	Permission-based model			
	Accuracy	F1	Precision	Recall
Logistic regression	96.8	95.5	96.2	94.8
Linear discriminant analysis	94.7	92.7	91.8	93.6
K-nearest neighbor	96.4	95	95.9	94
Classification and regression tree	96.8	95.5	96.6	94.4
Random forest	95.5	93.4	99.0	88.4

Table 3 Evaluation of intent-based classifiers

Classifiers	Intent-based model			
	Accuracy	F1	Precision	Recall
Logistic regression	88.3	82.2	90.2	75.4
Linear discriminant analysis	86.5	77.8	94.9	65.9
K-nearest neighbor	86.4	77.4	95.1	65.3
Classification and regression tree	88.5	82.3	92.4	74.2
Random forest	86.2	76.5	98.2	62.7

Table 4 Evaluation of API call-based classifiers

Classifiers	API Call-based model			
	Accuracy	F1	Precision	Recall
Logistic regression	94.1	91.5	94.8	88.4
Linear discriminant analysis	89.2	84.7	86.2	83.2
K-nearest neighbor	88.7	85.9	77.8	95.9
Classification and regression tree	94.0	91.4	94.3	88.6
Random forest	90.7	85.6	96.4	77.0

$$F1 - Score = \frac{2 * Recall * Precision}{Recall + Presision} \tag{5}$$

Table 2 provides the performance evaluation measures, namely accuracy, F1-score, precision and recall, of the classifiers present in the permission-based node of the consortium blockchain. Tables 3 and 4 show the values for the intent-based and API call-based nodes of the consortium network, respectively.

11 Results

Figure 6 visualizes the comparison between each individual node as an independent system and our proposed system that encompasses the nodes as components of the entire system. The measures used for comparison include accuracy, F1-score, precision and recall (or sensitivity). The proposed system proves to be

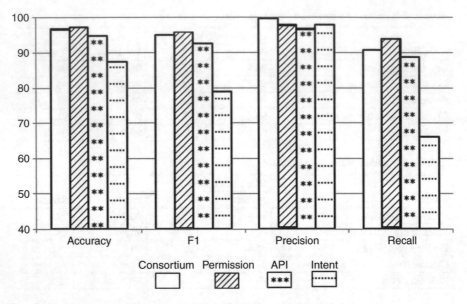

Fig. 6 Working and overall structure of the malware detecting blockchain system

comparatively better than the single nodes, in terms of performance. It should be noted that the frequency of unique API calls and intents were not included in the training of the respective nodes. However, the results indicate 99.6% precision provided by the proposed system; this indicates that our system performs better than the individual nodes.

12   Conclusion

To address the problem of detecting malicious codes in malware and extracting the corresponding evidence in mobile devices, the proposed system outperformed the individual malware detectors as it was determined to provide better precision. The results indicate that the label used by the majority of the participating nodes to classify performs better compared to when any single entity is used for the classification process. And since the proposed system requires the participation of mostly independent anti-malware agencies, it is important that a tamper-proof platform is set in place to ensure a healthy competition among the different participants. This proposed system will not only encourage anti-malware firms to partake in the classification process, which would provide better results, since the ledger itself can easily be made available, it would help the general public better understand the importance and working of cyber security. This proves that the blockchain network facilitated an environment for individual entities to work together without raising conflicts. Further, our system increases the awareness among the consortium members regarding the various malware signatures present in the current environment.

References

1. 9apps. 9apps: Free android apps. http://www.9apps.com/, accessed August 2021.
2. Yousra Aafer, Wenliang Du, and Heng Yin. Droidapiminer: Mining api-level features for robust malware detection in android. In *International conference on security and privacy in communication systems*, pages 86–103. Springer, 2013.
3. Daniel Arp, Michael Spreitzenbarth, Malte Hubner, Hugo Gascon, Konrad Rieck, and CERT Siemens. Drebin: Effective and explainable detection of android malware in your pocket. In *Ndss*, volume 14, pages 23–26, 2014.
4. Patrick PK Chan and Wen-Kai Song. Static detection of android malware by using permissions and api calls. In *2014 International Conference on Machine Learning and Cybernetics*, volume 1, pages 82–87. IEEE, 2014.
5. Gereon Dahmen and Volker Liermann. Hyperledger composer—syndicated loans. In *The Impact of Digital Transformation and FinTech on the Finance Professional*, pages 45–70. Springer, 2019.
6. Parvez Faruki, Ammar Bharmal, Vijay Laxmi, Vijay Ganmoor, Manoj Singh Gaur, Mauro Conti, and Muttukrishnan Rajarajan. Android security: a survey of issues, malware penetration, and defenses. *IEEE communications surveys & tutorials*, 17(2):998–1022, 2014.
7. Francois Gagnon and Frederic Massicotte. Revisiting static analysis of android malware. In *10th {USENIX} Workshop on Cyber Security Experimentation and Test ({CSET} 17)*, 2017.
8. Jingjing Gu, Binglin Sun, Xiaojiang Du, Jun Wang, Yi Zhuang, and Ziwang Wang. Consortium blockchain-based malware detection in mobile devices. *IEEE Access*, 6:12118–12128, 2018.
9. Hyunjae Kang, Jae-wook Jang, Aziz Mohaisen, and Huy Kang Kim. Detecting and classifying android malware using static analysis along with creator information. *International Journal of Distributed Sensor Networks*, 11(6):479174, 2015.
10. Weizhi Meng, Elmar Wolfgang Tischhauser, Qingju Wang, Yu Wang, and Jinguang Han. When intrusion detection meets blockchain technology: a review. *Ieee Access*, 6:10179–10188, 2018.
11. Michael Nofer, Peter Gomber, Oliver Hinz, and Dirk Schiereck. Blockchain. *Business & Information Systems Engineering*, 59(3):183–187, 2017.
12. Saurabh Raje, Shyamal Vaderia, Neil Wilson, and Rudrakh Panigrahi. Decentralised firewall for malware detection. In *2017 International Conference on Advances in Computing, Communication and Control (ICAC3)*, pages 1–5. IEEE, 2017.
13. Leonard Richardson. Beautiful soup documentation. *Dosegljivo*: https://www.crummy.com/software/BeautifulSoup/bs4/doc/.[*Dostopano: 7. 7. 2018*], 2007.
14. A-D Schmidt, Rainer Bye, H-G Schmidt, Jan Clausen, Osman Kiraz, Kamer A Yuksel, Seyit Ahmet Camtepe, and Sahin Albayrak. Static analysis of executables for collaborative malware detection on android. In *2009 IEEE International Conference on Communications*, pages 1–5. IEEE, 2009.
15. Guillermo Suarez-Tangil, Santanu Kumar Dash, Mansour Ahmadi, Johannes Kinder, Giorgio Giacinto, and Lorenzo Cavallaro. Droidsieve: Fast and accurate classification of obfuscated android malware. In *Proceedings of the Seventh ACM on Conference on Data and Application Security and Privacy*, pages 309–320, 2017.
16. Sushma Verma and SK Muttoo. An android malware detection framework-based on permissions and intents. *Defence Science Journal*, 66(6):618, 2016.
17. welivesecurity. Trends 2016: (in)security everywhere. https://www.welivesecurity.com/wp-content/uploads/2016/02/eset-trends-2016-insecurity-everywhere.pdf, accessed August 2021.

18. Dong-Jie Wu, Ching-Hao Mao, Te-En Wei, Hahn-Ming Lee, and Kuo-Ping Wu. Droidmat: Android malware detection through manifest and api calls tracing. In *2012 Seventh Asia Joint Conference on Information Security*, pages 62–69. IEEE, 2012.
19. Songyang Wu, Yong Zhang, Bo Jin, and Wei Cao. Practical static analysis of detecting intent-based permission leakage in android application. In *2017 IEEE 17th International Conference on Communication Technology (ICCT)*, pages 1953–1957. IEEE, 2017.
20. Ke Xu, Yingjiu Li, and Robert H Deng. Iccdetector: Icc-based malware detection on android. *IEEE Transactions on Information Forensics and Security*, 11(6):1252–1264, 2016.
21. Raima Zachariah, K Akash, Mohammed Sajmal Yousef, and Anu Mary Chacko. Android malware detection a survey. In *2017 IEEE international conference on circuits and systems (ICCS)*, pages 238–244. IEEE, 2017.
22. Win Zaw Zarni Aung. Permission-based android malware detection. *International Journal of Scientific & Technology Research*, 2(3):228–234, 2013.
23. Zibin Zheng, Shaoan Xie, Hongning Dai, Xiangping Chen, and Huaimin Wang. An overview of blockchain technology: Architecture, consensus, and future trends. In *2017 IEEE international congress on big data (BigData congress)*, pages 557–564. IEEE, 2017.
24. Yajin Zhou and Xuxian Jiang. Dissecting android malware: Characterization and evolution. In *2012 IEEE symposium on security and privacy*, pages 95–109. IEEE, 2012.
25. Jiawei Zhu, Zhengang Wu, Zhi Guan, and Zhong Chen. Api sequences based malware detection for android. In *2015 IEEE 12th Intl Conf on Ubiquitous Intelligence and Computing and 2015 IEEE 12th Intl Conf on Autonomic and Trusted Computing and 2015 IEEE 15th Intl Conf on Scalable Computing and Communications and Its Associated Workshops (UIC-ATC-ScalCom)*, pages 673–676. IEEE, 2015.

BERT for Malware Classification

Joel Alvares and Fabio Di Troia (iD)

Abstract In this paper, we aim to accomplish malware classification using word embeddings. Specifically, we trained machine learning models using word embeddings generated by BERT. We extract the "words" directly from the malware samples to achieve multi-class classification. In fact, the attention mechanism of a pre-trained BERT model can be used in malware classification by capturing information about the relation between each opcode and every other opcode belonging to a specific malware family. As means of comparison, we repeat the same experiments with Word2Vec. Differently than BERT, Word2Vec generates word embeddings where words with similar context are considered closer, being able to classify malware samples based on similarity. As classification algorithms, we used and compared Support Vector Machines (SVM), Logistic Regression, Random Forests, and Multi-Layer Perceptron (MLP). We found that the classification accuracy obtained by the word embeddings generated by BERT is effective in detecting malware samples, and superior in accuracy when compared to the ones created by Word2Vec.

1 Introduction

Malware is a computer program created with the intention to cause harm and damage to personal data, or gain unauthorized access to a user's system. Many are the techniques used by malware to conceal their malicious intent. One way is to masquerade itself as a legitimate program. This behavior has been observed, among others, in trojans and ransomware programs [2].

Identification and classification of malware is very critical to information security. According to the Sophos 2021 threat report [23], malware programs contributed to 34% of all the breaches in a survey consisting of 3500 IT professionals who worked on remote infrastructure and cloud-based infrastructure. Each malicious

J. Alvares · F. Di Troia (✉)
San Jose State University, San Jose, CA, USA
e-mail: fabio.ditroia@sjsu.edu

© The Author(s), under exclusive license to Springer Nature Switzerland AG 2022
M. Stamp et al. (eds.), *Artificial Intelligence for Cybersecurity*, Advances in
Information Security 54, https://doi.org/10.1007/978-3-030-97087-1_7

piece of code shares common characteristics within a certain family and tends to differ from malware samples belonging to a different family. It is necessary to identify these unique characteristics which would help classify malware codes belonging to numerous families [4]. Word embeddings can be used to quantify these unique characteristics of a malware sample, and they can be generated by *state-of-the-art* machine learning models, such as BERT [26] and Word2Vec [3]. The embeddings capture useful information that serves as training features for the classification models. In this paper, the focus is on the effectiveness of the word embeddings generated in the context of malware classification.

The remainder of the paper is organized in the following Sections. It starts with a survey of relevant work in Sect. 2. Then, the building blocks of the research are introduced in Sect. 3, that is, the background of the word embedding models and the applied classification models. Next, the dataset used, the applied methodology, and the accomplished experiments and the results are analyzed in Sect. 4. Finally, Sect. 5 contains the conclusions and suggestions for future work.

2 Related Work

Malware writers are constantly analyzing computer systems and their software in search of security faults that can be exploited by specific malware programs. To obfuscate their malicious intent, such programs implement sophisticated techniques to mask them as benign software and, thus, becoming invisible to malware recognition software [18]. This is the reason malware detection has become a challenging task. A lot of malware recognition techniques rely on signature-based detection. The antivirus program that relies on signature-based detection generally computes the hash of the files and compares it with the hash of known malware signatures [28]. However, modifying the code by inserting dead code within the malicious code is one easy way to avoid detection. Furthermore, this malware recognition technique is also inefficient, because all the files of a given user are scanned and compared with known available malicious signatures, which is a time consuming process. According to [25], a number of metamorphic malware families, such as, MetaPHOR, Zmist, Zperm, Regswap, and Evol morph after each new infection. Detecting these malware samples is challenging and it can defeat signature-based detection. Metamorphic malware morphs the code by using a combination of substitution, insertion, deletion, and transposition. However, the metamorphic malware can be identified by machine learning techniques because they are able to notice the subtle differences between malware and benign samples despite the use of morphing [30]. The effectiveness of the different machine learning techniques depends on the input features extracted from the dataset. Some possible features that can be used are signatures [28], API calls [27], and opcodes [5].

Natural Language Processing (NLP) techniques extract rich information, known as word embeddings, from sentences of a language, and are able to identify the meaning of a sentence, generate sentences with similar meaning, or fill the blanks within a sentence. The NLP models extract information of the relation of a word with every other word of a sentence. The model groups together words with similar meaning and maps them to a higher dimensional space where similar words in meaning are grouped together. This information helps NLP models accomplish several classification and prediction tasks. The NLP models can be used in the field of malware recognition to generate embeddings for malware samples. The malware samples that belong to the same family have features that are closely related. This information can be used by classifiers to group together malware samples that belong to the same family. BERT is one type of NLP model that can be used to generate word embeddings to capture information of every component of the input with respect to every other component. More details about the architecture of transformers and the attention mechanism of BERT can be found in [26], while an analysis of the attention heads of the BERT model can be found in [8]. The attention heads of BERT capture various patterns and linguistic notions.

Another example is Word2Vec, that was used in previous research to generate word embeddings for malware samples, with performance comparable to traditional machine learning techniques, such as, Hidden Markov Models and Principal Component Analysis [7]. The opcode sequences within malware samples are treated as a language in [1], and context is captured using Word2Vec. The classification is carried out using k-nearest neighbors (k-NN). The results derived by utilizing word embeddings generated by Word2Vec to achieve malware classification proves that NLP based models can extract rich features that assist with classification accuracy. This success induces to test newer NLP based models. Thus, differently than the previous work and in addition to it, we introduce the use of BERT in malware detection. BERT implements a transformer-based model that consists of encoders and decoders along with an attention mechanism [26]. The BERT model will be explained in further detail in Sect. 3. The experiments performed in this paper primarily focus on generating embeddings using BERT and comparing the classification accuracy with Word2Vec using a variety of classifiers.

3 Background

This section provides more details on the key components of this paper, that is, the NLP models and the implemented classifiers. The NLP models introduced are BERT and Word2Vec, while the classifiers are SVM, Random Forests, Logistic Regression, and MLP. The dataset, the results, and the experiments implemented using these building blocks are described in Sect. 4.

3.1 NLP Models

Natural Language Processing (NLP) is the subfield of Artificial Intelligence (AI) that enables machines to understand the language spoken by humans. The models that help achieve this result are known as NLP models. Training an NLP model from scratch is a tedious task and it requires a massive dataset and computational resources. For this reason, a pre-trained NLP model is often used to achieve the tasks related to NLP. Transfer learning is an example of technique used to transfer the knowledge gained by the model during the training phase to achieve other tasks on a different dataset to which it has never been exposed before. The tasks subject to NLP application are, for instance, sentiment analysis, next sentence prediction, word embedding generation, and more [21].

3.1.1 Word Embeddings

Word Embeddings are used in natural language processing as a representation of the words of a sentence in vector values such that words of similar meaning are grouped together in the vector space. This information can be used by classifiers to identify key features and efficiently accomplish classification. Features need to be extracted from the malware samples, which can be done by generating word embeddings from the malware samples. These word embeddings capture information and group together features that are unique to a specific malware family. They are generated using NLP based models such as Word2Vec and BERT. These word embeddings generated for every opcode in a malware sample can be represented as unit vectors and plotted in a circular heat map, as shown in Fig. 1 for, respectively, the malware families CeeInject, FakeRean, OnlineGames, Renos, and Winwebsec.

The circular heat map representation of the opcodes seem to differ for every malware family, even though the opcodes with higher frequencies across all the malware families are the opcodes *push*, *mov*, and *add*.

3.1.2 Word2Vec

Word2Vec is used to convert the input sequence of words to vectors, and map them to a higher dimensional space. The tutorial in [16] explains how Word2Vec uses neural networks to group together words with a similar meaning. For example, we can consider the following set of words:

$$w_0 = \text{"queen"}, w_1 = \text{"man"}, w_2 = \text{"woman"}, w_3 = \text{"king"}$$

In Fig. 2, we see how these words are mapped to a higher dimensional space by Word2Vec. Cosine similarity can then be used to identify words that are synonymous in nature. We can color the values using numbers, such that red

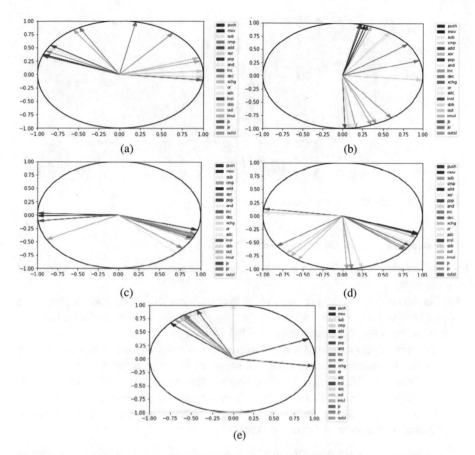

Fig. 1 Circular heatmaps for (**a**) CeeInject (**b**) FakeRean (**c**) OnlineGames (**d**) Renos (**e**) Winwebsec

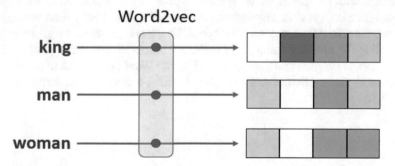

Fig. 2. Using Word2Vec to generate embeddings

represents a value close to 2, blue represents a value close to −2, and white represents a value close to 0.

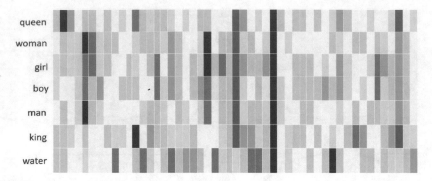

Fig. 3 Word embeddings represented as a color map

Based on Fig. 3, it can be observed that:

- The words "woman" and "girl" are considered similar to each other in several positions.
- The words "boy" and "girl" are similar in certain positions, but these positions are different from "woman" or "man". The algorithm could be capturing something similar between the words "boy" and "girl" such as "youth".
- The embeddings can be added and subtracted in order to form relations between words. For instance, in the case where the word embedding for the word "queen" is subtracted with the word embedding for the word "woman", and the word embedding for "man" is added, then the resultant word embedding is very close to the word embedding for the word "king". This can be represented as follows: "queen" − "woman" + "man" = "king"

Associating negative weights with frequently used words is another technique to improve the rate of training. In generating the output vectors, the positive weights associated with the model are all updated, while only a sample set of the negative weights are also updated. This reduces the impact of frequently used words while training the Word2Vec model. The Word2Vec model is used to generate word embedding for malware samples by using a window of size 6 and output size of 2 dimensions. We use the output generated by the Word2Vec model to generate unit vectors and plot a circular heat map which will be discussed in further detail in Sect. 3.1.1.

3.1.3 BERT

BERT is a transformer-based NLP model that is used to accomplish language-based tasks, such as, masked word prediction, sentiment classification, and more. The architecture consists of a stack of trained Transformer Encoders. BERT is able to generate the word embedding for a particular word by also taking into account the context in which it was used, known as contextualized word embeddings. The encoder uses attention to map the input to a set of vectors which store information of a given word with respect to every other word in the sentence. For instance, if we have the following input sentence: "*The boy drank water because he was thirsty*", the word "*he*" is associated with the word "*boy*", and the BERT model can identify this relation using attention. Attention helps BERT understand other relevant words in the sentence compared to the one that is currently being processed.

As shown in Fig. 4, the BERT model can accept at most 512 words as input. In general, a sentence in natural language does not exceed 512 words but the opcodes in a malware sample can exceed such value. In our experiments, the first 400 opcodes from each malware sample were sufficient to obtain good results. The BERT model used as a part of the experiments is DistilBERT, which is a smaller version of BERT that was open sourced by the HuggingFace team [9]. DistilBERT performs similarly to BERT but it is lightweight and, hence, more efficient. The DistilBERT model used is pre-trained on the English language. However, the model is neither trained nor fine-tuned to achieve malware classification. The classification token (CLS) from BERT, used to represent sentence-level classification output, captures the information about the entire sentence. In case of a malware sample, the CLS token captures the entire information of the sample. This information can be used in malware classification because the CLS token from the generated embedding collects information that helps with classification. For instance, if there are 2000 malware samples that BERT was trained on, and if 66 is the length of the tokens in the longest malware opcode sequence, as seen in Fig. 5, only the first column representing the CLS token is extracted from the 768 hidden units of BERT. A label is assigned to each of the 2000 sentences depending on the class of the malware sample.

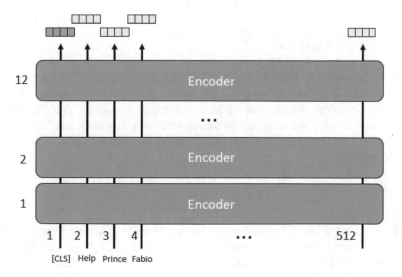

Fig. 4 Trained BERT components

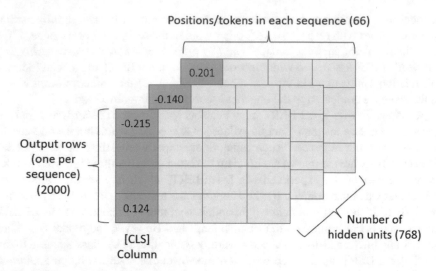

Fig. 5 Slicing BERT word embedding

3.2 *Classifiers*

Classification is the process of predicting the class or label of the input dataset. The input dataset is mapped to the desired output class depending on the features of the input data. The machine learning models, which enable the user to map the input data to its corresponding class, are known as classifiers. This Section will briefly introduce the classifiers used in our experiments.

3.2.1 Logistic Regression

Logistic regression is used to describe the input data and to find a correlation between them. The result of logistic regression is dichotomous in nature. A logistic regression model used to fit more than two classes is referred to as multinomial logistic regression. The model achieves classification using multinomial probability distribution. The assumption of logistic regression is a sigmoid function that can be defined as follows:

$$f(x) = \frac{1}{1 + e^{-(x)}} \tag{1}$$

The disadvantages of logistic regression are similar to linear regression. It is, in fact, prone to outliers, and assumption of linearity amongst dependent and independent variables. However, logistic regression model provides probabilities, and it is not just a classification model. It enables the user to identify the percentage with which a certain instance was assigned to a class. A detailed explanation and various strategies guidelines for logistic regression can be found in [19].

3.2.2 SVM

The main objective of SVM is to accomplish classification within the dataset by maximizing the distance between the separating hyperplane and the dataset.

As shown in Fig. 6, the hyperplane with the maximum distance from the dataset is chosen. The support vectors are the data points closest to the hyperplane. These are used by SVM to maximize the separation between the data points and the hyperplane. SVMs can be used to identify the subtle changes in malware samples belonging to a certain family as discussed in [29]. SVM identifies that the dataset may not be linearly separable by itself. Hence, the dataset is mapped to a higher dimensional space where a separating hyperplane can classify the dataset. This approach is often referenced to as kernel trick. For example, in Fig. 7 we see that the data on the left side is not linearly separable. However, the data can be easily separated by a hyperplane if the data is mapped as seen on the right side. One of the ways to achieve this is by using a polynomial kernel. There are many kernels that can be applied, and identifying the right one can be a challenging task, but it can significantly improve the classification accuracy without causing a major computation overhead. The classification process of SVMs and the mathematical proof can be found in [24].

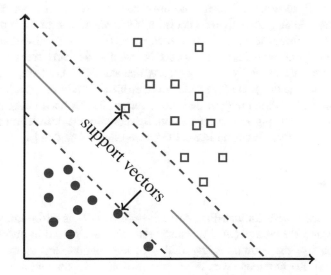

Fig. 6 SVM for binary classification

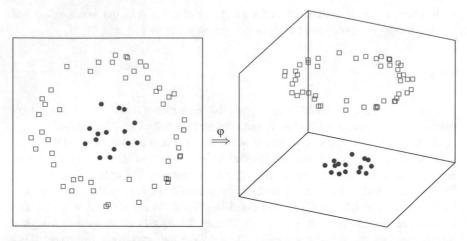

Fig. 7 Mapping input data to a higher dimension

3.2.3 Random Forests

Random Forests accomplish the classification of the dataset using an ensemble of decision trees. Every tree classifies the data independently from the others and votes for a specific class based on its prediction. The class with the highest number of votes is, then, selected as the classification output of the Random Forest. Basically, a large number of decision trees achieve the classification together as a committee, and the overall accuracy of such a committee outperforms the accuracy of an individual tree. In fact, an individual decision tree tends to overfit the input dataset, while a group of trees tends to protect each other from their individual errors. A problem that arises with Random Forests is that the decision trees may be too correlated with each other. Hence, bagging, which stands for bootstrap aggregation, is used to overcome this issue. The decision trees are formed using random samples of the training data which may or may not overlap. In this way, bagging prevents the Random Forest from overfitting the data by reducing the correlation among the decision trees. Further details on Random Forests can be found at [6].

3.2.4 MLP

A neuron, known as McCulloch-Pitts Artificial neuron [17], is the building block for a Multi Layered Perceptron (MLP). Multiple neurons are placed in different layers and the inputs of the neurons in the hidden or intermediate layers are outputs of the neurons in the previous layer. A neuron with three inputs and a single output is depicted in Fig. 8, where the inputs are X_0, X_1, and X_2, while the weights associated with these inputs are w_0, w_1, and w_2. The neuron generates an output $Y \in [0, 1]$ where 1 implies that the neuron was activated, while 0 implies that

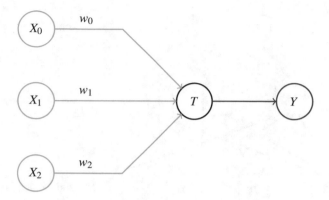

Fig. 8 Neuron of a neural network

the neuron remained inactive. The weights together with the input determine if the neuron should output or not. If the value $\sum w_i X_i$ is greater than the threshold T, then the neuron is activated. Equation 2 represents the function that a neuron of an MLP utilizes. An independent bias b is also introduced and updated during the training of the MLP.

$$f(x, y) = \sum_{i=0}(n - 1)w_i X_i + b \tag{2}$$

In case of binary classification, if the Eq. 2 generates a positive value, then we classify the input as class 1, or, if the function generates a negative value, the input is classified as class 2. The decision boundary of the binary classifier is represented by the Eq. 3. The decision boundary separates the inputs into the two classes in the output dimension space.

$$f(x, y) = w_0 x + w_1 y + b \tag{3}$$

An MLP consists of multiple layers of these perceptron's, as shown in Fig. 9 which consists of two hidden layers. Each edge of the MLP has a weight associated with it, and the definitive values of the weights are finalized after the training phase. More details on the MLP architecture can be found at [20].

4 Experiments and Results

In this section, we describe dataset used for the experiments, the parameters used for the machine learning models, and their classification results accomplished on the word embeddings generated by BERT and Word2Vec.

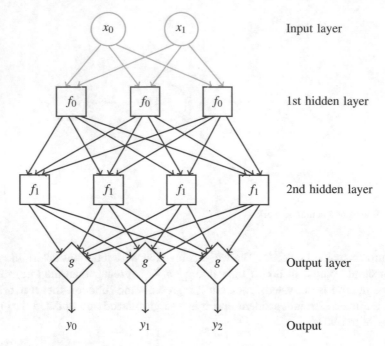

Fig. 9 Multi layer perceptron

Table 1 Malware dataset information

Malware family	Malware type	Nr. of samples
CeeInject	VirTool	899
FakeRean	Rogue	899
OnlineGames	Password stealer	900
Winwebsec	Rogue	897
Renos	Trojan downloader	900

4.1 Dataset

All our experiments were based on the malware families described in Table 1, along with the number of malware samples for each family [15].

A brief description of each malware family is given here.

1. CeeInject is malware that is generally used in combination with other malware families as it is used to conceal the other malware samples. The malware that CeeInject is used along with is installed in a user's machine without requesting any permissions [11].
2. FakeRean alerts the user for issues or viruses that do not exist on the system and asks for money in order to assist the user [14].
3. OnlineGames is used to track the login information of online games and keeps track of information of online gamers without consent [12].

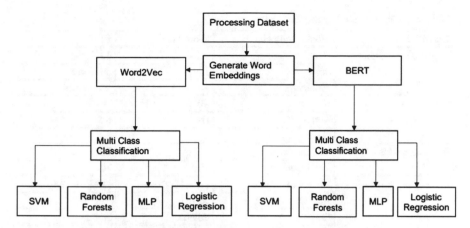

Fig. 10 Illustration of the organization of the project

4. Winwebsec is a trojan that pretends to be a legitimate antivirus software, informing the user that the system is corrupt and needs to be fixed. It tries to scare the user with the intention of extorting money [13].
5. Renos is a malware that shows to the user fake security warnings once it is downloaded and requests for payments to resolve the issues [10].

4.2 Methodology

In Fig. 10, we see an illustration of our approach. The input dataset of malware samples is processed and transformed into inputs for BERT and Word2Vec that generate the word embeddings. Then, these are directly used to train the machine learning models to achieve multi-class classification on the malware samples. The word embeddings generated are, thus, classified to their respective malware families with the help of the classifiers described in Sect. 3, that is, Support Vector Machines (SVMs), Random Forests, Multi Layer Perceptron (MLP), and Logistic Regression. In this way, the overall accuracy depends on the classification of the word embeddings which capture the essential characteristics of the malware samples.

4.3 Classifier Parameters

The parameters that were selected for the classification are shown in Table 2. We found these values to be the optimal ones by experimenting using GridSearchCV

Table 2 Parameters used by the classifiers

Classifier	Model parameter	Word2Vec	BERT
Logistic regression	C	42.1	42.1
	Solver	Lbfgs	newton-cg
	Multiclass	Auto	Multinomial
SVM	C	1000	1000
	Kernel	rbf	rbf
	Gamma	1	1
Random forests	max depth	20	20
	n estimators	100	100
MLP	Hidden layer size	(150,150,100)	(100,100,100)
	Activation function	ReLU	ReLU
	Solver	adam	adam
	Nr. of iterations	3000	10,000
	Learning rate	Constant	invscaling

from the scikit-learn library [22]. The parameters obtained are almost identical for the features generated by both BERT and Word2Vec.

4.4 Logistic Regression Results

Optimal results were obtained by the logistic regression model using the regularization parameter value $C = 42.1$. The different values for C were obtained using numpy's linspace function by dividing the range 0.0001 to 100 into 20 parts. The test accuracy of this model was 81.2% using the word embeddings generated by Word2Vec, and 83.54% using the word embeddings generated by BERT. The confusion matrices of the obtained results for BERT and Word2Vec are shown in Fig. 11.

The overall accuracy is unsatisfactory when compared with the other classifiers. One of the possible reasons is that the model is overfitting the decision boundary to the training dataset. This causes the model to perform poorly when exposed to new data.

4.5 SVM Results

Experiments were achieved on the SVM model and the ideal set of parameters that produced the maximum accuracy were selected. We tested different types of kernels, that is, radial basis function (rbf) kernel, linear, and polynomial, along with the regularization parameter C in the range 10 to 1000, and gamma value in the range 0.001 to 0.1. SVM maps the input features to a higher dimensional space in

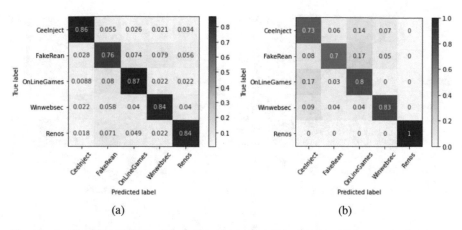

Fig. 11 Confusion matrix of logistic regression for (**a**) BERT features (**b**) Word2Vec features

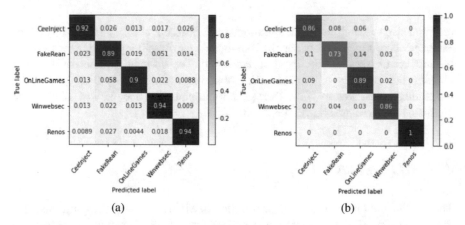

Fig. 12 Confusion matrix of SVM for (**a**) BERT features (**b**) Word2Vec features

order to form a decision boundary that separates the features into different classes. For this reason, SVM is able to successfully leverage the features in the word embeddings and group together malware samples with similar features obtaining a high classification accuracy of around 91.01% using the embedding generated by BERT. The embeddings generated by Word2Vec, instead, obtained a classification accuracy of 86.8%. The confusion matrices are shown in Fig. 12.

4.6 Random Forest Results

Random Forest is a neighborhood-based algorithm that classifies input features by grouping the ones that are closer to each other, and making decisions at

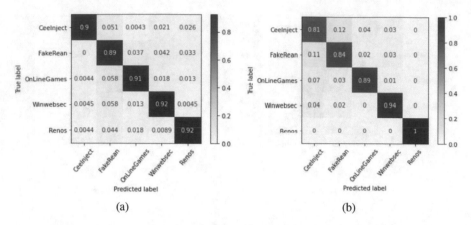

Fig. 13 Confusion matrix of random forest for (**a**) BERT features (**b**) Word2Vec features

different stages which segregate the inputs into different classes. The results of the experiments conducted show that the Random Forest classifier performs better when the number of trees and the depth is increased. The optimal parameters lead to a classification accuracy of 91.81% with embeddings generated by BERT, while the embeddings generated by Word2Vec gave a classification accuracy of 89.6%. The confusion matrices are shown in Fig. 13.

4.7   MLP Results

The multilayered perceptron performs quite closely as SVM by mapping the input features to a higher dimensional space, and accomplishing classification by forming a decision boundary to group together features that are closer to each other. A constant learning rate with a 30,30,30) hidden layer width and ReLU activation function provided the best results. The classifier converged and gave optimal outcome at around 10,000 iterations. The final accuracy obtained using the word embeddings generated by BERT was 86.83%, which is not surprisingly close to the accuracy obtained by SVM. The word embeddings generated by Word2Vec obtained a similar result with a final accuracy of around 86.6%. The confusion matrices are shown in Fig. 14.

4.8   Further Analysis

Random Forest is a neighborhood-based classification model. By our experiments, we noticed that the model performs poorly when the depth of the binary tree of the decision is shallow. It tended to overfit to the training data, as the training

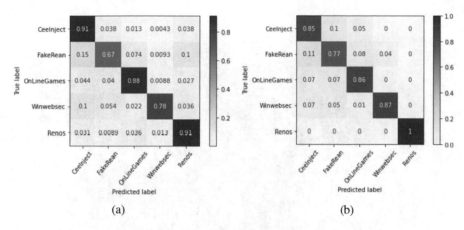

Fig. 14 Confusion matrix of MLP for (**a**) BERT features (**b**) Word2Vec features

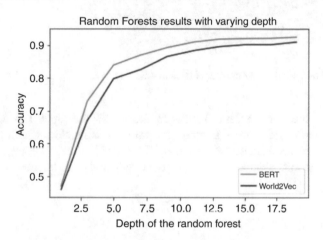

Fig. 15 Depth vs accuracy for RF using BERT and Word2Vec

accuracy was high, while the model performed insufficiently when tested on the test data. Figure 15 shows that the embeddings generated from both BERT and Word2Vec demonstrate improvement in the classification accuracy when the depth of the Random Forest was increased. The accuracy plateaus at depth 10 and gradually increases beyond this point. After further analysis, it was observed that such behavior was similar when both the depth and number of trees of the Random Forest were increased. It was also observed that a larger number of trees in the Random Forest classification model compensate for shallow depths. As seen in Fig. 16, the accuracy of the Random Forest model was high even when the depth of the decision trees was around 2.5. Beyond a depth of 2.5 there was a gradual increase in classification accuracy as the number of trees of the Random Forest classifier was increased. As described in Sect. 3.2.3, a larger number of decision trees can

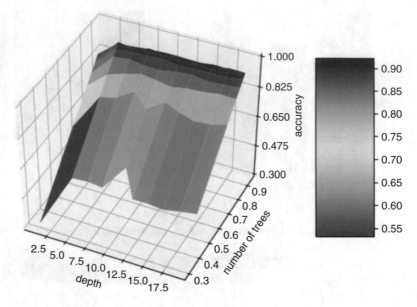

Fig. 16 Accuracy for depth vs number of tress in random forest

generalize to the training data. A class is chosen for the input data only when a majority of the decision trees generate the same classification, which protects the classification result from errors caused by the individual decision trees. This is in line with the results obtained as a part of the conducted experiments.

4.9 Summary

Word embeddings were generated by BERT and Word2Vec and they were classified using classifiers such as Logistic Regression, SVM, MLP, and Random Forests. Classification of malware samples by using word embedding generated with BERT performs better overall in comparison to Word2Vec, as shown in the Fig. 17 that summarizes the results. SVM, MLP, and Random Forests perform better overall in comparison to Logistic Regression, which is an expected outcome. MLP and SVM perform similarly as they try to find the decision boundary that best fits the data without overfitting it. Random Forests use an ensemble of decision trees to carry out the classification of the dataset and obtain a high classification accuracy. The results of this experiment prove that word embeddings generated by BERT can be used to accomplish multi-class malware classification of the dataset. When compared to Word2Vec, we see that is obtains better results. This is likely due to the attention mechanism that characterizes the BERT algorithm. In fact, by implementing the attention mechanism, BERT is able to maintain long-term dependencies as easily

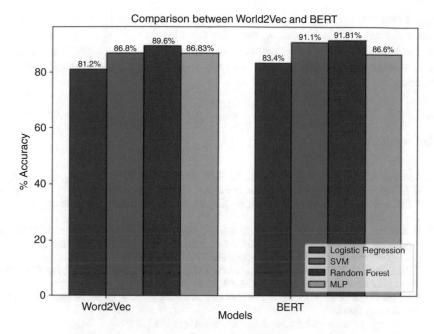

Fig. 17 Accuracies obtained after classification

as short-term dependencies. The number of opcodes selected per malware sample was 400, since BERT required a maximum of 512 words per sentence. Furthermore, while some of the malware samples had over 105 unique opcodes, the classification accuracy was not impacted, even when there was a smaller number of opcodes for some samples. This proves that word embeddings generated by BERT can capture rich features with even a limited subset of the opcodes for each input file.

5 Conclusions and Future Work

Based on the experiments conducted in this paper, it was observed that when the malware samples were mapped to word embeddings by capturing, grouping, and enriching the key components of the input features, it led to an improvement in classification accuracy while achieving malware classification. The promising results show that BERT was able to capture information that helps the classifier improving the classification accuracy. The results were superior to the ones obtained using Word2Vec on the same set of input parameters and the same set of classifiers. It proves that a transformer-based model such as BERT has applications beyond NLP. For future work, more research can be conducted in this area by using different versions of BERT. DistilBERT was used in these paper but further research can be accomplished using the other available versions of BERT. Morover, the BERT model was pre-trained on natural language input, but it may be able to generate richer features if it is trained on malware samples directly. Finally, research can be implemented using more malware families with more complex sets of data and observing how BERT captures the key information across multiple malware families.

References

1. Yara Awad, Mohamed Nassar, and Haidar Safa. Modeling malware as a language. In *2018 IEEE International Conference on Communications (ICC)*, pages 1–6, 2018.
2. P. Baldi and Y. Chavin. Smooth on-line learning algorithms for hidden markov models. *Neural Computation*, 6:307–318, 1994.
3. S. Banerjee. Word2vec — a baby step in deep learning but a giant leap towards natural language processing. https://laptrinhx.com/word2vec-a-baby-step-in-deep-learning-but-a-giant-leap-towards-natural-language-processing-3998188269/, 2018.
4. S. Basole, F. Di Troia, and M. Stamp. Multifamily malware models. *Journal of Computer Virology and Hacking Techniques*, 16:79–92, 2020.
5. D. Bilar. Opcodes as predictor for malware. *Int. J. Electron. Secur. Digit. Forensic*, 1(2):156–168, January 2007.
6. L. Breiman. Random forests. *Machine learning*, 45(1):5–32, 2001.
7. Aniket Chandak, Wendy Lee, and Mark Stamp. A comparison of word2vec, hmm2vec, and pca2vec for malware classification. https://arxiv.org/abs/2103.05763, 2021.
8. K. Clark, U. Khandelwal, O. Levy, and C. Manning. What does BERT look at? an analysis of BERT's attention. In *Proceedings of the 2019 ACL Workshop BlackboxNLP: Analyzing and Interpreting Neural Networks for NLP*, pages 276–286, Florence, Italy, August 2019. Association for Computational Linguistics.
9. HuggingFace. Distilbert. https://huggingface.co/transformers/model_doc/distilbert.html.
10. Microsoft Security Intelligence. Renos. https://www.microsoft.com/en-us/wdsi/threats/malware-encyclopedia-description?Name=TrojanDownloader:Win32/Renos&threatId=16054, 2006.
11. Microsoft Security Intelligence. Ceeinject. https://www.microsoft.com/en-us/wdsi/threats/malware-encyclopedia-description?Name=VirTool%3AWin32%2FCeeInject, 2007.
12. Microsoft Security Intelligence. Onlinegames. https://www.microsoft.com/en-us/wdsi/threats/malware-encyclopedia-description?Name=PWS%3AWin32%2FOnLineGames, 2008.
13. Microsoft Security Intelligence. Winwebsec. https://www.microsoft.com/security/portal/threat/encyclopedia/entry.aspx?Name=Win32%2fWinwebsec, 2010.
14. Microsoft Security Intelligence. Fakerean. https://www.microsoft.com/en-us/wdsi/threats/malware-encyclopedia-description?Name=Win32/FakeRean, 2011.
15. Samuel Kim. Pe header analysis for malware detection. Master's thesis, San Jose State University, Department of Computer Science, 2018.
16. C. McCormick. Word2vec tutorial - the skip-gram model. http://mccormickml.com/2016/04/19/word2vec-tutorial-the-skip-gram-model, 2016.
17. W. S. McCulloch and W. Pitts. A logical calculus of the ideas immanent in nervous activity. *The bulletin of mathematical biophysics*, 5(4):115–133, 1943.
18. C. Mihai and J. Somesh. Testing malware detectors. In *Proceedings of the 2004 ACM SIGSOFT International Symposium on Software Testing and Analysis*, ISSTA '04, page 34–44, New York, NY, USA, 2004. Association for Computing Machinery.
19. Fred C. Pampel. *Logistic Regression: A Primer*. SAGE Publications, Inc., 2000.
20. H. Ramchoun, M. A. J. Idrissi, Y. Ghanou, and M. Ettaouil. Multilayer perceptron: Architecture optimization and training. *Int. J. Interact. Multim. Artif. Intell.*, 4(1):26–30, 2016.

21. N. Ranjan, K. Mundada, K. Phaltane, and S. Ahmad. A survey on techniques in nlp. *International Journal of Computer Applications*, 134(8):6–9, 2016.
22. sklearn. Gridsearchcv. https://scikitlearn.org/stable/modules/generated/sklearn.model_selection.GridSearchCV.html.
23. SophosLabs. Sophos 2021 threat report. https://www.sophos.com/en-us/medialibrary/pdfs/technical-papers/sophos-2021-threat-report.pdf, 2021.
24. Mark Stamp. *Introduction to Machine Learning with Applications in Information Security*. Chapman and Hall/CRC, 2020.
25. Symantec. Internet security threat report: Malware. https://interactive.symantec.com/istr24-web, 2019.
26. Ashish Vaswani, Noam Shazeer, Niki Parmar, Jakob Uszkoreit, Llion Jones, Aidan N. Gomez, Lukasz Kaiser, and Illia Polosukhin. Attention is all you need. https://arxiv.org/abs/1706.03762, 2017.
27. S. Vemparala, F. Di Troia, C. Visaggio, T. Austin, and M. Stamp. Malware detection using dynamic birthmarks. In *Proceedings of the 2016 ACM on International Workshop on Security And Privacy Analytics*, IWSPA '16, page 41–46, New York, NY, USA, 2016. Association for Computing Machinery.
28. P. Vinod, R. Jaipur, R. Laxmi, and M. Gaur. Survey on malware detection methods. In *Proceedings of the 3rd Hackers Workshop on Computer and Internet Security*, pages 74–79, 2009.
29. M. Wadkar, F. Di Troia, and M. Stamp. Detecting malware evolution using support vector machines. *Expert Systems with Applications*, 143:113022, 2020.
30. W. Wong and M. Stamp. Hunting for metamorphic engines. *Journal of Computer Virology and Hacking Techniques*, 2:211–229, 2017.

Machine Learning for Malware Evolution Detection

Lolitha Sresta Tupadha and Mark Stamp

Abstract Malware evolves over time and antivirus must adapt to such evolution. Hence, it is critical to detect those points in time where malware has evolved so that appropriate countermeasures can be undertaken. In this research, we perform a variety of experiments on a significant number of malware families to determine when malware evolution is likely to have occurred. All of the evolution detection techniques that we consider are based on machine learning and can be fully automated—in particular, no reverse engineering or other labor-intensive manual analysis is required. Specifically, we consider analysis based on hidden Markov models (HMM) and the word embedding techniques HMM2Vec and Word2Vec.

1 Introduction

Malware is software that is intended to be malicious in its effect [1]. By one recent estimate, there are more than one billion malware programs in existence, with 560,000 new malware samples discovered every day [12]. Clearly, malware is a major cybersecurity threat, if not the most serious security threat today.

Since the creation of the ARPANET in 1969, there has been an exponential growth in the number of users of the Internet. The widespread use of computer systems along with continuous Internet connectivity of the "always on" paradigm makes modern computer systems prime targets for malware attacks. Malware comes in many forms, including viruses, worms, backdoors, trojans, adware, ransomware, and so on. Malware is a continuously evolving threat to information security.

In the field of malware detection, a signature typically consists of a string of bits that is present in a malware executable. Signature-based detection is the most popular method of malware detection used by anti-virus (AV) software [1]. But malware has become increasingly difficult to detect with standard signature-based

L. S. Tupadha · M. Stamp (✉)
San Jose State University, San Jose, CA, USA
e-mail: lolithasresta.tupadha@sjsu.edu; mark.stamp@sjsu.edu

© The Author(s), under exclusive license to Springer Nature Switzerland AG 2022
M. Stamp et al. (eds.), *Artificial Intelligence for Cybersecurity*, Advances in
Information Security 54, https://doi.org/10.1007/978-3-030-97087-1_8

approaches [34]. Virus writers have developed advanced metamorphic generators and obfuscation techniques that enable their malware to easily evade signature detection. For example, in [3], the authors prove that carefully constructed metamorphic malware can successfully evade signature detection.

Koobface is a recent example of an advanced form of malware. This malware was designed to target the users of social media, and its infection is spread via spam that is sent through social networking websites. Once a system is infected, Koobface gathers a user's sensitive information such as banking credentials, and it blocks the user from accessing anti-virus or other security-related websites [11].

Malware writers modify their code to deal with advances in detection, as well as to add new features to existing malware [2]. Hence, malware can be perceived as evolving over time. To date, most research into malware evolution has relied on software reverse engineering [7], which is labor intensive. Our goal is to detect malware evolution automatically, using machine learning techniques. We want to find points in time where it is likely that significant evolution has occurred within a given malware family. It is important to detect such evolution, as these points are precisely where modifications to existing detection strategies are urgently needed.

We note in passing that malware evolution detection can play an additional crucial role in malware research, beyond updating existing detection strategies to deal with new variants. Generally, in malware research, we consider samples from a specific family, without regard to any evolutionary changes that may have occurred over time. An adverse side effect of such an approach is that—with respect to any specific point in time—we are mixing together past, present, and future samples. Relying on training based on future samples to detect past (or present) samples is an impossibility in any real-world setting, yet it is seldom accounted for in research. By including an accurate evolutionary timeline, we can conduct far more realistic research. Thus, accurate information regarding malware evolution will also serve to make research results more realistic and trustworthy.

We consider several machine learning techniques to identify potential malware evolution, and our experiments are conducted using a significant number of malware families containing a large numbers of samples collected over an extended period of time. We extract the opcode sequence from each malware sample, and these sequences are used as features in our experiments. We group the available samples based on time periods and we train machine learning models on time windows. We compare the models to determine likely evolutionary points—substantial differences in models across a time boundary indicate significant change in the code base of the malware family under consideration. Specifically, we experiment with hidden Markov models (HMM) and word embedding techniques (Word2Vec and HMM2Vec). For comparison, we also consider logistic regression.

One limitation of this research is that we do not explicitly determine the importance of any evolution has actually occurred—we simply provide evidence of code modification within a specified time interval. To determine the significance of any evolution, we would need to carefully examine malware samples, most likely via labor-intensive reverse engineering techniques. A related issue is that a small modification to a malware family might result in a large change in the functionality of the code. Thus, the the importance of an evolutionary change may not be captured by the magnitude of the metrics that we use to quantify evolutionary change. Nevertheless, the work presented here can provide a first step in an overall malware evolution detection scheme. Specifically, our machine learning based approach allows anti-virus researchers to focus their efforts at specific points in time where evolution is most likely to have an impact on the performance of a malware family.

The remainder of this paper is organized as follows. In Sect. 2, we discuss a range of relevant background topics, including malware, related work, our dataset, and we introduce the learning techniques that we employ in our experiments. Section 3 contains our the experimental results, while Sect. 4 gives our conclusions along with a discussion of a few potential avenues for future work.

2 Background

In this section, we first give a brief introduction to malware. Then we consider related work in the area of malware evolution detection.

2.1 Malware

A computer worm is a kind of malware that spreads by itself over a network [1]. Examples of famous worms include Code Red, Blaster, Stuxnet, Santy, and, of course, the Morris Worm [33].

Viruses are the most common form of malware, and the word "virus" is often used interchangeably with "malware." A computer virus is similar to a worm but it requires outside assistance to transmit its infection from one system to another. Viruses are often considered to be parasitic, in the sense that they embed themselves in benign code. More advanced forms of viruses (and malware, in general) often use encryption, polymorphism, or metamorphism as means to evade detection [1]. These techniques are primarily aimed at defeating signature-based detection, although they can also be effective against more advanced detection strategies.

A trojan horse, or simple a trojan, is malicious software that appears to be innocent but carries a malicious payload. Trojans are particularly popular today, with the the vast majority of Android malware, for example, being trojans.

A trapdoor or backdoor is malware that allows unauthorized access to an infected system [33]. Such access allows an attacker to use the system in a denial of service (DoS) attack, for example.

Traditionally, malware detection has relied on static signatures, which typically consist of strings of bits found in specific malware samples. While effective, signatures can be defeated by a wide variety of obfuscation and morphing techniques, and the sheer number of malware samples today can make signature scanning infeasible.

Recently, machine learning and deep learning techniques have become the tools of choice for malware detection, classification, and analysis. We would argue that it is also critical to detect malware evolution, since we need to know when a malware family has evolved in a significant way so that we can update our detection techniques to account for such changes. As we see in the next section, this aspect of malware analysis has, thus far, received only limited attention from the research community.

2.2 Related Work

While there is a great deal of research involving applications of machine learning to malware detection, classification, and analysis, there are very few articles that consider malware evolution. In [10], analysis of malware based on code injection is considered. This works deals with shell code extracted from malware samples. The researchers used clustering techniques to analyze shell code to determine relationships between various samples. This work was successful in determining the similarities between samples, showing that a significant amount of code sharing had occurred. A drawback to the approach in this paper is that the authors only considered analysis of shell code. While shell code often serves as the attack vector for malware, other attack vectors are possible, and malware evolution is not restricted to the attack portion of the code. For example, a malware family might evolve to be more stealthy or obfuscated, without affecting the attack payload. Another limitation of this research is that it only considers software similarity, and not malware evolution, per se.

Malware evolution research is considered in [8]. One positive aspect of this research is that it considers a large dataset that spans two decades. The authors use techniques based on graph pruning and they claim to show specific properties of various families are inherited from other families. However, it is not clear whether these properties are inherited from other families, or were developed independently. In addition, this work relies on manual investigation. A primary goal of our research is to eliminate the need for such manual intervention.

The research presented in [29] is focused on detecting malware variants, which can be considered as a form of evolution detection. The authors apply semi-supervised learning techniques to malware samples that have been shown to evade machine learning based detection. In contrast, in our research, we use unsupervised learning techniques to detect significant evolutionary points in time which, again, serves to minimize the need for manual intervention.

The authors of [12] extract variety of features from Android malware samples, and then determine various trends based standard software quality metrics. These results are then compared to trends present in Android goodware. This work shows that the trends in the Android malware and goodware are similar, with changes in malware following a similar path as goodware. These results are not surprising, given that Android malware largely consists of trojans that, by necessity, would tend to have a great deal of overlap with goodware.

The work presented in [5] is focussed on malware taxonomy, which provides some insights into malware evolution, in the form of genealogical trajectories. This research is based on features extracted from malware encyclopedia entries, which have been developed by antivirus software vendors, such as TrendMicro. The authors use SVMs and language processing techniques to extract features on which their results are based.

In general, the features used in malware analysis can be considered to be either static or dynamic. Static features are those that can be collected without executing the code, whereas dynamic features require code execution or emulation. In general, static features are easier to collect, while dynamic features are more robust with respect to common obfuscation techniques [6].

The authors of [32] use multiple static features to perform malware classification among various families. The static features that are considered are byte n-grams, entropy, and image representations. In addition, hex-dump based features are also used, along with features extracted from disassembled files, including opcodes, API calls, and sectional information from portable executable (PE) files. This works provides interesting insights on a wide variety of static features.

The research that we present in this paper can be viewed as a continuation of work that originated in [36], where static PE file features of malware samples are used as the basis for malware evolution detection. This previous research employed linear support vector machine (SVM) techniques to train on samples from a specific family over sliding windows of time. The resulting SVM weights are compared based on a χ^2 measure, and observed differences in model weights are used to indicate potential evolutionary points in time.

The work in [30], which employs opcode sequences from malware samples to analyze malware evolution, is related to the research presented in [36]. In [30], the data is again divided into time windows, and support vector machine (SVM) techniques are used to observe evolutionary points in the malware samples. In addition, hidden Markov model (HMM) techniques are used as a secondary test to confirm suspected evolutionary points in time. Our research in this paper is a further extension to this previous work. We perform extensive experiments with HMMs and the word embedding techniques of Word2Vec and HMM2Vec to analyze malware evolution. We find that we can automatically detect significant evolution in malware families using these techniques.

2.3 Dataset

The dataset we use in this research consists of Windows portable executable files belonging to 15 malware families. Two families (Winwebsec and Zbot) are from the Malicia dataset [28], while the remaining families are from a larger dataset that was constructed using VirusShare [9]. Each malware family contains a a

Table 1 Number of samples
used in experiments

Family	Samples	Years
Adload	791	2009–2011
Bho	1116	2007–2011
Bifrose	577	2009–2011
CeeInject	742	2009–2012
DelfInject	401	2009–2012
Dorkbot	222	2005–2012
Hupigon	449	2009–2011
Ircbot	59	2009–2012
Obfuscator	670	2004–2017
Rbot	127	2001–2012
Vbinject	2331	2009–2018
Vobfus	700	2009–2011
Winwebsec	1511	2008–2012
Zbot	835	2009–2012
Zegost	506	2008–2011
Total	11,037	2001–2018

number of samples from an extended period of time. Samples belonging to a
malware family are assumed to have similar characteristics and to share a code
base. However, samples within the same family differ, as malware writers regularly
modify successful malware to perform slightly different functions, to make it harder
to detect, or for other purposes. The number of samples in each family in our dataset
is given in Table 1. The table also includes the time range over which the samples
were produced.

The malware families in our dataset encompass a wide variety of types, including
virus, trojan, backdoor, worms, and so on. Some of the families uses encryption and
other obfuscation techniques in an effort to evade detection. Next, we briefly discuss
each of the malware families listed in Table 1.

Bifrose   is a backdoor trojan [25]. As mentioned above, a trojan poses as innocent
software to trick the user into installing it, while a backdoor serves to give an
attacker unauthorized access to an infected system.

CeeInject   performs various malicious operations. CeeInject uses obfuscation
techniques to evade signature detection [16].

DelfInject   is a worm that resides on websites and is downloaded to a user's
machine when visiting an infected site. This malware is executed whenever the
system is restarted [17].

Dorkbot   is a worm that is used to steal credentials of users on an infected system.
It performs a denial of service (DoS) attack, and it is spread via messaging
applications [24].

Hotbar   is an adware virus that resides on websites and is downloaded onto
a user's system when visiting a site that hosts the malware. Hotbar is more

annoying than harmful, as it displays advertisements when the user browses the Internet [14].

Hupigon   is also a backdoor trojan, similar to Bifrose [15].

Obfuscator   evades signature detection using sophisticated obfuscation techniques. It can perform a variety of malicious activities [21].

Rbot   is a backdoor trojan that allows attackers into the system through an IRC channel. This is a relatively advanced malware that is typically used to launch denial of service (DoS) attacks [13].

VbInject   uses encryption techniques to evade signature detection. Its primary purpose is to disguise other malware that can be hidden inside of it. Its payload can vary from harmless to severe [18].

Vobfus   is a trapdoor that lets other malware into the system. It exploits the vulnerabilities of the Windows operating system autorun feature to spread on a network. This malware makes changes to the system configuration that cannot be easily undone [19].

Winwebsec   is a trojan that attempts to trick a user into paying money by portraying itself as anti-virus software. It gives deceptive messages claiming that the system has been infected [20].

Zbot   is a trojan that steals private user information from an infected system. It can target information such as system data and banking details, and it can be easily modified to acquire other types of data. This trojan is generally spread via spam [22].

Zegost   is another backdoor trojan that gives an attacker access to a compromised system [23].

We obtain Windows PE files for each sample in the families discussed above. All of our analysis is based on opcodes, so we first disassemble the files and extract the mnemonic opcode sequence from each, discarding labels, directives, and so on. Since opcodes encapsulate the function of the program we can expect opcode sequences to be useful in detecting code evolution. The resulting opcode sequence will serve as input to our machine learning techniques. In addition, we segregate the samples from each family according to their creation date. Next, we briefly describe each of the learning techniques considered in this paper.

2.4   Learning Techniques

In this section, we discuss the learning techniques that are used in our experiments. Specifically, we introduce hidden Markov models, HMM2Vec, Word2Vec, and logistic regression.

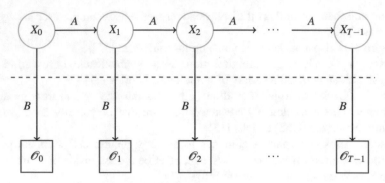

Fig. 1 Hidden Markov model [35]

2.4.1 Hidden Markov Models

As the name suggests, a hidden Markov model (HMM) includes a Markov process that is "hidden" in the sense that it cannot be directly observed. We do have access to a series of observations that are probabilistically related to the underlying (hidden) Markov process. We can train a model to fit a given observation sequence and, given a model, we can score an observation sequence to determine how closely it fits the model. A generic HMM is illustrated in Fig. 1.

The number of hidden states in an HMM is denoted as N, and hence A in Fig. 1 is an $N \times N$ row stochastic matrix that drives the hidden Markov process. The number of distinct observation symbols is denoted as M. The B matrix in Fig. 1 is $N \times M$, with each row representing a discrete probability distribution on the symbols, relative to a given (hidden) state. The B matrix serves to (probabilistically) relate the hidden states to the observations. Note that the B matrix is also row stochastic. An HMM is specified as $\lambda = (A, B, \pi)$, where π is a $1 \times N$ initial state distribution matrix.

2.4.2 Word2Vec

Word2Vec is a technique for embedding terms in a high-dimensional space, where the term embeddings are obtained by training a shallow neural network. After the training process, words that are more similar in context will tend to be closer together in the Word2Vec space.

Perhaps surprisingly, meaningful algebraic properties hold for Word2Vec embeddings. For example, according to [26], if we let

$$w_0 = \text{``king''}, \ w_1 = \text{``man''}, \ w_2 = \text{``woman''}, \ w_3 = \text{``queen''}$$

and $V(w_i)$ is the Word2Vec embedding of word w_i, then $V(w_3)$ is the vector that is closest—in terms of cosine similarity—to

Table 2 Training data

Offset	Training pairs
" one small step ..."	(one,small), (one,step)
"one small step for ..."	(small,one), (small,step), (small,for)
"one small step for man ..."	(step,one), (step,small), (step,for), (step,man)
"... small step for man one ..."	(for,small), (for,step), (for,man), (for,one)
"... step for man one giant ..."	(man,step), (man,for), (man,one), (man,giant)
"... for man one giant leap ..."	(one,for), (one,man), (one,giant), (one,leap)
"... man one giant leap for ..."	(giant,man), (giant,one), (giant,leap), (giant,for)
"... one giant leap for mankind"	(leap,one), (leap,giant), (leap,for), (leap,mankind)
"... giant leap for mankind"	(for,giant), (for,leap), (for,mankind)
"... leap for mankind "	(mankind,leap), (mankind,for)

$$V(w_0) - V(w_1) + V(w_2)$$

Suppose that we have a vocabulary of size M. We can encode each word as a "one-hot" vector of length M. For example, suppose that our vocabulary consists of the set of $M = 8$ words

$$W = (w_0, w_1, w_2, w_3, w_4, w_5, w_6, w_7)$$

$$= (\text{"for"}, \text{"giant"}, \text{"leap"}, \text{"man"}, \text{"mankind"}, \text{"one"}, \text{"small"}, \text{"step"})$$

Then we encode "for" and "man" as

$$E(w_0) = E(\text{"for"}) = 10000000 \text{ and } E(w_3) = E(\text{"man"}) = 00010000$$

respectively.

Now, suppose that our training data consists of the phrase

$$\text{"one small step for man one giant leap for mankind"} \tag{1}$$

To obtain training samples, we specify a window size, and for each offset we use all pairs of words within the specified window. For example, if we select a window size of two, then from (1), we obtain the training pairs in Table 2.

Consider the pair "(for,man)" from the fourth row in Table 2. As one-hot vectors, this training pair corresponds to input 10000000 and output 00010000.

A neural network similar to that in Fig. 2 is used to generate Word2Vec embeddings. The input is a one-hot vector of length M representing the first element of a training pair, such as those in Table 2, and the network is trained to output the second element of the ordered pair. The hidden layer consists of N linear neurons

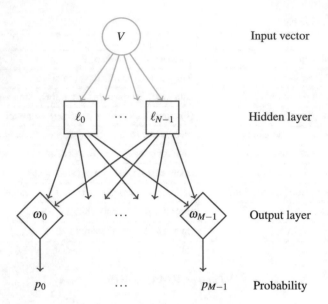

Fig. 2 Neural network for generating Word2Vec embeddings

and the output layer uses a softmax function to generate M probabilities, where p_i is the probability of the output vector corresponding to w_i for the given input.

Observe that the Word2Vec network in Fig. 2 has NM weights that are to be determined, as represented by the blue lines from the hidden layer to the output layer. For each output node ω_i, there are N edges (i.e., weights) from the hidden layer. The N weights that connect to output node ω_i form the Word2Vec embedding $V(w_i)$ of the word w_i.

A Word2Vec model can be trained using either a continuous bag-of-words (CBOW) or a skip-gram model. The model discussed in this section uses the CBOW approach, and that is what we employ in our experiments in this paper. Note that in our implementation, we use opcodes as the "words."

Several tricks are used to speed up the training of Word2Vec models. Such details are beyond the scope of this paper; see [27] for more information.

2.4.3 HMM2Vec

Analogous to Word2Vec, we can use the B matrix of a trained HMM to specify vector embeddings corresponding to the observations. More precisely, each column of the B matrix is associated with a specific observation, and hence we obtain vector embeddings of length N directly from the B matrix—we refer to the resulting embedding as HMM2Vec. Since HMM2Vec is not a standard vector embedding technique, in this section, we illustrate the process using a simple English text example.

Recall that an HMM is defined by the three matrices A, B, and π, and is denoted as $\lambda = (A, B, \pi)$. The π matrix contains the initial state probabilities, A contains the hidden state transition probabilities, and B consists of the observation probability distributions corresponding to the hidden states. Each of these matrices is row stochastic, that is, each row satisfies the requirements of a discrete probability distribution. Notation-wise, we let N be the number of hidden states, M is the number of distinct observation symbols, and T is the length of the observation (i.e., training) sequence. Note that M and T are determined by the training data, while N is a user-defined parameter. For more details in HMMs, see [35] or Rabiner's fine tutorial [31].

Suppose that we train an HMM on a sequence of letters extracted from English text, where we convert all upper-case letters to lower-case, and we discard any character that is not an alphabetic letter or word-space. Then $M = 27$, and we select $N = 2$ hidden states, and suppose we use $T = 50,000$ observations for training. Note that each observation is one of the $M = 27$ symbols (letters, together with word-space). For the example discussed below, the sequence of $T = 50,000$ observations was obtained from the Brown corpus of English [4], but any source of English text could be used.

For one specific case, an HMM trained with the parameters listed in the previous paragraph yields the B matrix in Table 3. Observe that this B matrix gives us two probability distributions over the observation symbols—one for each of the hidden states. We observe that one hidden state essentially corresponds to vowels, while the other corresponds to consonants. This simple example nicely illustrates the machine learning aspect of HMMs, as no a priori assumption was made concerning consonants and vowels, and the only parameter we selected was the number of hidden states N. The training process enabled the model to learn a crucial aspect of English directly from the data.

Suppose that for a given letter ℓ, we define its HMM2Vec representation $V(\ell)$ to be the corresponding row of the matrix B^{T} in Table 3. Then, for example,

$$V(a) = (0.13537 \ 0.00364) \quad V(e) = (0.21176 \ 0.00223)$$
$$V(s) = (0.00032 \ 0.11069) \quad V(t) = (0.00158 \ 0.15238) \tag{2}$$

Next, we consider the distance between these HMM2Vec representations. Instead of using Euclidean distance, we measure the cosine similarity.[1]

The cosine similarity of vectors X and Y is the cosine of the angle between the two vectors. Let $S(X, Y)$ denote the cosine similarity between vectors X and Y. Then for $X = (X_0, X_1, \ldots, X_{n-1})$ and $Y = (Y_0, Y_1, \ldots, Y_{n-1})$,

---

[1] Cosine similarity is not a true metric, since it does not, in general, satisfy the triangle inequality.

Table 3 Final B^{T} for HMM

Letter	State 0	1	Letter	State 0	1
a	0.13537	0.00364	n	0.00035	0.11429
b	0.00023	0.02307	o	0.13081	0.00143
c	0.00039	0.05605	p	0.00073	0.03637
d	0.00025	0.06873	q	0.00019	0.00134
e	0.21176	0.00223	r	0.00041	0.10128
f	0.00018	0.03556	s	0.00032	0.11069
g	0.00041	0.02751	t	0.00158	0.15238
h	0.00526	0.06808	u	0.04352	0.00098
i	0.12193	0.00077	v	0.00019	0.01608
j	0.00014	0.00326	w	0.00017	0.02301
k	0.00112	0.00759	x	0.00030	0.00426
l	0.00143	0.07227	y	0.00028	0.02542
m	0.00027	0.03897	z	0.00017	0.00100
Space	0.34226	0.00375	–	–	–

$$S(X, Y) = \frac{\sum_{i=0}^{n-1} X_i Y_i}{\sqrt{\sum_{i=0}^{n-1} X_i^2} \sqrt{\sum_{i=0}^{n-1} Y_i^2}}$$

In general, we have $-1 \le S(X, Y) \le 1$, but since our HMM2Vec encoding vectors consist of probabilities—and hence are non-negative values—in this case, we always have $0 \le S(X, Y) \le 1$.

When considering cosine similarity, the length of the vectors is irrelevant, as we are only considering the angle between vectors. Consequently, we might want to normalize all vectors to be of length one, say, $\widetilde{X} = X/\|X\|$ and $\widetilde{Y} = Y/\|Y\|$, in which case the cosine similarity simplifies to the dot product

$$S(X, Y) = S(\widetilde{X}, \widetilde{Y}) = \sum_{i=0}^{n-1} \widetilde{X}_i \widetilde{Y}_i$$

Henceforth, we use the notation $\widetilde{X}$ to indicate a vector X that has been normalized to be of length one.

For the vector encodings in (2), we find that for the vowels "a" and "e", the cosine similarity is $S(V(a), V(e)) = 0.9999$. In contrast, the cosine similarity of the vowel "a" and the consonant "t" is $S(V(a), V(t)) = 0.0372$. The normalized vectors $V(a)$ and $V(t)$ are illustrated in Fig. 3. Using the notation in this figure, cosine similarity is $S(V(a), V(t)) = \cos(\theta)$

Fig. 3 Normalized
vectors $\widetilde{V}(a)$ and $\widetilde{V}(t)$

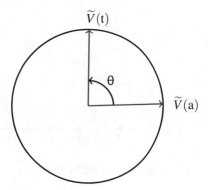

These results indicate that our HMM2Vec encodings—which are derived from a trained HMM—provide useful information on the similarity (or not) of pairs of letters. Note that we could obtain a vector encoding of any dimension by simply training an HMM with the number of hidden states N equal to the desired vector length.

In our experiments below, we consider HMM2Vec embeddings. However, in this research, models are trained on opcodes instead of letters, and hence the embeddings are relative to individual opcodes.

2.4.4 Logistic Regression

Logistic regression is used widely for classification problems. This relatively simple technique relies on the sigmoid function, which is also knows as the logistic function, and hence the name. The sigmoid function is defined as

$$S(x) = \frac{1}{1 + e^{-x}}.$$

Logistic regression can be viewed as a modification of linear regression. As with linear regression, logistic regression models the probability that observations take one of two (binary) values. Linear regression makes unbounded predictions whereas logistic regression converts the probability into the range 0 to 1 due to the use of the sigmoid function. The graph of the sigmoid function is given in Fig. 4, from which we can see that the output must be between 0 and 1.

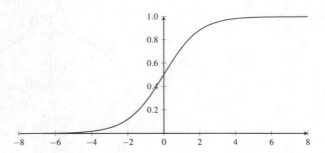

Fig. 4 Graph of sigmoid function

3 Experiments and Results

In this section, we discuss our evolution detection experiments and results. We divide this section into four subsections, one for each technique considered, namely, logistic regression, HMM, HMM2Vec and Word2Vec.

3.1 Logistic Regression Experiments

As mentioned above, in [30] the authors use linear SVMs to detect potential malware evolution. Logistic regression is a simpler technique that, like SVM, is widely used for classification. Hence, we train logistic regression models over time-windows, analogous to the SVM approach in [30]. Specifically, we divide our data into overlapping time windows of one year, with a slide length of one month. All of the samples from the most recent one year time window are taken as the +1 class, while samples from the current month are considered as the −1 class, and we train our logistic regression models on the resulting data. Each such model is represented by its weights, and we calculate the Euclidian distances between these weight vectors to measure the similarity of the models. We then plot these distances on a timeline—spikes in the graph indicate that the model has changed and hence evolution may have occurred. Figures 5 and 6 in the Appendix show the results of our logistic regression experiments for Winwebsec and Zegost, respectively.

The results in Figs. 5 and 6 are inconclusive. Although our logistic regression model achieves high accuracy in classifying samples, the weights of the hidden layer do not appear to provide clear information regarding changes in the malware samples. Apparently, the noise inherent in these weights overwhelms the relevant information.

3.2   Hidden Markov Model Experiments

All experiments in this section are based on the top thirty most frequent opcodes per family, with all other opcodes grouped into a single "other" category. Thus, our HMMs are all based on $M = 31$ distinct symbols. We use $N = 2$ hidden states in all experiments. We conduct two sets of experiments based on hidden Markov models (HMM). In both of these approaches, we train models, and we then score samples with the resulting models.

For our first set of experiments, we reserve the data from the first one-month time period to test our models, and hence we do not train a model on this data. For each subsequent one-month time window, we train a model, and then score the samples from the first one-month time period versus each of these models. We refer to this as HMM approach 1.

Consider two distinct one-month time periods, say time period X and Y. Suppose that we train an HMM on the data from time period X and another on the data from time period Y, which we denote as λ_X and λ_Y, respectively. If the samples from X and Y are similar, then we expect the HMMs λ_X and λ_Y to be similar, and hence they should produce similar scores on the reserved (first month) data. On the other hand, if the the samples from time periods X and Y differ significantly, then we expect the models λ_X and λ_Y to differ, and hence the scores on the reserved first-month test set should differ significantly. Figures 7, 8, 9 in the Appendix show results for three families based on this HMM approach 1.

In Figs. 7, 8, 9, we observe spikes in the graphs at various points in time, with relative stability over extended periods of time. Thus, this approach seems to have the potential to detect malware evolution.

Next, we consider another application of HMMs to our data. In this case, for each one-month time window, we use 75% of the available samples for training and reserve 25% for testing. Next, we train an HMM for each month—as above, we use $N = 2$, and we have $M = 31$ in each case.

Suppose we have data from consecutive months that we label as X and Y. We train model λ_X on the training data from time period X and we train a model λ_Y on the training data from time period Y. We then score each test sample from X with both λ_X and model λ_Y, giving us two score vectors. Since an HMM score depends on the length of the observation sequence, and since the observation sequence lengths vary between malware samples, each scores is normalized by dividing by the length of the observation sequence. As a result, each score is in the form of a log likelihood per opcode (LLPO). Note that If we have, say, m test samples in X, the score vector obtained from λ_X and the score vector obtained from λ_Y will both be of length m.

Once we generate these two vectors, we compute the Euclidean distance between the vectors, which we denote as d_X. We repeat this scoring process using the test samples from Y to obtain a distance d_Y, and we define the distance between time windows X and Y to be the average, that is,

$$d = \frac{d_X + d_Y}{2}.$$

We plot the graph of these distances—small changes in the distance from one month to the next suggests minimal change, whereas larger distances indicate potential evolution points. Figures 10, 11, 12, 13 in the Appendix give results for four malware families using this HMM-based technique, which we refer to as HMM approach 2.

The results in Figs. 10, 11, 12, 13 indicate that we see significant evolutionary change points when considering this second HMM technique. Together with the results for HMM approach 1, these results provide strong evidence that HMM-based techniques are a powerful tool for malware evolution detection.

3.3 HMM2Vec Experiments

In this section we present our experimental results using HMM2Vec. Recall that we discussed the HMM2Vec word embedding technique in Sect. 2.4.3. In these experiments, we select $N = 2$ and we have $M = 31$. Recall that the HMM2Vec embeddings are determined by the columns of the B matrix from our trained HMM, and that each embedding vector is of length N.

A technical difficulty arises when considering HMM2Vec embeddings. That is, the order of the hidden states can vary between models—even when training on the same data, different random initializations can cause the hidden states to differ in the resulting trained models. Since we only consider models with $N = 2$ hidden states, we account for this possibility in our HMM2Vec experiments by computing the distance between B matrices twice, once with the order of the rows flipped in one of the models. More precisely, suppose that we want to compare the two HMMs $\lambda = (A, B, \pi)$ and $\widetilde{\lambda} = (\widetilde{A}, \widetilde{B}, \widetilde{\pi})$, where $N = \widetilde{N} = 2$ and $M = \widetilde{M}$. We first compute the distance based on the HMM2Vec embeddings determined by the matrices B and $\widetilde{B}$ (we ignore A and $\widetilde{A}$, as well as π and $\widetilde{\pi}$). Denote the rows of B as B_1 and B_2 and, similarly, let $\widetilde{B}_1$ and $\widetilde{B}_2$ be the rows of $\widetilde{B}$. Compute

$$d_1 = d(B_1 \| B_2, \widetilde{B}_1 \| \widetilde{B}_2) \text{ and } d_2 = d(B_1 \| B_2, \widetilde{B}_2 \| \widetilde{B}_1)$$

where "$\|$" is the concatenation operator, and $d(x, y)$ is the Euclidean distance between vectors x and y. We define the HMM2Vec distance between λ and $\widetilde{\lambda}$ as

$$d(\lambda, \widetilde{\lambda}) = \min\{d_1, d_2\}.$$

We divide the dataset into overlapping windows of one year, with a slide length of one month and we train an HMM (with $N = 2$ and $M = 31$) on each window. We compute the distance between adjacent windows using the method described in the previous paragraph, and we graph the resulting distances. The graphs obtained for three families are given in Figs. 14, 15, 16 in the Appendix.

The results in Figs. 14, 15, 16 indicate that HMM2Vec is successful in identifying potential evolution in these particular families. We observe significant spikes (i.e., evolutionary points) in most families using this technique.

3.4   Word2Vec Experiments

In this set of experiments, we use Word2Vec to generate vector embeddings of opcodes. We compare the resulting models by concatenating the embedding vectors, and computing the distance between the resulting vectors. As above, we divide the dataset into overlapping time windows of one year, with a slide length of one month. The Word2Vec models are trained as outlined in Sect. 2.4.2.

When training Word2Vec, the window size W refers to the length of the window used to determine training pairs, while the vector length V is the number of components in each embedding vector. We experimented with different window sizes and found that $W = 5$ works best. We also experimented with different vector sizes—in Figs. 17, 18, 19 in the Appendix, we give results for the Zbot family for $V = 2$, $V = 3$, and $V = 5$, respectively. In general, we do not find any improvement for larger values of V, and hence we use $V = 2$ in all of our subsequent Word2Vec experiments.

Results from our Word2Vec experiments for three families are given in Figs. 20, 21, 22 in the Appendix. These results show potential evolutionary points in almost all the malware families and we conclude that Word2Vec is also a useful technique for detect potential malware evolution points.

3.5   Discussion

Here, we first discuss the results given by each technique considered in this section. Then we compare our results to the most closely related previous work.

The two HMM scoring techniques that we first considered provide different models and scores, yet the results are similar. This provides evidence of correctness and consistency, and also some evidence of actual evolution.

Our HMM2Vec and Word2Vec experiments were somewhat different, since they focus on longer time windows of one year, whereas the HMM techniques both are based on one-month time intervals. In any case, both HMM2Vec and Word2Vec performed well and consistently with each other. Again, this consistency is evidence of correctness of the implementations, and of evolution detection.

Combining either HMM2Vec or Word2Vec with either of the HMM scoring techniques provides a two-step strategy for detecting evolution. That is, we can use a year-based technique (either HMM2Vec or Word2Vec) to see if there is any indication of evolution over such a time window. If so, we can then use one of the HMM scoring techniques to determine where within that one-year window the strongest evolutionary points occur. In this way, we could rapidly filter out time periods that are unlikely to be of interest, and then in the secondary phase, detect precise times at which interesting evolutionary changes have most likely occurred. For example, both Word2Vec and HMM2Vec indicate that evolution in the Winwebsec family took place during the time period November 2010–June 2011. Then experimenting with the first HMM scoring approach on the Winwebsec family during the November 2010–June 2011 time period indicates that the precise point of evolution was June 2011.

Next, we compare our work to that in [30], which considered the same malware evolution problem and used the same dataset as in our research. In [30], linear SVM models are trained over on year time windows, with a slide of one month. The resulting linear SVM model weights are compared using a χ^2 distance computation. Furthermore, Word2Vec feature vectors (derived from opcode sequences) were used as input features to their SVM models. Once a χ^2 similarity graph has been generated, an HMM-based approach is used on either sides of a spike to confirm that evolution has occurred.

Comparing our results with those given in [30], our techniques are more efficient, as we omit the SVM training and our work factor is less than their secondary test. In spite of these simplifications, we find that our detect strategy is at least as sensitive as that in [30]. For example, we see clear spikes in some families (e.g., DelfInject, Dorkbot, and Zbot) for which previous work found, at best, ambiguous results.

Next we briefly summarize our results per family. We refer to previous work in [30] in some of these cases.

Adload: For this family, both HMM2Vec and Word2Vec did not result in any significant spikes in the graphs, and hence we do not see indications of evolutionary change. On the other hand, the results given by our HMM techniques show significant spikes for this family.

Bho: The results generated by Word2Vec for this family indicate that malware evolution occurred during the September 2009–December 2010 timeframe. Using our HMM approach, we are able to see that malware evolution happened during October 2010, which is consistent with the Word2Vec results. This is also consistent with results given in [30], based on SVM analysis.

Bifrose: The results generated by HMM2Vec did not provide any major spikes in the graph, but we can see indications of slower change over time. The graph generated by Word2Vec gives us a better understanding of changes in this malware family, since we could see that significant evolution occurred during the November 2009–March 2011 time period. Again, the results given by the HMM approaches narrow down the evolution point—in this case, to March 2011. A similar graph is given in [30], indicating evolution during November 2010–May 2011, which is consistent with our results.

CeeInject: For CeeInject, we obtain clear results from all experiments we performed. The results given by HMM2Vec and Word2Vec shows significant evolution during the August 2010–July 2011 time window, and we identify the month of clearest change as November 2010 based on our HMM approaches. The results for this family given in [30] show similar evolution during September 2010–May 2011.

DelfInject: We obtained significant results for this family using our HMM-based approaches. This family shows evolution occurring during January 2011. In this case, we do not observe significant spikes for Word2Vec or HMM2vec with their longer time windows. In [30], no evolutionary points are detected for this family.

Dorkbot: Similar to CeeInject, we obtain strong results on Dorkbot from all of our experiments. Specifically, the evidence strongly points to malware evolution during 2011.

Hupigon: The results received from Word2Vec technique show that significant malware evolution in this family happened during the July 2010–April 2011 period. Results from the HMM approaches narrow the time period to February 2011. Results given by the SVM approach in [30] are consistent with these results.

Ircbot: The results generated by Word2Vec indicates that malware evolution occurred in this family slowly throughout 2011. That is, there is no major spikes observed, but the graph shows a slow changing trend.

Obfuscator: We could not derive significant information from this family. Graphs plotted on this family had many spikes which we could not interpret regarding malware evolution.

Rbot: Graphs generated based on Word2Vec show significant evolution in this malware family. Significant results were not observed for this family in any previous research.

VbInject: We could not observe a significant spike in this malware family in any of our experiments.

Vobfus: The results generated by our experiments shows that evolution in this family occurred during the December 2009–January 2011 timeframe. The results given in [30] indicate evolution during November 2010–May 2011.

Winwebsec: We observe evolution in this malware family using Word2Vec, where a spike appears in December 2010–July 2011. The previous research in [30] did not indicate evolution for this family.

Zbot: Experiments conducted on this family inidcate significant changes. Specifically, we observe a spike between April 2011–November 2011.

Zegost: From our Word2Vec experiments, we see significant spikes in the August 2010–September 2011 and July 2010–July 2011 timeframes.

Our experiments indicate significant evolution in almost all the malware families considered. By comparing the results given by our two HMM techniques, HMM2Vec, and Word2Vec, we can see that there are clear similarities in the results for most families. When we observe such similar evolution points across different experiments, it increases our confidence in the results. As further evidence, we found that the evolutionary points generated in previous research in [30] matches with our experiments, and we detect additional points of interested, as compared to previous research, indicating that our techniques may be somewhat more sensitive.

In some cases, we found potential evolutionary points with the HMM techniques, but not with HMM2Vec or Word2Vec. We conjecture that this is a result of the longer time windows (one year) used in the latter two approaches, while the HMM techniques are based on monthly time windows. These longer time windows may not be as sensitive in cases where a changes are less pronounced or transient.

4 Conclusion and Future Work

In previous research—first in [36] and subsequently in [30]—it has been shown that malware evolution can be detected using machine learning techniques. In this paper, we extend this previous work by exploring additional learning techniques. We find that various HMM-based techniques and Word2Vec provide powerful tools for automatically detecting malware evolution.

Here, we conducted all of our experiments based on mnemonic opcodes derived from the malware samples. For future work, it would be useful to consider experiments with other features extracted from the malware samples. While mnemonic opcodes perform well, extracting such opcodes is relatively expensive. It is possible that other, less costly features can be used. Also, by considering dynamic features, we might gain more information about evolution within a malware family. Finally, the use of additional neural networking and deep learning techniques should be considered. Word2Vec performed well, and it is likely that more sophisticated techniques would result in more discriminative ability, which would enable more fine grained analysis of evolutionary trends.

Appendix

See Figs. 5–22.

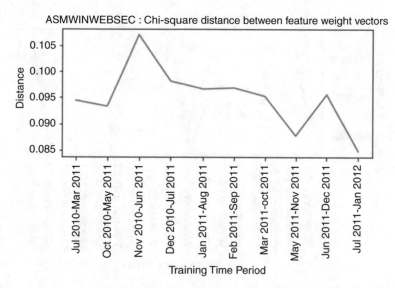

Fig. 5 Logistic regression results for Winwebsec

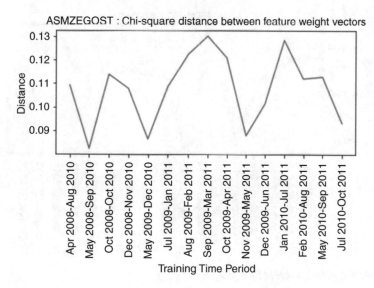

Fig. 6 Logistic regression results for Zegost

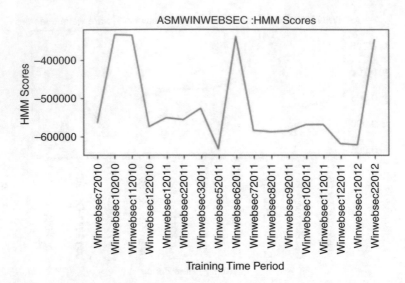

Fig. 7 HMM approach 1 results for Winwebsec

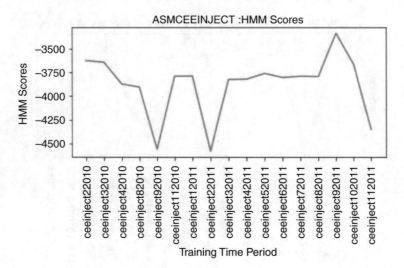

Fig. 8 HMM approach 1 results for CeeInject

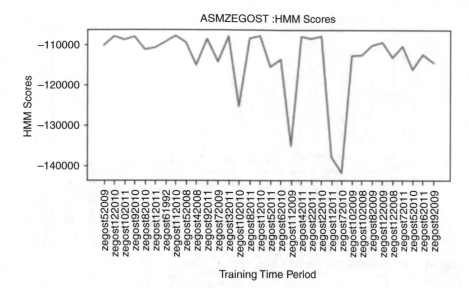

Fig. 9 HMM approach 1 results for Zegost

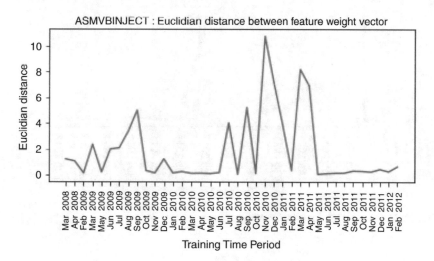

Fig. 10 HMM approach 2 results for VbInject

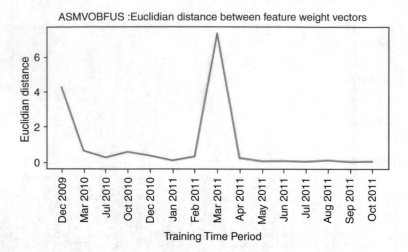

Fig. 11 HMM approach 2 results for Vobfus

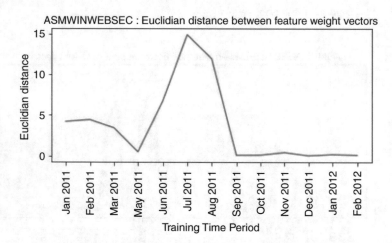

Fig. 12 HMM approach 2 results for Winwebsec

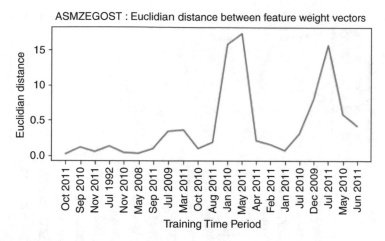

Fig. 13 HMM approach 2 results for Zegost

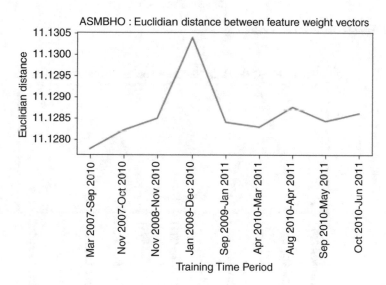

Fig. 14 HMM2Vec results for Bho

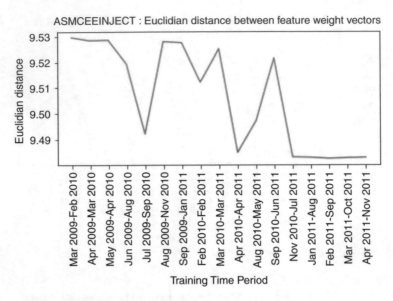

Fig. 15 HMM2Vec results for CeeInject

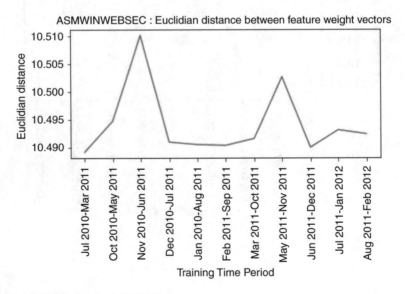

Fig. 16 HMM2Vec results for Winwebsec

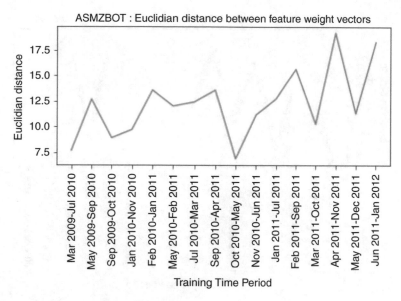

Fig. 17 Word2Vec for Zbot with $V = 2$

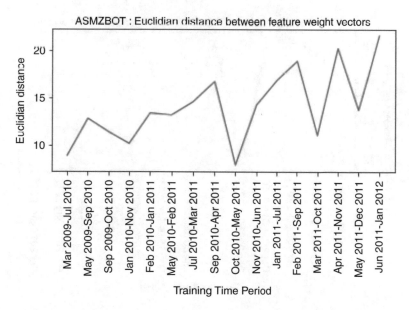

Fig. 18 Word2Vec for Zbot with $V = 3$

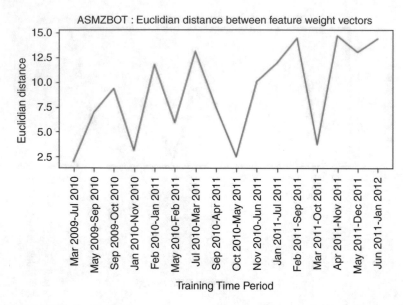

Fig. 19 Word2Vec for Zbot with $V = 5$

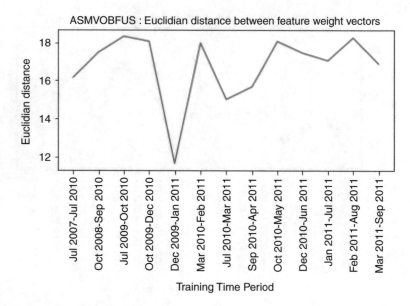

Fig. 20 Word2Vec results for Vobfus

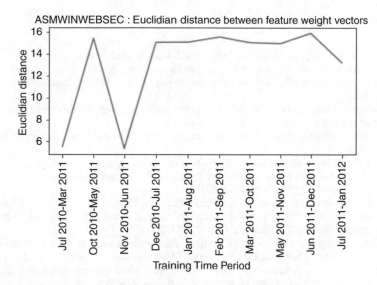

Fig. 21 Word2Vec results for Winwebsec

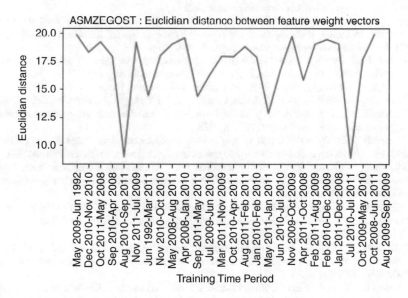

Fig. 22 Word2Vec results for Zegost

References

1. John Aycock. *Computer Viruses and Malware*. Springer, 2006.
2. Marius Barat, Dumitru-Bogdan Prelipcean, and Dragoş Gavriluţ. A study on common malware families evolution in 2012. *Journal of Computer Virology and Hacking Techniques*, 9(4):171–178, 2013.
3. Jean-Marie Borello and Ludovic Mé. Code obfuscation techniques for metamorphic viruses. *Journal in Computer Virology*, 4(3):211–220, 2008.
4. The Brown corpus of standard American English. http://www.cs.toronto.edu/~gpenn/csc401/a1res.html.
5. Zhongqiang Chen, Mema Roussopoulos, Zhanyan Liang, Yuan Zhang, Zhongrong Chen, and Alex Delis. Malware characteristics and threats on the Internet ecosystem. *The Journal of Systems & Software*, 85(7):1650–1672, 2012.
6. Anusha Damodaran, Fabio Di Troia, Corrado Visaggio, Thomas Austin, and Mark Stamp. A comparison of static, dynamic, and hybrid analysis for malware detection. *Journal of Computer Virology and Hacking Techniques*, 13(1):1–12, 2017.
7. A Gupta, P Kuppili, A Akella, and P Barford. An empirical study of malware evolution. In *First International Communication Systems and Networks and Workshops*, pages 1–10, 2009.
8. Archit Gupta, Pavan Kuppili, Aditya Akella, and Paul Barford. An empirical study of malware evolution. In *First International Communication Systems and Networks and Workshops*, pages 1–10, 2009.
9. Samuel Kim. PE header analysis for malware detection. Master's thesis, San Jose State University, Department of Computer Science, 2018.
10. Justin Ma, John Dunagan, Helen J Wang, Stefan Savage, and Geoffrey M Voelker. Finding diversity in remote code injection exploits. In *Proceedings of the 6th ACM SIGCOMM Conference on Internet Measurement*, pages 53–64, 2006.
11. Robert McMillan. COMPUTERWORLD: Researchers take down Koobface servers — Criminals behind the botnet made more than $2 million in one year, 2010. https://www.computerworld.com/article/2750985/researchers-take-down-koobface-servers.html.
12. Francesco Mercaldo, Andrea Di Sorbo, Corrado Aaron Visaggio, Aniello Cimitile, and Fabio Martinelli. An exploratory study on the evolution of Android malware quality. *Journal of Software: Evolution and Process*, 30(11), 2018.
13. Microsoft. Win32 Rbot detected with Windows Defender antivirus. https://www.microsoft.com/en-us/wdsi/threats/malware-encyclopedia-description?Name=Win32%2FRbot, 2005.
14. Microsoft. Adware: Win32 Hotbar detected with Windows Defender antivirus. https://www.microsoft.com/en-us/wdsi/threats/malware-encyclopedia-description?Name=Adware%3AWin32%2FHotbar, 2006.
15. Microsoft. Backdoor: Win32 Hupigon detected with Windows Defender antivirus. https://www.microsoft.com/en-us/wdsi/threats/malware-encyclopedia-description?Name=Backdoor%3AWin32%2FHupigon, 2006.
16. Microsoft. Virtool: Win32 CeeInject detected with Windows Defender antivirus. https://www.microsoft.com/en-us/wdsi/threats/malware-encyclopedia-description?Name=VirTool%3AWin32%2FCeeInject, 2007.
17. Microsoft. Virtool: Win32 DelfInject detected with Windows Defender antivirus. https://www.microsoft.com/en-us/wdsi/threats/malware-encyclopedia-description?Name=VirTool:Win32/DelfInject&ThreatID=-2147369465, 2007.
18. Microsoft. Virtool: Win32 VBInject detected with Windows Defender antivirus. https://www.microsoft.com/en-us/wdsi/threats/malware-encyclopedia-description?Name=VirTool:Win32/VBInject&ThreatID=-2147367171, 2010.
19. Microsoft. Win32 Vobfus detected with Windows Defender antivirus. https://www.microsoft.com/en-us/wdsi/threats/malware-encyclopedia-description?name=win32%2Fvobfus, 2010.
20. Microsoft. Win32 Winwebsec detected with Windows Defender antivirus. https://www.microsoft.com/security/portal/threat/encyclopedia/entry.aspx?Name=Win32%2fWinwebsec, 2010.

21. Microsoft. Win32 Obfuscator detected with Windows Defender antivirus. https://www.microsoft.com/en-us/wdsi/threats/malware-encyclopedia-description?Name=Win32%2FObfuscator, 2011.
22. Microsoft. Win32 Zbot detected with Windows Defender antivirus. http://www.symantec.com/securityresponse/writeup.jsp?docid=2010-011016-3514-99, 2011.
23. Microsoft. Win32 Zegost detected with Windows Defender antivirus. https://www.symantec.com/security-center/writeup/2011-060215-2826-99, 2011.
24. Microsoft. Worm: Win32 Dorkbot detected with Windows Defender antivirus. https://www.microsoft.com/en-us/wdsi/threats/malware-encyclopedia-description?Name=Worm%3AWin32/Dorkbot, 2011.
25. Microsoft. Win32 Bifrose detected with Windows Defender antivirus. https://www.trendmicro.com/vinfo/us/threat-encyclopedia/malware/bifrose, 2012.
26. Tomas Mikolov, Kai Chen, Greg Corrado, and Jeffrey Dean. Efficient estimation of word representations in vector space. https://arxiv.org/abs/1301.3781, 2013.
27. Tomas Mikolov, Ilya Sutskever, Kai Chen, Greg Corrado, and Jeffrey Dean. Distributed representations of words and phrases and their compositionality. https://papers.nips.cc/paper/5021-distributed-representations-of-words-and-phrases-and-their-compositionality.pdf, 2013.
28. Antonio Nappa, M Zubair Rafique, and Juan Caballero. The MALICIA dataset: Identification and analysis of drive-by download operations. *International Journal of Information Security*, 14(1):15–33, 2015.
29. Jacob Ouellette, Avi Pfeffer, and Arun Lakhotia. Countering malware evolution using cloud-based learning. In *8th International Conference on Malicious and Unwanted Software*, pages 85–94, 2013.
30. Sunhera Paul and Mark Stamp. Word embedding techniques for malware evolution detection. In *Malware Analysis Using Artificial Intelligence and Deep Learning*, pages 321–343. Springer, 2021.
31. Lawrence R. Rabiner. A tutorial on hidden Markov models and selected applications in speech recognition. *Proceedings of the IEEE*, 77(2):257–286, 1989. https://www.cs.sjsu.edu/~stamp/RUA/Rabiner.pdf.
32. Sacid Rezaci, Ali Afraz, Fereidoon Rezaei, and Mohammad Reza Shamani. Malware detection using opcodes statistical features. In *8th International Symposium on Telecommunications*, IST, pages 151–155, 2016.
33. Mark Stamp. *Information Security: Principles and Practice*. Wiley, 2011.
34. Mark Stamp. *Introduction to Machine Learning with Applications in Information Security*. CRC Press, Boca Raton, 2018.
35. Mark Stamp. A revealing introduction to hidden Markov models. https://www.cs.sjsu.edu/~stamp/RUA/HMM.pdf, 2018.
36. Mayuri Wadkar, Fabio Di Troia, and Mark Stamp. Detecting malware evolution using support vector machines. *Expert Systems with Applications*, 143, 2020.

Part II
Other Security Topics

Gambling for Success: The Lottery Ticket Hypothesis in Deep Learning-Based Side-Channel Analysis

Guilherme Perin, Lichao Wu, and Stjepan Picek

Abstract Deep learning-based side-channel analysis (SCA) represents a strong approach for profiling attacks. Still, this does not mean it is trivial to find neural networks that perform well for any setting. Based on the developed neural network architectures, we can distinguish between small neural networks that are easier to tune and less prone to overfitting but could have insufficient capacity to model the data. On the other hand, large neural networks have sufficient capacity but can overfit and are more difficult to tune. This brings an interesting trade-off between simplicity and performance.

This work proposes to use a pruning strategy and recently proposed Lottery Ticket Hypothesis (LTH) as an efficient method to tune deep neural networks for profiling SCA. Pruning provides a regularization effect on deep neural networks and reduces the overfitting posed by overparameterized models. We demonstrate that we can find pruned neural networks that perform on the level of larger networks, where we manage to reduce the number of weights by more than 90% on average. This way, pruning and LTH approaches become alternatives to costly and difficult hyperparameter tuning in profiling SCA. Our analysis is conducted over different masked AES datasets and for different neural network topologies. Our results indicate that pruning, and more specifically LTH, can result in competitive deep learning models.

G. Perin · L. Wu
Delft University of Technology, Delft, The Netherlands
e-mail: G.Perin@tudelft.nl; L.Wu-4@tudelft.nl

S. Picek (✉)
Delft University of Technology, Delft, The Netherlands
Radboud University, Nijmegen, The Netherlands
e-mail: S.Picek@tudelft.nl

M. Stamp et al. (eds.), *Artificial Intelligence for Cybersecurity*, Advances in Information Security 54, https://doi.org/10.1007/978-3-030-97087-1_9

1   Introduction

Several side-channel analysis (SCA) approaches exploit various sources of information leakage from electronic devices. Common examples of side channels are timing [15], power [16], and electromagnetic (EM) emanation [26]. Besides a division based on side channels, it is possible to divide SCA based on the attacker's capabilities into non-profiling and profiling attacks. Non-profiling attacks require fewer assumptions but often require thousands to millions of measurements (traces) to break a target, especially if protected with countermeasures. Profiling attacks are considered one of the strongest possible attacks as the attacker has control over a clone device to build its complete profile [6]. This profile is given by a parametric statistical model used by the attacker to generalize to side-channel information collected from similar devices to recover the secret information.

The history of profiling side-channel analysis (SCA) spans around 20 years, and it is possible to distinguish among several research directions. The first direction used techniques like (pooled) template attack [6, 8] or stochastic models [28] and managed to improve the attack performance over non-profiling attacks significantly. Then, the second direction moved toward machine learning in SCA, and again, a plethora of results [12, 17, 24] indicated that machine learning could outperform other profiling SCA methods. More recently, as the third direction, we see a change of focus to deep learning techniques. Intuitively, we can find at least two reasons for this: (1) deep learning show superior practical results in breaking targets protected with countermeasures [19], and (2) deep learning does not require pre-processing like feature selection [22] or dimensionality reduction [1]. While the SCA community progressed quite far in the deep learning-based SCA in just a few years, there are many knowledge gaps. One example would be how to successfully and systematically find neural networks that manage to break various targets.

Thus, we still need to find approaches that allow designing neural networks that perform well for various targets. We aim to have an approach that transforms a good-performing architecture for one scenario into a good-performing architecture for a different scenario. Finally, it would be ideal if the top-performing architectures could be small (so they are more computationally efficient, and hopefully, easier to understand). Unfortunately, this is not easy as the search space turns into infinite neural network configuration possibilities. There are no general guidelines on how to construct a neural network that will break a target. Current efforts mainly concentrate on finding better hyperparameters by defining modest and optimal ranges for hyperparameters, resulting in large and exhaustive search spaces. Examples of applied hyperparameters search in profiling SCA are random search [21], Bayesian optimization [32], reinforcement learning [27], or approaches following a specific methodology [31, 35]. Still, there are alternatives to how to provide neural networks that are small and perform well. Note that larger neural networks can also perform well for profiling SCA. However, they suffer more from overfitting, and tuning large models becomes more difficult due to the increased hyperparameter search spaces.

Regularization techniques are indicated to correct large models by limiting their capacity, although constructing efficient regularizers can be a highly complex task.

In the machine learning domain, there is a technique called *pruning* (or *sparsification*) that refers to a systematical removal of parameters from neural networks. Commonly, pruning is used on large neural networks that show good performance. The goal is to produce a smaller network with similar performance to be deployed in memory-constrained devices. Also, pruning offers an alternative and cheap solution for regularizing large deep learning models. While pruning [4, 13] is a rather standard technique in deep learning, it has not been investigated before in the SCA domain to the best of our knowledge. Similarly, the Lottery Ticket Hypothesis [9] attracted quite some attention in the machine learning community, but none (as far as we know) in the SCA community.

This chapter applies the recent *Lottery Ticket Hypothesis* (LTH) in the profiling side-channel analysis. After training a (relatively large) neural network, we apply the pruning process by removing the activity of small weights from the neural network. We then re-initialize the pruned neural network with the same initial weights set for the original large neural network. The pruned and re-initialized network shows equal or, most of the time, superior performance compared to the baseline trained network. We emphasize:

- Pruning is convenient for deep neural networks that overfit. Finding efficient and small networks is more difficult than starting with a large model and then pruning it. In this chapter, we consider neural network architectures with up to one million trainable parameters.
- Pruning has two main advantages for SCA: (1) When the baseline model is not carefully tuned and overfits or underfits, pruning (and specially LTH process) may "tune" the model size. (2) Pruning acts as a strong regularizer, which is important for noisy and small SCA datasets. Moreover, techniques such as explainability and interpretability can be used to define pruning strategies efficiently.

The results demonstrate that when the large network cannot reach a successful attack (low guessing entropy), applying the Lottery Ticket Hypothesis leads to a successful key recovery, even when the number of profiling traces is low. More importantly, we verify that when training a large deep neural network provides guessing entropy close to a random guess, a pruned and re-initialized neural network can reduce the entropy of the target key. Our main contributions are:

1. We introduce the pruning approach into profiling SCA, enabling us to propose a procedure that can work on top of other approaches. Our approach can be applied to any neural network, regardless of whether it is selected randomly or obtained through some other methodology. Naturally, depending on how good is the original network, the results from our approach can differ.
2. We demonstrate that the Lottery Ticket Hypothesis holds for SCA, which is a significant finding due to different metrics used in SCA. The original publication [9] measures LTH efficiency through test accuracy. Here, our metric

is guessing entropy from the attack traces. As reported in this chapter, we can find smaller, better, and stable networks by using the pruning and weight initialization based on LTH, even when the original network does not return successful attack results.

2   Background

2.1   Notation

Let calligraphic letters like $\mathcal{X}$ denote sets, and the corresponding upper-case letters X denote random variables and random vectors $\mathbf{X}$ over $\mathcal{X}$. The corresponding lower-case letters x and $\mathbf{x}$ denote realizations of X and $\mathbf{X}$, respectively. Next, let k be a key candidate that takes its value from the keyspace $\mathcal{K}$, and k^* the correct key. We define a dataset as a collection of traces $\mathbf{T}$, where each trace $\mathbf{t}_i$ is associated with an input value (plaintext or ciphertext) $\mathbf{d}_i$ and a key $\mathbf{k}_i$. When considering only a specific key byte j, we denote it as $k_{i,j}$, and input byte as $d_{i,j}$.

The dataset consists of $|T|$ traces. From $|T|$ traces, we use N traces for the profiling set, V traces for the validation set, and Q traces for the attack set. Finally, θ denotes the vector of parameters to be learned in a profiling model, and $\mathcal{H}$ denotes the hyperparameters defining the profiling model.

2.2   Supervised Machine Learning in Profiling SCA

Supervised machine learning considers the machine learning task of learning a function f mapping an input X to the output Y ($f : \mathcal{X} \rightarrow Y$) based on input-output pairs. The function f is parameterized by $\theta \in \mathbb{R}^n$, where n represents the number of trainable parameters.

Supervised learning happens in two phases: training and test, corresponding to SCA's profiling and attack phases. Thus, in the rest of this chapter, we use the terms profiling/training and attack/testing interchangeably. As the function f, we consider a deep neural network with the *Softmax* output layer.

The goal of the training phase is to learn parameters θ' that minimize the empirical risk represented by a loss function L on a dataset T of size N.

In the attack phase, the goal is to make predictions about the classes

$$y(t_1, k^*), \ldots, y(t_Q, k^*),$$

where k^* represents the secret (unknown) key on the device under the attack (or the key byte). The outcome of predicting with a model f on the attack set is a two-dimensional matrix P with dimensions equal to $Q \times c$ (the number of classes c

depends on the leakage model as the class label v is derived from the key and input through a cryptographic function and a leakage model). To reach the probability that a certain key k is the correct one, we use the maximum log-likelihood approach:

$$S(k) = \sum_{i=1}^{Q} \log(\mathbf{p}_{i,v}).$$ (1)

The value $\mathbf{p}_{i,v}$ denotes the probability that for a key k and input d_i, we obtain the class v.

We are interested in reaching good generalization with machine learning algorithms, denoting how well the concepts learned by a machine learning model apply to previously unseen examples. At the same time, we aim to avoid underfitting and overfitting. Overfitting happens when a model learns the detail and noise in the training data, negatively impacting the model's performance on unseen data. Underfitting happens with a model that cannot model the training data or generalize to unseen data.

In SCA, an adversary is not interested in predicting the classes in the attack phase but in obtaining the secret key k^*. To estimate the effort required to obtain the key, we will use the guessing entropy (GE) or success rate metrics [29]. An attack outputs a key guessing vector $\mathbf{g} = [g_1, g_2, \ldots, g_{|\mathcal{K}|}]$ in decreasing order of probability, which means that g_1 is the most likely key candidate and $g_{|\mathcal{K}|}$ the least likely key candidate. The success rate is the average probability that the secret key k^* is the first element of the key guessing vector $\mathbf{g}$. Guessing entropy is the average position of k^* in $\mathbf{g}$. Commonly, averaging is done over 100 independent experiments to obtain statistically significant results. As common in the deep learning-based SCA, we consider multilayer perceptron (MLP) and convolutional neural networks (CNNs).

2.3 Leakage Models and Datasets

During the execution of the cryptographic algorithm, the processing of sensitive information produces a specific leakage. In this chapter, we consider the Hamming weight leakage model since the considered datasets leak significantly in this model. There, the attacker assumes the leakage is proportional to the sensitive variable's Hamming weight. This leakage model results in nine classes when considering a cipher that uses an 8-bit S-box ($c = 9$).

ASCAD Datasets The first target platform we consider is an 8-bit AVR microcontroller running a masked AES-128 implementation [3]. There are two versions of the ASCAD dataset. The first version of the ASCAD dataset has a fixed key and 50,000 traces for profiling and 10,000 for testing. The second version of the ASCAD dataset has random keys, and it consists of 200,000 traces for profiling and 100,000 for testing. For both versions, we attack the key byte 3 unless specified differently.

For the ASCAD dataset, the third key byte is the first masked byte. For ASCAD with the fixed key, we use a pre-selected window of 700 features, while for ASCAD with random keys, the window size equals 1400 features. These datasets are available at [2].

CHES CTF 2018 Dataset This dataset refers to the CHES Capture-the-flag (CTF) AES-128 dataset, released in 2018 for the Conference on Cryptographic Hardware and Embedded Systems (CHES). The traces consist of masked AES-128 encryption running on a 32-bit STM microcontroller. We use 45,000 traces for the training set (CHES CTF Device C), containing a fixed key. The attack set consists of 5000 traces (CHES CTF Device D). The key used in the training and validation set is different from the key configured for the test set. CHES CTF 2018 trace sets contain the power consumption of the full AES-128 encryption, with a total number of 650,000 features per trace. The raw traces were pre-processed in the following way. First, a window resampling is performed. Later, we concatenated the trace intervals representing the processing of the masks (beginning of the trace) with the samples indicating the processing of S-boxes located after an interval without any particular activity (flat power consumption profile). The resulting traces have 2200 features. The original dataset is available at [7], and the processed traces are provided at [25].

3   Related Works

The goal of finding neural networks that perform well in SCA is probably the most explored direction in machine learning-based SCA. The first works commonly considered multilayer perceptron and reported good results even though there were not many available details about hyperparameter tuning or the best-obtained architectures [10, 11, 20, 33]. In 2016, Maghrebi et al. made a significant step forward in the profiling SCA as they investigated the performance of convolutional neural networks [19]. Since the results were promising, this paper started a series of works where deep learning techniques (most dominantly MLP and CNNs) were used to break various targets efficiently.

Soon after, works from Cagli et al. [5], Picek et al. [23], and Kim et al. [14] demonstrated that deep learning could efficiently break implementations protected with countermeasures. While those works also discuss hyperparameter tuning, it was still not straightforward to understand the effort required to find the neural networks that performed well. This effort became somewhat clearer after Benadjila et al. investigated hyperparameter tuning for the ASCAD dataset [3]. Indeed, while considering only a subset of possible hyperparameters, the tuning process was far from trivial.

Zaid et al. proposed a methodology for CNNs for profiling SCA [35]. While the methodology has limitations, the results obtained are significant as they reached top performance with never smaller deep learning architectures. This direction is further investigated by Wouters et al. [31] who reported some issues with [35] but managed to find even smaller neural networks that perform similarly well. Still, the proposed methodologies have some issues. First, it is not easy to use those methodologies and generalize for other datasets or neural network architectures. Second, the conflicting results among those methodologies indicate it is difficult to find a single approach that works the best for everything.

Perin et al. conducted a random search in pre-defined ranges to build deep learning models to form ensembles [21]. Their findings showed that even random search (when working on some reasonable range of hyperparameters) could find neural networks that perform extremely well. Finally, van ver Valk et al. used a technique called mimicking to find smaller neural networks that perform like the larger ones [30]. Still, the authors did not use pruning but ran experiments until they found a smaller network that outputs the same results as the larger one. Thus, the approaches are significantly different.

Thus, while the approaches mentioned work as evident from the excellent attack performance, there are still unanswered questions. What is clear is that we can reach good results with (relatively) small neural networks. What remains to be answered is how to adapt those methodologies for different datasets, or can we find even smaller neural network architectures that perform as well (or better). We aim to provide the answers to those questions in this work.

4   The Lottery Ticket Hypothesis (LTH)

The Lottery Ticket Hypothesis (LTH) was originally proposed by Franke and Carbin in [9] as a technique to improve pruned neural network performances. The main goal of pruning is to remove unnecessary weights to achieve the smallest neural network by keeping the original baseline performance. The baseline model refers to the trained neural network architecture that is not pruned. This way, pruned neural networks are suitable for memory-constrained devices and can deliver faster inference. With the LTH, authors verified that re-initializing the pruned neural network with the same initial weights from the baseline neural network shows equivalent or superior performance to the baseline model. In short, authors define the following: *"Lottery Ticket Hypothesis: a randomly initialized dense neural network contains a sub-network that is initialized such that - when trained in isolation - it can match the test accuracy of the original network after training for at most the same number of iterations"*.

A fixed sparsity level gives the amount of pruned weights and denotes the percentage of the removed network (e.g., 90% sparsity on an MLP would remove 90% of weight connections). The top-performing sub-networks are then called the *winning tickets*. There are two main ways to deploy LTH: one-shot pruning and iterative pruning. The latter defines the next sparsity level according to the results obtained from the previously evaluated sparsity level amount. This way, instead of defining a fixed sparsity level as in the case of one-shot pruning, the process iteratively finds the maximum possible sparsity level that delivers satisfactory results.

4.1 Pruning Strategy

To find an efficient pruned network, the large overparameterized baseline model must be trained before applying pruning to remove unnecessary weights. The pruning process applied in this work removes the smallest weights from the trained weights obtained from training the baseline model for a fixed amount of epochs. The activations in the forward propagation are mostly affected by larger weight values. Therefore, pruning the smallest weights remove those weights that are not significantly impacting the predictions. Different pruning strategies could be considered. Here we show that even the most simple method based on weight magnitude already delivers efficient results.

As shown in the experimental results section, we apply the LTH process on public (and protected) AES datasets, which also works when considering other than accuracy performance metrics (e.g., success rate, guessing entropy). The process starts by training an overparameterized neural network model for a single target AES key byte (note that our baseline models are assumed to be overparameterized in comparison to state-of-the-art works, e.g., [27, 31, 35]). Afterward, the model is pruned by removing smallest weights, and this pruned model is re-initialized and retrained (with more efficiency, e.g., fewer epochs) for all AES key bytes. Therefore, LTH reduces the complexity of deep neural network tuning in profiling SCA. We give the pruning strategy procedure for LTH in Algorithm 1. Although our process iterates over all sparsity levels s (in our case, from 1 to 99%), we do not consider it as iterative pruning because we do not set a metric to stop the process. Our main goal is to evaluate the profiling attack performance for all evaluated sparsity levels.

Algorithm 1 Pruning strategy

1: **procedure** PRUNING STRATEGY(original neural network f, original dataset x, random initial
    weights θ_0, training epoch θ_j, trained weight θ_j, pruning ratio $P_\%$, mask m)
2:      **for** $s = 1$ to 99 **do**
3:          $\theta_j \leftarrow$ Pretrain Model $f(x, \theta_0)$ for j epochs
4:          $m \leftarrow$ Prune $s\%$ of the smallest weights from θ_j
5:          **for** $i = 1$ to j **do**
6:              Train $f(x, \theta_0 \odot m)$
7:          **end for**
8:      **end for**
9: **end procedure**

In Fig. 1, we depict an one-shot pruning procedure. The first part of the figure displays the reference training procedure with no pruning where the weights at the beginning of the training process are different from those at epochs A and B. The

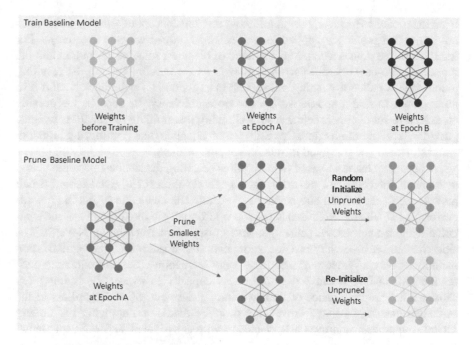

Fig. 1 One-shot pruning procedure for LTH

lower figure shows the setup when we prune the smallest weights and are left to choose whether we randomly initialize the remaining weights or re-initialize them from the original weights.

4.2   Winning Tickets in Profiling SCA

In [9], a *winning ticket* is defined as a sub-network that, when trained in isolation (after being re-initialized with the same baseline model initial weights), provides classification accuracy equivalent or superior to the baseline model. For profiling SCA, we define winning ticket as a sub-network that provides a test guessing entropy lower than or equivalent to the guessing entropy obtained from the original baseline model. Note that hyperparameters defined for the baseline model (which affects the total number of training parameters) and the number of profiling traces directly affect the chances to identify a winning ticket as demonstrated in Sect. 5.

Recall, pruning refers to removing neurons (neuron-based pruning) or weight connections (weight-based pruning) from the neural network activity. The most popular pruning technique consists of keeping a number of weight connections based on their weight value. This means that the smallest weights are pruned out from the model.[1] However, one should note that the concept of *winning ticket* does not imply that pruning is applied to well-selected pruned weights or neurons. For instance, if one prunes a certain percentage of elements selected at random and the remaining sub-network still performs as well as the baseline model, the resulting model is still called a winning ticket. Obviously, pruning techniques should also be explored to find a sub-network with more efficiency. In Sect. 5, we provide an extensive set of experimental results showing that pruning the smallest weights provides excellent results for SCA. Still, we do not claim that pruning, e.g., random weights, would not give good results for specific settings.

Ideally, deep learning-based SCA requires selecting the smallest possible neural network architecture that provides good generalization for a given target. Small models are faster to train and easier to interpret. The challenge of finding a well-performing small architecture may grow proportionally to the difficulty of the evaluated side-channel dataset (misalignment, noise, countermeasures). Nevertheless, side-channel traces usually provide a low signal-to-noise ratio, and regularization techniques play an important role in leakage learnability. Small models are self-regularized, mainly because they offer less capacity to overfit the training set. This justifies the importance of finding winning tickets in SCA. Regardless of the evaluated dataset, starting from a large baseline model and applying the Lottery Ticket Hypothesis improves the chances to create a small and efficient neural network model.

5   Experimental Results

5.1   Baseline Neural Networks

In our experiments, we define six different baseline models: three MLPs and three CNNs. Here, the main idea is to demonstrate how pruning and weight re-initialization (the Lottery Ticket Hypothesis) provide different SCA results if the baseline model varies in size or capacity. The MLP models are selected based on the sizes of commonly used architectures from the related works [3, 19, 21]. CNN models contain relatively fewer trainable parameters, and we define them based on efficient results obtained with smaller models as presented in [21, 31, 35].

Table 1 lists the hyperparameter configurations for MLP4, MLP6, and MLP8 models. The main idea is to verify how pruning and re-initialization work for MLP architectures with different numbers of dense layers and, consequently, different number of trainable parameters. Note that we have not selected very large neural network models. All of them contain less than one million trainable parameters. Here, the goal is to demonstrate that even a moderately-sized model can be significantly reduced according to the Lottery Ticket Hypothesis procedure presented in

---

[1] Similarly, pruning can be considered as keeping the largest weights in model.

Table 1 MLP architectures (batch size 400, learning rate 0.001, ADAM, *selu* activation functions). Number of parameters vary for different datasets due to different input layer dimensions

Layer	MLP4	MLP6	MLP8
Dense_1	200 neurons	200 neurons	200 neurons
Dense_2	200 neurons	200 neurons	200 neurons
Dense_3	200 neurons	200 neurons	200 neurons
Dense_4	200 neurons	200 neurons	200 neurons
Dense_5	–	200 neurons	200 neurons
Dense_6	–	200 neurons	200 neurons
Dense_7	–	–	200 neurons
Dense_8	–	–	200 neurons
Softmax	9 neurons	9 neurons	9 neurons
Parameters (ASCAD Random Keys)	402,609	483,009	563,409
Parameters (ASCAD Fixed Key)	262,609	343,009	423,409
Parameters (CHES CTF 2018)	562,609	643,009	723,409

Algorithm 1 and still keep or provide improved profiling SCA results. While it could be said that neural networks with up to one million trainable parameters are small, we note that the state-of-the-art results report significantly smaller architectures (even significantly fewer than 100,000 trainable parameters) [27, 31, 35].

The principle also holds for the chosen CNN models. Table 2 shows three CNN architectures, denoted as CNN3, CNN4, and CNN4-2. We defined relatively small CNNs (but still larger than state-of-the-art in, e.g., [35]), which are sufficient to break the evaluated datasets. CNN3 has only one convolution layer, while CNN4 and CNN4-2 contain two convolution layers each. In particular, CNN4-2 has larger dense layers than CNN4 to allow more complex relations between the input-output data pairs to be found (and allow more overfitting to happen). It is important to note that we define the same models for three different datasets. It is expected that for baseline models (without pruning), the performance might not be optimal for all cases. Although it is out of this chapter's scope to identify one model that generalizes well for all scenarios, we demonstrate that applying the Lottery Ticket Hypothesis procedure is a step forward in this important deep learning-based profiling SCA research direction.

We also provide experimental results demonstrating that the procedure described in Sect. 4 depends on several aspects such as the number of profiling traces and the sparsity level in the pruning process. By identifying the optimal sparsity level for pruning, we can drastically improve the performance of re-initialized sub-networks. Moreover, in some scenarios, we show that even when a large baseline model cannot recover the key, the pruned and re-initialized sub-network succeeds, especially when the number of profiling traces is small.

Interpreting Plots
This section's results are given in terms of guessing entropy for different baseline models, datasets, and sparsity levels. The sparsity level is provided in the x-axis,

Table 2 CNN architectures (batch size 400, learning rate 0.001, ADAM, *selu* activation function). Number of parameters vary for different datasets due to different input layer dimensions

Layer	CNN3	CNN4	CNN4-2
Conv1D_1	16 filters ks=10, stride=5	16 filters ks=10, stride=5	16 filters ks=10, stride=5
MaxPool1D_1	ks=2, stride=2	ks=2, stride=2	ks=2, stride=2
–	BatchNorm	BatchNorm	BatchNorm
Conv1D_2	–	16 filters ks=10, stride=5	16 filters ks=10, stride=5
MaxPool1D_2	–	ks=2, stride=2	ks=2, stride=2
–	–	BatchNorm	BatchNorm
Dense_1	128 neurons	128 neurons	256 neurons
Dense_2	128 neurons	128 neurons	256 neurons
Softmax	9 neurons	9 neurons	9 neurons
Parameters (ASCAD Random Keys)	302,713	47,305	124,489
Parameters (ASCAD Fixed Key)	159,353	32,969	95,817
Parameters (CHES CTF 2018)	466,553	63,689	157,257

where we apply pruning to the trained baseline neural network from 1% up to 99%. In each plot, there is a dashed green line that represents the average resulting guessing entropy for the baseline model *without pruning*. Thus, the green line is shown together with the plots to indicate the obtained guessing entropy when baseline models are trained for 300 epochs without any pruning. We consider 300 epochs to skip possible underfitting scenarios. The models we consider range from 32,969 to 723,409 trainable parameters, and with 300 epochs, there are no extreme overfitting cases.

The dashed red line is the resulting average guessing entropy after the trained baseline model is pruned according to the indicated sparsity level (*x*-axis) and initialized with *random weights* and trained for 50 epochs. Finally, the blue line is the resulting average guessing entropy from the same previous pruned model and re-initialized with *initial weights* from the baseline model according to LTH and trained for 50 epochs. For each sparsity level, each experiment is repeated ten times. Therefore, each plot results from training $98 \times 2 \times 10 = 1960$ pruned models. The plots also present the margin variation obtained with ten experiments (depicted as the area in the respective color).

We briefly discuss the limits that pruning and the Lottery Ticket Hypothesis offer regarding the results and their explainability:

1. Pruning allows smaller neural networks that perform on the level or even better than larger ones. This results from the regularization effect provided by pruning out small weights according to some strategy (random pruning or LTH).
2. The Lottery Ticket Hypothesis assumes there will be smaller, good performing sub-networks, so-called winning tickets. Winning tickets in profiling SCA allow reaching small sub-networks with good attack performance, as measured with GE. This provides an alternative solution for hyperparameter tuning in which pruning is used to extract the best possible performance from a model by disabling unnecessary weight connections.
3. In profiling SCA, finding an efficient model is also characterized by determining a good balance between the model's fitting capacity (i.e., its number of trainable parameters) and its generalization. Regularization is the method that provides this balance if one chooses to avoid tuning the model's hyperparameters. However, finding good regularizers might also pose critical difficulties, especially when there are more hyperparameters to be tuned due to the regularizer choice. Therefore, the pruning, and LTH process, offer a cheap and easy-to-deploy alternative to regularize a large model. However, not all neural network sizes will necessarily be converted from a baseline model that performs poorly into an optimal one just by applying pruning strategies as regularizers. Other aspects, such as dataset nature and the number of profiling traces, also will affect the pruned model's performance.
4. Pruning and LTH are not methods to provide explainability. However, explainability and interpretability can be used to improve the pruning strategy. This can be done by analysing, e.g., gradients [18], neuron relevance to classification [34] or simply weight magnitude [9].

5.2 ASCAD with a Fixed Key

Figures 2, 3, and 4 provide results for the ASCAD Fixed Key dataset when MLP4, MLP6, and MLP8 are used as baseline models, respectively, for different number

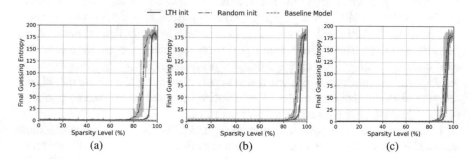

Fig. 2 ASCAD Fixed Key, MLP4. (**a**) 30,000 profiling traces. (**b**) 40,000 profiling traces. (**c**) 50,000 profiling traces

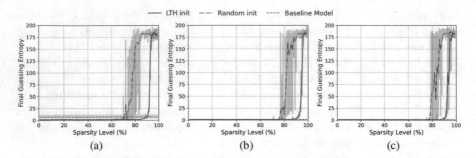

Fig. 3 ASCAD Fixed Key, MLP6. (**a**) 30,000 profiling traces. (**b**) 40,000 profiling traces. (**c**) 50,000 profiling traces

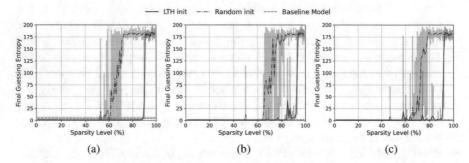

Fig. 4 ASCAD Fixed Key, MLP8. (**a**) 30,000 profiling traces. (**b**) 40,000 profiling traces. (**c**) 50,000 profiling traces

of profiling traces. There, we can immediately conclude that random initialization (*random init* in figure legends) and LTH initialization (*LTH init* in figure legends) provide different final guessing entropy results for different MLP sizes and the number of profiling traces. For the LTH case, the model size and the number of profiling traces have a small impact, and we can observe that, for all scenarios, pruning up to 90% of the weights show similar key recovery results.

On the other hand, if the pruned models are initialized with random weights, the model's performance is directly related to model size and the number of profiling traces. Adding more profiling traces improves the behavior of the model that is randomly initialized, approaching the model's behavior that is re-initialized according to LTH. The baseline model performs better than the pruned model that uses random initialization if the percentage of pruned weights is larger than 50% (for MLP8), and the number of profiling traces is sufficient to build a strong model. For the pruned model that follows the LTH initialization, the baseline model performs better only if we prune more than 90% of weights.

Interestingly, we can observe that pruning and LTH weight initialization show very stable results. Repeating the experiments ten times for each sparsity level tends to provide similar final guessing entropy values. Random weight initialization after

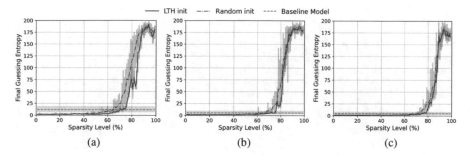

Fig. 5 ASCAD Fixed Key, CNN3. (**a**) 30,000 profiling traces. (**b**) 40,000 profiling traces. (**c**) 50,000 profiling traces

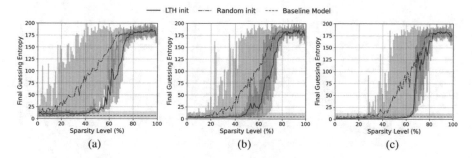

Fig. 6 ASCAD Fixed Key, CNN4. (**a**) 30,000 profiling traces. (**b**) 40,000 profiling traces. (**c**) 50,000 profiling traces

pruning clearly shows different final guessing entropy results, which is an obvious consequence of the randomness of weight initialization.

Next, we give results for the three different CNN architectures. Figures 5 and 6 indicate that for 30,000 training traces, as the dataset is small, the baseline model generally performs well but shows signs of overfitting. Then, pruning up to 60% of weights improves the performance regardless of the weight initialization procedure, although the LTH approach shows more stable and superior results. Increasing the number of traces shows improved behavior for the baseline model. Still, carefully selected sub-networks are sufficient to break the target, even when pruning 80% of weights.

Going to a more complex architecture (CNN4), the baseline model performs well and can reach a guessing entropy of one. However, this baseline model shows more variation from the ten repeated experiments. Simultaneously, pruning enables similar performance where the larger the training set, the smaller the differences between weight initialization procedures (LTH initialization or random). In Fig. 7, we consider the most complex CNN architecture. Interestingly, for 40,000 and 50,000 traces, we observe an even better performance of pruned networks when compared to the baseline model for up to 60% pruned weights. Again, LTH initialization tends to provide more stable results.

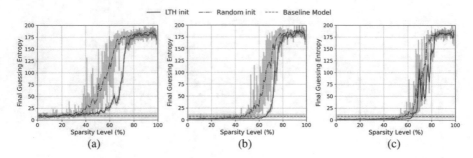

Fig. 7 ASCAD Fixed Key, CNN4-2. (**a**) 30,000 profiling traces. (**b**) 40,000 profiling traces. (**c**) 50,000 profiling traces

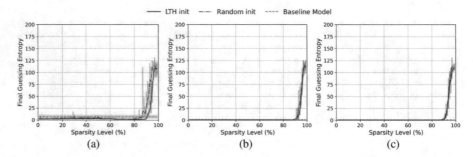

Fig. 8 ASCAD Random Keys, MLP4. (**a**) 60,000 profiling traces. (**b**) 100,000 profiling traces. (**c**) 200,000 profiling traces

5.3 ASCAD with Random Keys

In this section, we provide results for the ASCAD Random Keys dataset, as introduced in Sect. 2.3. Again, we apply the LTH procedure for a different number of profiling traces (60,000, 100,000, and 200,000) on the six different baseline models (MLP4, MLP6, MLP8, CNN3, CNN4, and CNN4-2).

Figure 8 shows results for different number of profiling traces and the MLP4 baseline model. With four dense layers, this MLP can be considered a small model, which is sufficient to break the ASCAD dataset for a large number of profiling traces (above 100,000), as indicated by the baseline model guessing entropy results. However, if the number of profiling traces is reduced (60,000), the guessing entropy result for the baseline model trained for 300 epochs is worse due to overfitting. On the other hand, applying the LTH process on this MLP4 baseline model shows good results even when the number of profiling traces is reduced. A natural alternative to fix the baseline model training would be to reduce the number of epochs to limit the overfitting. However, we expose this result (Fig. 8a) to demonstrate how pruning (even from 1% of weights) already regularizes the model and delivers successful attack results (we also must mention that pruned model is trained for fewer epochs, also reducing overfitting).

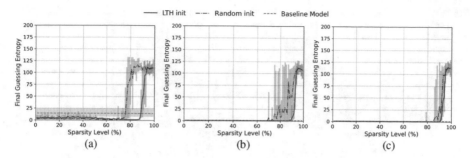

Fig. 9 ASCAD Random Keys, MLP6. (**a**) 60,000 profiling traces. (**b**) 100,000 profiling traces. (**c**) 200,000 profiling traces

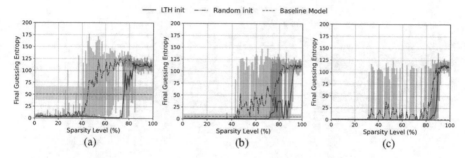

Fig. 10 ASCAD Random Keys, MLP8. (**a**) 60,000 profiling traces. (**b**) 100,000 profiling traces. (**c**) 200,000 profiling traces

The observations are confirmed in Figs. 9 and 10 for MLP models with more capacity (MLP6 and MLP8). Indeed, profiling sets that are too small cause overfitting for the baseline model, which can be easily resolved following the pruning method. Notice that random initialization always works worse than LTH initialization, and it also gives more irregular behavior due to the randomness in the process. This is even more evident in Fig. 10 where the variation of random initialization after pruning is very significant. This confirms that the LTH is valid in the profiling SCA context. As shown in Fig. 9, pruning approximately 90% of the weights from the baseline model results in a successful attack when weights are initialized with the LTH process.

Comparing Figs. 9 and 10, the larger baseline models tend to provide less successful results when the LTH procedure is applied. Larger baseline models may overfit training data more easily, and, as a consequence, the pruning process is applied to a model that might overfit. The solution for this problem is to consider early stopping for the baseline model training. This way, pruning would be applied to the baseline model weights when they reach the best training epoch. To confirm our hypothesis, we can consider Fig. 10c. The baseline model (MLP8) is trained on 200,000 profiling traces for 300 epochs and does not overfit, as seen in the final baseline model's guessing entropy. In this case, the pruned model performance with

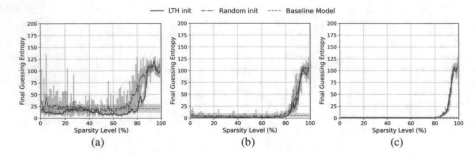

Fig. 11 ASCAD Random Keys, CNN3. (**a**) 60,000 profiling traces. (**b**) 100,000 profiling traces. (**c**) 200,000 profiling traces

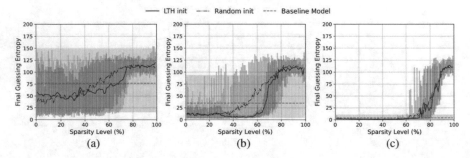

Fig. 12 ASCAD Random Keys, CNN4. (**a**) 60,000 profiling traces. (**b**) 100,000 profiling traces. (**c**) 200,000 profiling traces

LTH initialization is as good as for smaller baseline models trained on the same number of profiling traces (see, e.g., Fig. 9c).

The CNN architectures selected for this analysis show better guessing entropy results for the baseline model when more profiling traces are used, as shown in Figs. 11, 12, and 13. However, when less profiling traces are used, as is the case of results provided in Figs. 11b, 12b, and 13b, the baseline guessing entropy is not reaching one on average. Adding more profiling traces helps, but the number of profiling traces should align with the model complexity. The evaluated CNN models worked well for the ASCAD Fixed Key dataset, as shown in the last section. However, these models (especially CNN4 and CNN4-2) appear less appropriate for the ASCAD Random Keys dataset. In such cases, pruning plays an important role in (partially) overcoming this. After pruning, it is possible to reach very low GE values (under 5) for a specific percentage of pruned weights. In particular, results show that pruning plus LTH initialization is better than pruning plus random initialization. For all cases, we can prune up to around 50% of weights and still reach good performance even though we use (relatively) simple CNN architectures.

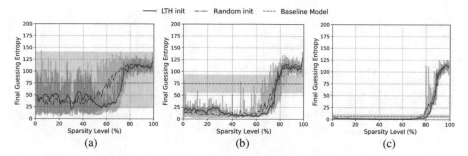

Fig. 13 ASCAD Random Keys, CNN4-2. (**a**) 60,000 profiling traces. (**b**) 100,000 profiling traces. (**c**) 200,000 profiling traces

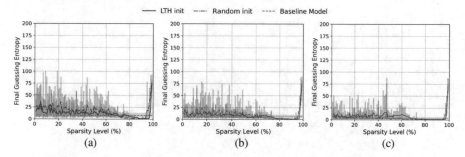

Fig. 14 CHES CTF 2018, MLP4. (**a**) 20,000 profiling traces. (**b**) 30,000 profiling traces. (**c**) 40,000 profiling traces

5.4 CHES CTF 2018

For the CHES CTF 2018 dataset, we repeated the experiments on the same neural network architectures defined in Tables 1 and 2. In this case, we observed much better results for the three selected MLPs and CNN3 than results obtained for CNN4 and CNN4-2. These results again confirm the practical advantage of the LTH procedure in profiling SCA.

Figure 14 shows the guessing entropy for different sparsity levels on three different number of profiling traces: 20,000, 30,000, and 40,000. As indicated by the dashed green line in Figs. 14a, 14b, and 14c, the baseline guessing entropy cannot reach one for MLP4 trained on 300 epochs. Adding more profiling traces helps, but still, GE stays slightly above one on average. When the network is pruned, we can immediately see how GE improves, especially for sparsity levels around 80–95%. The LTH initialization shows better (at least more stable) results than random initialization. Figures 15 and 16 confirm our observations as more profiling traces is required for good attack performance for the baseline model, especially as the architecture becomes more complex. On the other hand, we can prune up to 95% of weights if we follow the LTH initialization and still reach superior attack performance.

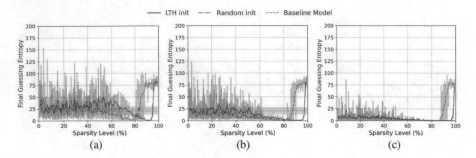

Fig. 15 CHES CTF 2018, MLP6. (**a**) 20,000 profiling traces. (**b**) 30,000 profiling traces. (**c**) 40,000 profiling traces

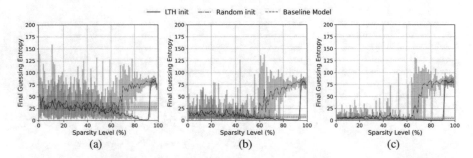

Fig. 16 CHES CTF 2018, MLP8. (**a**) 20,000 profiling traces. (**b**) 30,000 profiling traces. (**c**) 40,000 profiling traces

Results for CNNs on the CHES CTF 2018 dataset are acceptable (i.e., converging to GE close to one) for the CNN3 architecture only, as shown in Fig. 17. There, we see the benefit of adding more profiling traces as the baseline model overfits. Still, some sub-networks are providing better attack performance. For CNN4 and CNN4-2 (Figs. 18 and 19), the baseline model provides poor performances when trained on 300 epochs. We postulate this happens as the baseline model has a significantly larger capacity than needed, so it either overfits or underfits, becoming similar to random guessing. In other words, CNN4 and CNN4-2 on smaller profiling sets (lower than 30,000 traces) show no generalization for the baseline model, indicating that these two models are not compatible with the target dataset. We can observe how the LTH procedure reduces guessing entropy for specific sparsity level ranges even with those models. Observing Figs. 18 and 19, for sparsity levels around 70%, LTH initialization reach significantly lower guessing entropy values (GE ≤ 70) after training for 50 epochs. Increasing the number of attack traces (we consider only 2000 attack traces) could lead to successful key recovery, which is particularly interesting if a baseline model provided performance close to random guessing. When the number of profiling traces is increased to 40,000 traces (Figs. 18c and 19c), the baseline model shows slightly better results and the LTH initialization still improves the attack performance. In this case, we can verify that

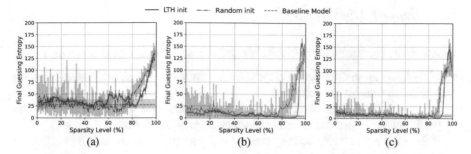

Fig. 17 CHES CTF 2018, CNN3. (**a**) 20,000 profiling traces. (**b**) 30,000 profiling traces. (**c**) 40,000 profiling traces

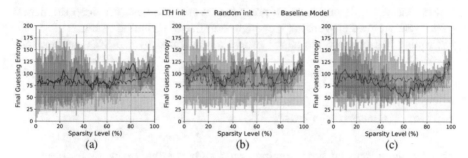

Fig. 18 CHES CTF 2018, CNN4. (**a**) 20,000 profiling traces. (**b**) 30,000 profiling traces. (**c**) 40,000 profiling traces

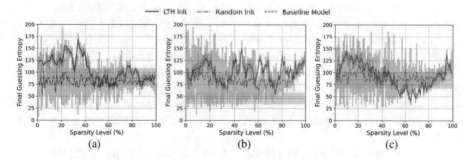

Fig. 19 CHES CTF 2018, CNN4-2. (**a**) 20,000 profiling traces. (**b**) 30,000 profiling traces. (**c**) 40,000 profiling traces

random initialization might not be a good procedure, as the guessing entropy results are inferior to the baseline model results.

5.5 General Observations

Based on the conducted experiments, we provide several general observations:

- If the baseline model works poorly for a limited set of attack traces, pruning might still improve performance.
- If the baseline works well and does not overfit, then pruning maintains the performance but produces smaller and regularized networks.
- If there are not enough profiling traces for the model capacity, it will overfit, and pruning can help avoid that.
- More profiling traces improve pruning results, but it also reduces differences between weight initialization techniques.
- Pruning and LTH initialization procedure works the best, provided the neural network architectures are large enough to utilize the winning tickets.
- Pruning can improve the attack results as indicated by the SCA performance metrics.

6 Conclusions and Future Work

This chapter discussed how pruning could improve the attack performance for deep learning-based side-channel analysis. We considered the recently proposed Lottery Ticket Hypothesis that assumes there are small sub-networks in the original network that perform on the same level as the original network. To the best of our knowledge, both of those concepts were never before investigated in profiling SCA. Our experimental investigation confirms this hypothesis for profiling SCA, which allows us to prune up to 90% of weights and still reach good attack performance. Thus, we manage to reach the same attack performance for significantly smaller networks (easier to tune and faster to train). What is more, we show how pruning helps when a large network overfits or has issues due to imbalanced data. In such cases, pruning, besides resulting in smaller architectures, enables improved attack performance.

As future work, we plan to consider more sophisticated pruning techniques and different leakage models. Finally, as discussed, pruning allows smaller neural networks and good performance but does not provide insights into neural networks' explainability. It could be interesting to consider various feature visualization techniques to evaluate the important features before and after the pruning. Also, explainability and interpretability techniques could be efficiently applied here to select weights to be pruned.

Acknowledgments This work was supported in part by the Netherlands Organization for Scientific Research NWO project DISTANT (CS.019) and project PROACT (NWA.1215.18.014).

References

1. C. Archambeau, E. Peeters, F. X. Standaert, and J. J. Quisquater. Template attacks in principal subspaces. In Louis Goubin and Mitsuru Matsui, editors, *Cryptographic Hardware and Embedded Systems - CHES 2006*, pages 1–14, Berlin, Heidelberg, 2006. Springer Berlin Heidelberg.
2. ASCAD GitHub Repository. Website, 2018. https://github.com/ANSSI-FR/ASCAD.
3. Ryad Benadjila, Emmanuel Prouff, Rémi Strullu, Eleonora Cagli, and Cécile Dumas. Deep learning for side-channel analysis and introduction to ASCAD database. *J. Cryptographic Engineering*, 10(2):163–188, 2020.
4. Davis Blalock, Jose Javier Gonzalez Ortiz, Jonathan Frankle, and John Guttag. What is the state of neural network pruning?, 2020.
5. Eleonora Cagli, Cécile Dumas, and Emmanuel Prouff. Convolutional neural networks with data augmentation against jitter-based countermeasures. In Wieland Fischer and Naofumi Homma, editors, *Cryptographic Hardware and Embedded Systems – CHES 2017*, pages 45–68, Cham, 2017. Springer International Publishing.
6. Suresh Chari, Josyula R. Rao, and Pankaj Rohatgi. Template attacks. In Burton S. Kaliski Jr., Çetin Kaya Koç, and Christof Paar, editors, *Cryptographic Hardware and Embedded Systems - CHES 2002, 4th International Workshop, Redwood Shores, CA, USA, August 13–15, 2002, Revised Papers*, volume 2523 of *Lecture Notes in Computer Science*, pages 13–28. Springer, 2002.
7. CHES CTF 2018. Website, 2018. https://chesctf.riscure.com/2018/news.
8. Omar Choudary and Markus G. Kuhn. Efficient template attacks. In Aurélien Francillon and Pankaj Rohatgi, editors, *Smart Card Research and Advanced Applications*, pages 253–270, Cham, 2014. Springer International Publishing.
9. Jonathan Frankle and Michael Carbin. The lottery ticket hypothesis: Training pruned neural networks. *CoRR*, abs/1803.03635, 2018.
10. R. Gilmore, N. Hanley, and M. O'Neill. Neural network based attack on a masked implementation of AES. In *2015 IEEE International Symposium on Hardware Oriented Security and Trust (HOST)*, pages 106–111, May 2015.
11. Annelie Heuser, Stjepan Picek, Sylvain Guilley, and Nele Mentens. Side-channel analysis of lightweight ciphers: Does lightweight equal easy? In Gerhard P. Hancke and Konstantinos Markantonakis, editors, *Radio Frequency Identification and IoT Security - 12th International Workshop, RFIDSec 2016, Hong Kong, China, November 30 - December 2, 2016, Revised Selected Papers*, volume 10155 of *Lecture Notes in Computer Science*, pages 91–104. Springer, 2016.
12. Annelie Heuser and Michael Zohner. Intelligent Machine Homicide - Breaking Cryptographic Devices Using Support Vector Machines. In Werner Schindler and Sorin A. Huss, editors, *COSADE*, volume 7275 of *LNCS*, pages 249–264. Springer, 2012.
13. Steven A. Janowsky. Pruning versus clipping in neural networks. *Phys. Rev. A*, 39:6600–6603, Jun 1989.
14. Jaehun Kim, Stjepan Picek, Annelie Heuser, Shivam Bhasin, and Alan Hanjalic. Make some noise. unleashing the power of convolutional neural networks for profiled side-channel analysis. *IACR Transactions on Cryptographic Hardware and Embedded Systems*, pages 148–179, 2019.
15. Paul C. Kocher. Timing Attacks on Implementations of Diffie-Hellman, RSA, DSS, and Other Systems. In *Proceedings of CRYPTO'96*, volume 1109 of *LNCS*, pages 104–113. Springer-Verlag, 1996.
16. Paul C. Kocher, Joshua Jaffe, and Benjamin Jun. Differential power analysis. In *Proceedings of the 19th Annual International Cryptology Conference on Advances in Cryptology*, CRYPTO '99, pages 388–397, London, UK, UK, 1999. Springer-Verlag.
17. Liran Lerman, Romain Poussier, Gianluca Bontempi, Olivier Markowitch, and François-Xavier Standaert. Template attacks vs. machine learning revisited (and the curse of

dimensionality in side-channel analysis). In *International Workshop on Constructive Side-Channel Analysis and Secure Design*, pages 20–33. Springer, 2015.

18. Congcong Liu and Huaming Wu. Channel pruning based on mean gradient for accelerating convolutional neural networks. *Signal Processing*, 156:84–91, 10 2018.

19. Houssem Maghrebi, Thibault Portigliatti, and Emmanuel Prouff. Breaking cryptographic implementations using deep learning techniques. In *International Conference on Security, Privacy, and Applied Cryptography Engineering*, pages 3–26. Springer, 2016.

20. Zdenek Martinasek, Jan Hajny, and Lukas Malina. Optimization of power analysis using neural network. In Aurélien Francillon and Pankaj Rohatgi, editors, *Smart Card Research and Advanced Applications*, pages 94–107, Cham, 2014. Springer International Publishing.

21. Guilherme Perin, Lukasz Chmielewski, and Stjepan Picek. Strength in numbers: Improving generalization with ensembles in machine learning-based profiled side-channel analysis. *IACR Transactions on Cryptographic Hardware and Embedded Systems*, 2020(4):337–364, Aug. 2020.

22. S. Picek, A. Heuser, A. Jovic, and L. Batina. A systematic evaluation of profiling through focused feature selection. *IEEE Transactions on Very Large Scale Integration (VLSI) Systems*, 27(12):2802–2815, 2019.

23. Stjepan Picek, Annelie Heuser, Alan Jovic, Shivam Bhasin, and Francesco Regazzoni. The curse of class imbalance and conflicting metrics with machine learning for side-channel evaluations. *IACR Transactions on Cryptographic Hardware and Embedded Systems*, 2019(1):209–237, Nov. 2018.

24. Stjepan Picek, Annelie Heuser, Alan Jovic, Simone A. Ludwig, Sylvain Guilley, Domagoj Jakobovic, and Nele Mentens. Side-channel analysis and machine learning: A practical perspective. In *2017 International Joint Conference on Neural Networks, IJCNN 2017, Anchorage, AK, USA, May 14–19, 2017*, pages 4095–4102, 2017.

25. Preprocessed CHES CTF 2018 dataset. Website, 2021. http://aisylabdatasets.ewi.tudelft.nl/.

26. Jean-Jacques Quisquater and David Samyde. Electromagnetic analysis (EMA): Measures and counter-measures for smart cards. In Isabelle Attali and Thomas Jensen, editors, *Smart Card Programming and Security*, pages 200–210, Berlin, Heidelberg, 2001. Springer Berlin Heidelberg.

27. Jorai Rijsdijk, Lichao Wu, Guilherme Perin, and Stjepan Picek. Reinforcement learning for hyperparameter tuning in deep learning-based side-channel analysis. *IACR Trans. Cryptogr. Hardw. Embed. Syst.*, 2021(3):677–707, 2021.

28. Werner Schindler, Kerstin Lemke, and Christof Paar. A stochastic model for differential side channel cryptanalysis. In Josyula R. Rao and Berk Sunar, editors, *Cryptographic Hardware and Embedded Systems – CHES 2005*, pages 30–46, Berlin, Heidelberg, 2005. Springer Berlin Heidelberg.

29. François-Xavier Standaert, Tal G. Malkin, and Moti Yung. A unified framework for the analysis of side-channel key recovery attacks. In Antoine Joux, editor, *Advances in Cryptology - EUROCRYPT 2009*, pages 443–461, Berlin, Heidelberg, 2009. Springer Berlin Heidelberg.

30. D. van der Valk, M. Krcek, S. Picek, and S. Bhasin. Learning from a big brother - mimicking neural networks in profiled side-channel analysis. In *2020 57th ACM/IEEE Design Automation Conference (DAC)*, pages 1–6, 2020.

31. Lennert Wouters, Victor Arribas, Benedikt Gierlichs, and Bart Preneel. Revisiting a methodology for efficient CNN architectures in profiling attacks. *IACR Transactions on Cryptographic Hardware and Embedded Systems*, 2020(3):147–168, Jun. 2020.

32. Lichao Wu, Guilherme Perin, and Stjepan Picek. I choose you: Automated hyperparameter tuning for deep learning-based side-channel analysis. Cryptology ePrint Archive, Report 2020/1293, 2020. https://eprint.iacr.org/2020/1293.

33. Shuguo Yang, Yongbin Zhou, Jiye Liu, and Danyang Chen. Back propagation neural network based leakage characterization for practical security analysis of cryptographic implementations. In Howon Kim, editor, *Information Security and Cryptology - ICISC 2011*, pages 169–185, Berlin, Heidelberg, 2012. Springer Berlin Heidelberg.

34. Seul-Ki Yeom, Philipp Seegerer, Sebastian Lapuschkin, Simon Wiedemann, Klaus-Robert Müller, and Wojciech Samek. Pruning by explaining: A novel criterion for deep neural network pruning. *CoRR*, abs/1912.08881, 2019.
35. Gabriel Zaid, Lilian Bossuet, Amaury Habrard, and Alexandre Venelli. Methodology for efficient CNN architectures in profiling attacks. *IACR Transactions on Cryptographic Hardware and Embedded Systems*, 2020(1):1–36, Nov. 2019.

Evaluating Deep Learning Models and Adversarial Attacks on Accelerometer-Based Gesture Authentication

Elliu Huang, Fabio Di Troia (iD)**, and Mark Stamp**

Abstract Gesture-based authentication has emerged as a non-intrusive, effective means of authenticating users on mobile devices. Typically, such authentication techniques have relied on classical machine learning techniques, but recently, deep learning techniques have been applied this problem. Although prior research has shown that deep learning models are vulnerable to adversarial attacks, relatively little research has been done in the adversarial domain for behavioral biometrics. In this research, we collect tri-axial accelerometer gesture data (TAGD) from 46 users and perform classification experiments with both classical machine learning and deep learning models. Specifically, we train and test support vector machines (SVM) and convolutional neural networks (CNN). We then consider a realistic adversarial attack, where we assume the attacker has access to real users' TAGD data, but not the authentication model. We use a deep convolutional generative adversarial network (DC-GAN) to create adversarial samples, and we show that our deep learning model is surprisingly robust to such an attack scenario.

1 Introduction

With the ubiquity of technology, authentication has become an essential part of everyday life. Passwords and PINs are the most common forms of authentication, but biometrics are also popular. Biometric authentication includes physiological (e.g., facial recognition and fingerprint) and behavioral (e.g., gait and keystroke dynamics) approaches [5].

While physiological biometric authentication has proven to be highly effective, sensors and equipment required for such approaches are usually costly. Additionally, attackers can sometimes bypass such a system if they have access to a copy of the required features [31]. On the other hand, behavioral biometrics not only have the potential to be cost effective, they may also be more secure, at least in

E. Huang · F. Di Troia · M. Stamp (✉)
San Jose State University, San Jose, CA, USA
e-mail: ehuang817@gmail.com; fabio.ditroia@sjsu.edu; mark.stamp@sjsu.edu

© The Author(s), under exclusive license to Springer Nature Switzerland AG 2022
M. Stamp et al. (eds.), *Artificial Intelligence for Cybersecurity*, Advances in Information Security 54, https://doi.org/10.1007/978-3-030-97087-1_10

243

cases where attackers have difficulty imitating the relevant features. Furthermore, the non-intrusive nature of behavioral biometrics may be considered desirable, in comparison to physiological biometric authentication.

Gesture-based authentication is a relatively recent behavioral biometric that has achieved promising results. There are various techniques for analyzing gestures, including acceleration, angular motion, 3D motion, and a mix of the three. Several machine learning techniques, including those we discuss in Sect. 3, have been applied to the gesture-based authentication problem.

In this research, we explore the effectiveness of deep learning techniques on gesture-based authentication. Our research is based on a new dataset that we have collected. Given that our tri-axial accelerometer gesture data (TAGD) are time series, we consider two time series classification (TSC) techniques: support vector machines (SVM) and one-dimensional convolutional neural networks (1D-CNN). When combined with feature extraction techniques, our SVM model provides for rudimentary analysis of our TAGD data, as well as a basis for comparison to our 1D-CNN model. We also generate adversarial samples using generative adversarial networks (GAN) and use these samples to explore the robustness of our 1D-CNN model against a realistic adversarial attack.

The remainder of this paper is structured as follows. Sect. 2 discusses relevant work in the field of gesture-based authentication and adversarial attacks on biometric authentication. Relevant background information on the machine learning techniques we consider is presented in Sect. 3. Section 4 provides an overview of our dataset, including specific steps of the data collection process and data preprocessing techniques. Details of our adversarial strategy are introduced in Sect. 5. We present our experimental results for our classification models and an adversarial attack in Sect. 6. Our conclusion and a brief discussion of future research directions are provided in Sect. 7.

2 Related Work

Relative to the vast research literature on behavioral biometrics, there is comparatively little work on gesture-based authentication. In this section, we provide an overview of research on gesture-based authentication and adversarial attacks on such security systems.

Most gesture-based authentication techniques can be categorized into two main methods, namely, touchscreen and motion gestures [11]. There are, however, studies that combine both into a single authentication system [8].

Touchscreen-based gesture authentication methods typically analyze touch dynamics, including various inputs recorded from a touchscreen interface such as finger size and pressure. One approach consisted of collecting finger behavior and position data and authenticated users via SVMs [3]. Another study employed particle swarm optimization to find patterns in touchscreen dynamics [24].

Motion gestures generally rely on accelerometer and gyroscope data to analyze the acceleration and angular motion of the mobile device. Prior research in this domain has applied dynamic time warping (DTW) [22], SVMs [23], and hidden Markov models (HMM) [15] to authenticate users. One gesture-based approach employed a more sophisticated method that involved the "leap motion" controller that collects 3D motion data and applied similarity thresholds to authenticate users [19]. Another approach analyzed full-body and hand-gestures in 3D space using two-stream CNNs [32].

Adversarial attacks on gesture-based authentication is not a well-researched area, but there are a handful of relevant studies in the general field of behavioral biometrics. One study analyzed behavioral mouse dynamics and found that deep learning authentication models were susceptible to adversarial attacks [28]. Adversarial samples have been generated using a fast gradient sign method (FGSM) to create perturbations in the data, with a gated recurrent unit (GRU) then used to generate adversarial samples. Another study [1] analyzed the resilience of continuous touch-based authentication systems (TCAS) to adversarial attacks. These researchers found that their TCAS trained with the help of generative adversarial networks (GAN) had a lower false acceptance rate than that of vanilla TCAS. The paper [2] reports on experiments with randomization attacks on gesture-based security systems that use SVMs, and finds that their models are highly vulnerable to adversarial attacks.

Similarly, adversarial learning on time series classification has not seen much research. Most adversarial attacks involve small perturbations of the original data using state-of-the-art FGSM or basic iterative method (BIM) in order to "trick" a classification or regression model. A study of adversarial attacks on multivariate time series regression found that three of the most popular deep learning models— CNNs, GRUs, and long short-term memory (LSTM)—were highly susceptible to such attacks [25]. Another study also used FGSM and BIM to create small perturbations in time series data, which significantly lowered the classification accuracy of deep learning models for vehicle sensors and electricity consumption data [13].

3 Background

In this section, we discuss the machine learning techniques that form the basis of our experiments. These techniques, namely, support vector machines, convolutional neural networks, and generative adversarial networks, will serve as the basis for our classification experiments in Sect. 6. We also introduce our strategies for adversarial attacks.

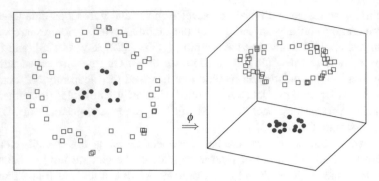

Fig. 1 The kernel trick

3.1 Support Vector Machines

Support vector machines (SVMs) are one of the most popular supervised learning techniques for classification and regression. SVMs attempt to find the optimal separating hyperplane between two labeled sets of training data [27]. However, a dataset need not be linearly separable, in which case we can employ the "kernel trick." As depicted in Fig. 1, the kernel trick maps the input data into a higher-dimensional space where it is more likely to be linearly separable. The kernel trick, together with "soft margin" calculations that allow for classification errors, makes an SVM an extremely powerful and flexible tool in the field of machine learning.

A classification problem with a small training sample size and high dimensionality is prone to overfitting [9]. Feature selection techniques can help to prevent this problem by discarding features, with minimal loss—or even improvements—in performance. In our SVM experiments, we used support vector machine recursive feature elimination (SVM-RFE) for feature selection. SVM-RFE consists of eliminating the least significant feature (based on linear SVM weights), then training a model on the reduced feature set. This process is repeated until the desired number of features is reached, or the performance degrades beyond acceptable limits.

3.2 1D Convolutional Neural Networks

Typically, convolutional neural networks (CNN) are associated with feature extraction and classification for images, which generally involves two-dimensional convolutional neural networks (2D-CNN). In such a model, 2D data (e.g., images) are fed into a CNN and classified via a final fully-connected layer.

In this paper, we do not consider images; instead, we have temporal sequences of fixed length. Such data is suitable for one-dimensional convolutional neural networks (1D-CNN). While not as common as 2D-CNNs, 1D-CNNs are used for signal processing and sequence classification, with numerous applications in biomedical and civil engineering [21].

The architecture of a 1D-CNN is analogous to that of a 2D-CNN, with the key differences being the dimensionality of the input data and the convolution operation. 2D-CNNs typically use a rectangular kernel that slides from left to right, top to bottom. In contrast, 1D-CNNs employ a kernel that spans some number of variables and slides along a vector.

3.3 Adversarial Strategy

We also consider how our deep learning models perform under adversarial attacks. While several studies analyze adversarial attacks involving small perturbations of the original data [20], we explore a scenario where we assume the intruder has access to a real users' gesture data. We test both poisoning and evasion attacks using learning models to generative adversarial samples; specifically, we use a type of generative adversarial network (GAN) to produce adversarial samples.

3.3.1 Deep Convolutional Generative Adversarial Networks

Two competing neural networks are trained in a GAN—a generative network and a discriminative network—with the generative network creating fake data that is designed to defeat the discriminative network. The two networks are trained simultaneously following a game-theoretic approach. In this way, both networks improve, with the ultimate objective being a model (discriminative, generative, or both) that is stronger than it would have been if it was trained only on the real training data; see [12] for additional details.

An overview of GAN structure is depicted in Fig. 2. Among many other uses, GANs have been used to generate realistic adversarial samples.

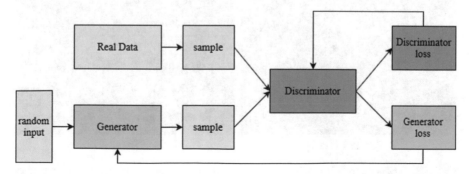

Fig. 2 Overview of GAN structure

4 Dataset

In this section, we give an overview of the data collected and specify the steps in the data collection process. We also include a discussion of data preprocessing and the feature engineering techniques that we have employed.

4.1 Data Collection

In this research, we collect users' tri-axial accelerometer gesture data (TAGD) while the user holds a smartphone and writes their "signature" in the air, similar to the process in [18]. The accelerometer sensor of the Physics Toolbox Sensor Suite [10] was used to collect data. A screenshot of the application is shown in Fig. 3. The tri-axial acceleration is represented as separate curves—the red curve represents

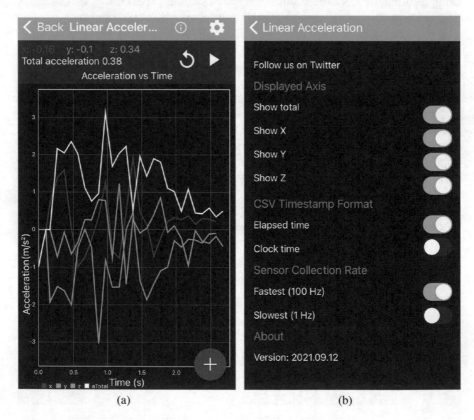

Fig. 3 Screenshots of physics toolbox sensor suite app. (**a**) Accelerometer sensor. (**b**) App settings

acceleration along the x-axis, the green curve represents acceleration along the y-axis, the blue curve represents acceleration along the z-axis, and the white curve represents the total magnitude of acceleration. Data is collected at a frequency of 100 Hz, i.e., 100 data points per second. We performed data collection solely on the Apple iOS platforms in order to reduce the possibility of smartphone type being a confounding variable. We note that iPhone models varied from iPhone 8 to iPhone X.

After installing the app, he user performs the following steps to collect signature data.

1. Tap the red button to start recording accelerometer data
2. Move the smartphone in the air to draw a signature
3. Tap the red button again to stop recording data
4. Upload data in the form of a CSV file into Google Drive

These steps were repeated 50 times to collect 50 signatures for each user.

In total, we collected 50 signatures from each of 46 different users who volunteered to provide such data. Users typically chose their initials as their signature, but they were free to create their own unique signature. Generally, the time to write each signature varied between 3 and 7 seconds, and the entire data collection process for one individual required about 20 minutes. Our dataset is freely available for use by other researchers at [17]. A sample of our TAGD data is shown in Fig. 4.

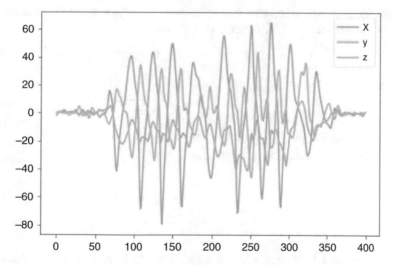

Fig. 4 Sample tri-axial acceleration time sequence

4.2 Data Preprocesssing

The raw tri-axial data is of variable length. As discussed below, we resize all signatures to the same size, as most traditional classification techniques require input data of a fixed size. We have applied several feature engineering techniques to the resulting time series.

4.2.1 Feature Engineering

As noted above, each signature is a temporal sequence of tri-axial accelerometer data. First, we extract statistical features based on the acceleration for each axis, ignoring the sequential nature of the data. The resulting distributions vary—we compute the following statistical measures of shape, center, and spread for each of the three axes.

Length (L) The number of data points in the signature.
Mean (μ) The center of the distribution is the mean

$$\mu = \frac{1}{n} \sum_{i=1}^{n} x_i = \frac{1}{n}(x_1 + x_2 + \cdots + x_i)$$

Median (m) The median is another measure of the center of the distribution.
Standard deviation (σ) The standard deviation

$$\sigma = \sqrt{\frac{\sum_{i=1}^{n}(x_i - \mu)}{n - 1}}$$

measures the variability in the signature data.
Kurtosis (k) The kurtosis is computed as

$$k = \frac{\sum_{i=1}^{n}(x_i - \mu)^4}{n\sigma^4}$$

and it measures the weight of the tails relative to the center of the distribution and provides additional information related to the signature motion.
Skewness (s) The symmetry of the distribution, which is computed as

$$s = \frac{\sum_{i=1}^{n}(x_i - \mu)^3}{n\sigma^3}$$

can help us understand the "smoothness" of motion in a signature.

All of these features provide some information about underlying patterns in users' signatures.

For each signature, we calculate a feature vector of the form

$$(L, \mu_x, \mu_y, \mu_z, m_x, m_y, m_z, \sigma_x, \sigma_y, \sigma_z, k_x, k_y, k_z, s_x, s_y, s_z)$$

consisting of the measures of shape, center, and spread of the distribution, as discussed above. This feature vector of 16 elements is utilized only in our SVM experiments, below.

4.2.2 Time Series Resampling

Since 1D-CNNs and GANs require feature vectors of fixed length, we use `tslearn` [30], we resize all the TAGDs to length 400. The resizing function in `tslearn` interpolates for arrays less than the target size. We chose to resample all the time series to length 400 since the median length of the sequences is 380, while the mean length is 400.

5 Implementation

In this section, we discuss specific details of the models that we use in relation to our TAGD data. We also outline the adversarial attacks that we consider.

5.1 DC-GAN Structure

We use deep convolutional GANs (DC-GAN) to replicate our time series data in a form that will serve as adversarial samples [4]. The generator and discriminator models are both based on 1D-CNN models. The generator essentially performs the functions of a convolutional layer in reverse—the input is an arbitrary sequence of values and it uses transposed convolution layers to shape the data into a desired form. The discriminator can be based on a traditional convolutional neural network. The fully-connected layer outputs a value between -1 and $+1$, with a negative output indicating a fake sample and positive output indicating a real sample. The generator and discriminator are connected by a loss function, which provides

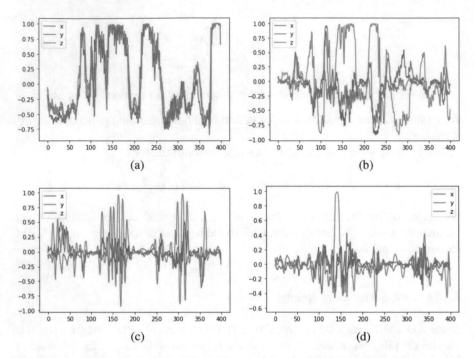

Fig. 5 DC-GAN generated acceleration sequences. (**a**) Acceleration sequence after 10 epochs. (**b**) Acceleration sequence after 25 epochs. (**c**) Acceleration sequence after 50 epochs. (**d**) Acceleration sequence after 100 epochs

feedback to both models. Over several epochs of training, the generator should becomes better at generating adversarial samples while the discriminator should become better at distinguishing between real and fake samples.

In our GAN model, the generator has an input consisting of a sequence of 100 values sampled from a normal distribution. After the data passes through three transposed convolution layers, the sequence of 100 values are transformed into a sequence of the same size as the TAGD, that is, 400×3. The discriminator model closely follows the architecture of the 1D-CNN classification model outlined above, with the major difference being the binary output of the fully-connected layer. The generator and discriminator are connected by a binary cross entropy loss function. In our experiments, we vary the number of training epochs to see how effective the adversarial samples are in breaking down our model.

In Fig. 5, we have examples of TAGD generated using DC-GANs with different training epochs. The number of training epochs is directly related to how well the DC-GAN model can replicate data. As we can see in Fig. 5a and b, the data is quite random doesn't resemble the real TAGD in Fig. 4, whereas Fig. 5c and d appear closer to the real TAGD. As we train the DC-GAN more epochs, the adversarial samples resemble the real data more.

5.2   Adversarial Attack

Similar to [26], we "poison" our training dataset with adversarial samples generated from DC-GANs, meaning that we mix in adversarial samples with our real training dataset. Then, we train our 1D-CNN on the poisoned dataset and try to classify real data. The accuracy of classifying with the poisoned training dataset suggests how well our 1D-CNN can survive a poisoning attack while simultaneously indicating how well our DC-GAN can generate adversarial samples.

6   Experiments and Results

In this section, we present and analyze the results of the experiments outlined in the previous section. For our first experiment, we provide the results of a multiclass classification problem using SVMs, based on the statistical features discussed in Sect. 4.2.1. Then we apply a deep learning technique, 1D-CNNs, in the multiclass classification problem. Both of these techniques, SVMs and 1D-CNNs, produce strong results. Then we move on to adversarial learning where we use GANs to generate adversarial samples. With the GAN-generated adversarial samples, we show that our deep learning model is robust under a poisoning type of adversarial attack.

The authentication problem is inherently a binary classification problem. For the technique considered in this paper, the confusion matrix is of the form given in Fig. 6, where

$$TP = \text{true positives}$$

$$FP = \text{false positives}$$

$$TN = \text{true negatives}$$

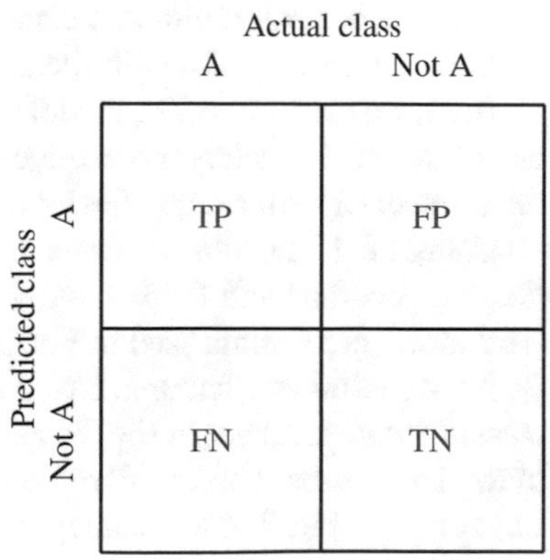

Fig. 6 Generic confusion matrix

$$FN = \text{false negatives}$$

Note that appending an "R" to TP, FP, TN, or FN represents the corresponding rate.

We use several metrics in our multiclass classification problem. The most basic metric we consider is the accuracy, which we calculate from a 46 by 46 confusion matrix. Of course, higher accuracy indicates a more successful model.

We also measure the false acceptance rate (FAR), which is the rate at which a different user is classified as the actual user, and the false reject rate (FRR), which is the rate at which real users are mis-classified as other users, that is,

$$FAR = FPR = \frac{FP}{FP + TN}$$

$$FRR = FNR = \frac{FN}{TP + FN}$$

Colloquially, the FAR is sometimes referred to as the fraud rate, while the FRR is known as the insult rate.

Since we are dealing with multiple users, we calculate FAR and FRR for each individual user and report the average for the 46 different users [14]. Intuitively, lower FAR and FRR indicates greater success in the classification model.

In our SVM experiments, we use SVM and RFE libraries from `scikit-learn`. In our 1D-CNN and GAN experiments, we implemented `Keras` libraries [6] to develop our models. For all the experiments, we use an 80–20 train-test split, i.e. 80% of the data is used to train the model, while the remaining 20% is used for testing.

6.1   SVM Results

We first considered rudimentary experiments with various kernels and values of parameters and found that a linear kernel with regularization parameter $C = 1000$ worked the best for multiclass classification. As a result, for all experiments in this section, we use an SVM with linear kernel and $C = 1000$.

Here, we train our SVM model on the feature vector described in Sect. 4.2.1 and use SVM-RFE to select the strongest features. We analyze the relationship between the number of features and the FAR, FPR, and accuracy.

Using all 16 features discussed in Sect. 4.2.1, the SVM achieved a 95% classification accuracy and 0.0014 and 0.057 FAR and FRR, respectively. These results (and more) are summarized in Fig. 7.

As we eliminate features based on the rankings determined by SVM-RFE, the classification accuracy in Fig. 7c generally decreases, dropping to about 89% with 9 of the 16 features. This is still quite strong, considering the number of classes. As we can see from Fig. 7, the accuracy generally decreases slightly as we eliminate more

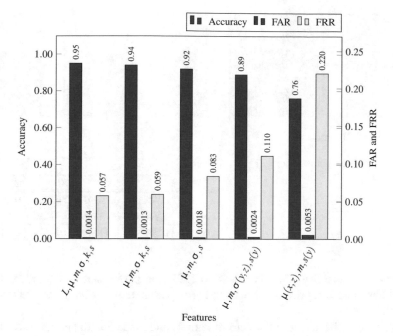

Fig. 7 Linear SVM results for selected combinations of features

features, although the accuracy does not drop below 90% until we have eliminated 7
features. Similarly, the FAR is exceptionally low even as features are eliminated,
staying below 1% for every combination of features. However, the FRR starts at
around 5% and increases to more than 20% once we have eliminated 10 features.

6.2 1D-CNN Results

We also experiment with a 1D-CNN as our classification model, where the model
follows the architecture in [7]. The 1D-CNN model is trained on temporal sequences
of fixed length, as described in Sect. 5. The model performs one-dimensional convolutions along the time axis with RELU activation functions after each convolution
layer. The first convolution layer produces 128 filters, while the second convolution
layer produces 256 filters. A dropout layer follows the convolution layers to prevent
overfitting [16]. Then we have a 1D max-pooling layer to downsample the data and
highlight the key features. The output of the max-pooling layer is passed through
a flatten layer, which is then passed through two fully-connected layers. The last
fully-connected layer produces the value that corresponds to the classification. This
model is illustrated in Fig. 8.

After fine-tuning hyper-parameters (dropout rate and number of filters), we
moved on to experiment with kernel size and stride length. We found the accuracy

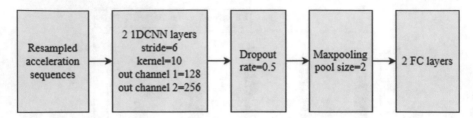

Fig. 8 1D-CNN architecture

Table 1 Stride length versus kernel size

Stride	Kernel			
	3	5	10	25
1	0.90457	0.89391	0.91276	0.90870
3	0.92696	0.93348	0.93630	0.93587
6	0.93087	0.94000	0.94109	0.94022

hovered around 89–91%. According to [29], the performance of our 1D-CNN should be more receptive to changes in hyper-parameters involving the convolution layers.

We tested different combinations of kernels and stride lengths in the convolution layers. These results are summarized in Table 1. We see that most of the results are fairly strong, ranging from a low accuracy of 90% to a high of 94%. Generally, as the kernel size increases, the accuracy increases. Similarly, as the stride length increases, the accuracy tends to increase. This is most likely due to the fact that larger kernels and stride lengths produce more refined features for the fully-connected layers.

6.3 Adversarial Results

We trained a GAN to generate adversarial samples, and using these samples to determine whether our deep learning model is robust under adversarial attack. First, we determine how close our GAN-generated data is to real data through a simulated poisoning attack. Then we test how well the adversarial samples can evade an authenticator model.

As outlined in Sect. 5.2, we poison our training dataset with an increasing percentage of fake data. All real samples are always included, so any changes in classification accuracy should be caused by our adversarial (fake) samples. We trained our DC-GAN for different numbers of epochs and different numbers of adversarial samples. Generally, a higher number of epochs results in better imitations of the original data. Increasing the number of adversarial samples in the training dataset gauges how well our 1D-CNN can resist large-scale poisoning.

Table 2 Adversarial attack results (1840 real samples)

Adversarial samples	epochs			
	10	25	50	100
100	0.93609	0.93130	0.94761	0.94565
250	0.94587	0.94848	0.93783	0.94196
600	0.94087	0.93978	0.94002	0.94152
1840	0.94000	0.94152	0.92239	0.93435

The results of our poisoning attacks are given in Table 2. The accuracy remains relatively high at more than 90%, even for high training epochs and up to a 1:1 ratio of real to fake data. Classification accuracy does decrease slightly when there are more adversarial samples in the training dataset, but the loss in accuracy is not large. These results show that our 1D-CNN is highly resistant to poisoning attacks.

We conjecture that the reason for the limited success of our adversarial attack is that it is exceedingly difficult to generate realistic signatures of the type considered in this paper. While the real signatures in our dataset vary wildly, those that we generated using DC-GAN appear to exhibit more homogeneity. We plan to investigate this issue further in future work.

7 Conclusion and Future Work

Previous research has shown that SVMs are a viable techniques for accelerometer-based gesture authentication [18]. In this paper, we expanded on and improved upon previous work. First, we refined the feature selection process with SVM-RFE to select the best features, while maintaining a high classification accuracy. Then, we used deep learning models, specifically 1D-CNNs, for classification. We obtained strong results, with greater than 90% classification accuracy, slightly surpassing the accuracy of our SVM model. Lastly, we experimented with adversarial attacks on our 1D-CNN model, namely, poisoning and evasion attacks. These simulate realistic attacks, assuming an intruder has access to the real data, but not the model itself. Our results indicate that our 1D-CNN is robust under such attacks, achieving greater than 90% accuracy for poisoning attacks and near perfect accuracy for evasion attacks.

For future work, additional machine learning techniques could be considered. For example, long-short term memory (LSTM) models could be used instead of 1D-CNNs, since LSTMs generally perform well on sequential data. Additionally, as mentioned in Sect. 6.3, we would like to explore alternative methods of generating adversarial samples to determine whether we can improve on the limited adversarial attack results obtained with DC-GANs.

References

1. Mohit Agrawal, Pragyan Mehrotra, Rajesh Kumar, and Rajiv Ratn Shah. Defending touch-based continuous authentication systems from active adversaries using generative adversarial networks. *arXiv preprint arXiv:2106.07867*, 2021.
2. Mohammad Al-Rubaie and J Morris Chang. Reconstruction attacks against mobile-based continuous authentication systems in the cloud. *IEEE Transactions on Information Forensics and Security*, 11(12):2648–2663, 2016.
3. Ala Abdulhakim Alariki and Azizah Abdul Manaf. Touch gesture authentication framework for touch screen mobile devices. *Journal of Theoretical & Applied Information Technology*, 62(2), 2014.
4. Sukarna Barua, Sarah Monazam Erfani, and James Bailey. FCC-GAN: A fully connected and convolutional net architecture for GANs. *arXiv preprint arXiv:1905.02417*, 2019.
5. Debnath Bhattacharyya, Rahul Ranjan, Farkhod Alisherov, Minkyu Choi, et al. Biometric authentication: A review. *International Journal of u-and e-Service, Science and Technology*, 2(3):13–28, 2009.
6. Jason Brownlee. *Deep learning with Python: Develop Deep Learning Models on Theano and TensorFlow using Keras*. Machine Learning Mastery, 2016.
7. Jason Brownlee. 1d convolutional neural network models for human activity recognition. https://machinelearningmastery.com/cnn-models-for-human-activity-recognition-time-series-classification/, 2020.
8. Attaullah Buriro, Bruno Crispo, Filippo Delfrari, and Konrad Wrona. Hold and sign: A novel behavioral biometrics for smartphone user authentication. In *2016 IEEE security and privacy workshops*, SPW, pages 276–285, 2016.
9. Xue-wen Chen and Jong Cheol Jeong. Enhanced recursive feature elimination. In *Sixth International Conference on Machine Learning and Applications*, ICMLA 2007, pages 429–435, 2007.
10. Chrystian Vieyra. Physics toolbox sensor suite. https://apps.apple.com/us/app/physics-toolbox-sensor-suite/id1128914250, 2016.
11. Gradeigh D. Clark and Janne Lindqvist. Engineering gesture-based authentication systems. *IEEE Pervasive Computing*, 14(1):18–25, 2015.
12. Antonia Creswell, Tom White, Vincent Dumoulin, Kai Arulkumaran, Biswa Sengupta, and Anil A. Bharath. Generative adversarial networks: An overview. *IEEE Signal Processing Magazine*, 35(1):53–65, 2018.
13. Hassan Ismail Fawaz, Germain Forestier, Jonathan Weber, Lhassane Idoumghar, and Pierre-Alain Muller. Adversarial attacks on deep neural networks for time series classification. In *2019 International Joint Conference on Neural Networks*, IJCNN, pages 1–8, 2019.
14. Margherita Grandini, Enrico Bagli, and Giorgio Visani. Metrics for multi-class classification: an overview. *arXiv preprint arXiv:2008.05756*, 2020.
15. Dennis Guse. Gesture-based user authentication on mobile devices using accelerometer and gyroscope. Master's thesis, Technische Universität Berlin, 2017.
16. Geoffrey E. Hinton, Nitish Srivastava, Alex Krizhevsky, Ilya Sutskever, and Ruslan R. Salakhutdinov. Improving neural networks by preventing co-adaptation of feature detectors. *arXiv preprint arXiv:1207.0580*, 2012.
17. Elliu Huang. Gesture dataset, 2021. Available from the authors upon request.
18. Elliu Huang, Fabio Di Troia, Mark Stamp, and Preethi Sundaravaradhan. A new dataset for smartphone gesture-based authentication. https://www.scitepress.org/Papers/2021/104258/104258.pdf, 2021.
19. Satoru Imura and Hiroshi Hosobe. A hand gesture-based method for biometric authentication. In *International Conference on Human-Computer Interaction*, pages 554–566, 2018.
20. Hassan Ismail Fawaz, Germain Forestier, Jonathan Weber, Lhassane Idoumghar, and Pierre-Alain Muller. Adversarial attacks on deep neural networks for time series classification. *arXiv e-prints*, pages arXiv–1903, 2019.

21. Serkan Kiranyaz, Onur Avci, Osama Abdeljaber, Turker Ince, Moncef Gabbouj, and Daniel J Inman. 1d convolutional neural networks and applications: A survey. *Mechanical Systems and Signal Processing*, 151, 2021.
22. Jiayang Liu, Lin Zhong, Jehan Wickramasuriya, and Venu Vasudevan. uWave: Accelerometer-based personalized gesture recognition and its applications. *Pervasive and Mobile Computing*, 5(6):657–675, 2009.
23. Duo Lu, Kai Xu, and Dijiang Huang. A data driven in-air-handwriting biometric authentication system. In *2017 IEEE International Joint Conference on Biometrics*, IJCB, pages 531–537, 2017.
24. Yuxin Meng, Duncan S. Wong, Roman Schlegel, et al. Touch gestures based biometric authentication scheme for touchscreen mobile phones. In *International Conference on Information Security and Cryptology*, pages 331–350, 2012.
25. Gautam Raj Mode and Khaza Anuarul Hoque. Adversarial examples in deep learning for multivariate time series regression. In *2020 IEEE Applied Imagery Pattern Recognition Workshop*, AIPR, pages 1–10, 2020.
26. Luis Muñoz-González, Bjarne Pfitzner, Matteo Russo, Javier Carnerero-Cano, and Emil C Lupu. Poisoning attacks with generative adversarial nets. *arXiv preprint arXiv:1906.07773*, 2019.
27. Mark Stamp. *Introduction to machine learning with applications in information security*. CRC Press, Taylor & Francis Group, Boca Raton, FL, 2018.
28. Yi Xiang Marcus Tan, Alfonso Iacovazzi, Ivan Homoliak, Yuval Elovici, and Alexander Binder. Adversarial attacks on remote user authentication using behavioural mouse dynamics. In *2019 International Joint Conference on Neural Networks*, IJCNN, pages 1–10, 2019.
29. Wensi Tang, Guodong Long, Lu Liu, Tianyi Zhou, Jing Jiang, and Michael Blumenstein. Rethinking 1d-CNN for time series classification: A stronger baseline. *arXiv preprint arXiv:2002.10061*, 2020.
30. Romain Tavenard. `tslearn` documentation. https://tslearn.readthedocs.io/en/stable/, 2021.
31. Cong Wu, Kun He, Jing Chen, Ziming Zhao, and Ruiying Du. Liveness is not enough: Enhancing fingerprint authentication with behavioral biometrics to defeat puppet attacks. In *29th USENIX Security Symposium*, USENIX Security 20, pages 2219–2236, 2020.
32. Jonathan Wu, Prakash Ishwar, and Janusz Konrad. Two-stream CNNs for gesture-based verification and identification: Learning user style. In *Proceedings of the IEEE Conference on Computer Vision and Pattern Recognition Workshops*, pages 42–50, 2016.

Clickbait Detection for YouTube Videos

Ruchira Gothankar, Fabio Di Troia (iD), and Mark Stamp

Abstract YouTube videos often include captivating descriptions and intriguing thumbnails designed to increase the number of views, and thereby increase the revenue for the person who posted the video. This creates an incentive for people to post clickbait videos, in which the content might deviate significantly from the title, description, or thumbnail. In effect, users are tricked into clicking on clickbait videos. In this research, we consider the challenging problem of detecting clickbait YouTube videos. We experiment with multiple state-of-the-art machine learning techniques using a variety of textual features.

1 Introduction

Today, web content is increasingly popular and people rely on information obtained from the internet. Furthermore, with the diversity of available resources, the amount of time spent on the internet has increased. Many platforms provide a medium where virtually anyone can publish information that is accessible to a large number of people. However, the credibility of such information is not guaranteed.

Online sources of information include blogs, video sharing platforms, and social media, among others. Many of these applications have been developed with the main intent to generate revenue. Hence, unscrupulous people can use false information to increase their viewership and increase their revenue. Clickbait is false and deceptive information that lures users to click a link, watch a video, or read an article. It aims to exploit the user's curiosity by providing misleading—though captivating—information. Clickbait has become a marketing tool in many sectors to entice users and thereby to generate revenue. Publishing eye-catching information to manipulate and trick users is a common practice to increase the viewership and spread brand awareness. A clickbait can be an image, a sensational headline, or a misleading video or audio content. While clickbait sources help in gaining attention, there are

R. Gothankar · F. Di Troia · M. Stamp (✉)
San Jose State University, San Jose, CA, USA
e-mail: ruchira.gothankar@sjsu.edu; fabio.ditroia@sjsu.edu; mark.stamp@sjsu.edu

© The Author(s), under exclusive license to Springer Nature Switzerland AG 2022   261
M. Stamp et al. (eds.), *Artificial Intelligence for Cybersecurity*, Advances in
Information Security 54, https://doi.org/10.1007/978-3-030-97087-1_11

many disadvantages and negative ramifications. In fact, clickbait not only wastes the time of viewers, but also affects the trustworthiness of the underlying platform [25].

YouTube is a video publishing platforms where users upload videos and share them with others. When uploading a video, the user adds a title, a description, and a thumbnail. The other users then view the title and thumbnail before deciding whether to view the video. Hence, this data become crucial parameters on which the users can base their decision to watch a video or not. For this reason, many YouTube content creators (aka YouTubers) use clickbait title and thumbnails that might deviate from the actual content to increase viewership for a video, and thereby generate more revenue.

A recent example includes the COVID-19 pandemic, where individuals have posted misleading health-related content, including some fake cures for COVID-19. Some other common examples of clickbait are video titles such as "You'll Never Believe What Happened Next. . .", "The 10 documentaries you should watch before you die", "You Can Now Travel Abroad Without Having to. . .","You Won't Believe. . ." and so on [13]. Figure 1 shows an example of clickbait video on YouTube.

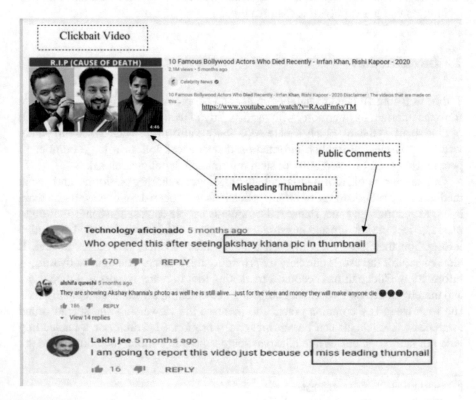

Fig. 1 Clickbait video example [30]

The clickbait problem is somewhat similar to that of spam detection. Spam, which is unsolicited emails, often includes misleading messages that are sent to deceive users by redirecting them to websites for the purpose of advertising or attack. Therefore, considerable research has been focused on detecting spam. In this research, we are concerned with detecting clickbait YouTube videos. The YouTube platform relies on users to manually flag suspected malicious or clickbait content. However, a more automated approach would clearly be desirable. We consider machine learning and deep learning based solutions to the clickbait detection problem.

The remainder of this paper is organized as follows. Section 2 considers relevant previous work and background topics related to natural language processing (NLP). In Sect. 3, we discuss our experimental setup, including the datasets used. Section 4 contains our experimental results and our analysis of these results. In Sect. 5, we give our conclusions and we discuss possible directions for future work.

2 Background

This Section discusses relevant work done in this field. We mainly focuses on clickbait detection, fake news detection, image forgery detection, and hoax detection. Apart from these topics, we also discuss advancements in natural language processing (NLP).

2.1 Related Work

Clickbait is a way to attract the attention of the users by luring them to access specific contents. However, misleading information is present on the internet in multiple forms and is often used interchangeably in different contexts. For example, a hoax is spreading false stories of, say, a celebrity death [30], while an example of a forgery is an image that suggests false information. We now discuss and analyze the performance of previous works on clickbait, fake news, forgery, and hoax detection.

2.1.1 Clickbait Detection

In 2016 [4], Chakraborty et al. implemented an ML classifier to detect clickbait. They also created a browser extension to help readers navigate around clickbait. They used the headlines from the Wiki-news corpus and used 18,513 articles as legitimate posts. For the clickbait posts, they used articles from popular domains containing illegitimate content. To train their classifier, they used a set of 14 features spanning linguistic analysis, word patterns, and N-gram. They achieved an accuracy of about 89% using a support vector machine (SVM) classifier.

Elyashar et al. [8] developed an approach based on feature engineering. Their work focused on detecting clickbait posts in online social media. They performed linguistic analysis using a machine learning classifier which could differentiate between legitimate and illegitimate posts. The dataset used for analysis was provided by the 2017 Clickbait Challenge [20]. The results of their experiments suggest that malicious content tends to be longer than the benign content. They also concluded that the title of the post played an important role to identify a clickbait.

Glenski et al. [11] developed a network model which is a linguistically infused network to detect fake tweets. This model, which is based on long short term memory (LSTM) and convolutional neural networks (CNN), used the text of tweets, images, and description for training. Furthermore, the pretrained embedding model GloVe was used as the embedding layer. They achieved an accuracy of 82%. Zhou [36] proposed a self-attentive neural network model using gated recurrent units (GRU) for predicting fake tweets. They performed multi classification using the annotation scheme. As proof of the success of their approach, they ranked first in the Clickbait Challenge 2017 with an F-score of 0.683.

2.1.2 Fake News Detection

Fake news is a type of misinformation that has received considerable attention in recent years. The main idea is to analyze the text content of a news item to check if the statements are valid or not. Ahmad et al. [2] implemented an ensemble model based on the linguistic features of the text which involved a combination of multiple machine learning algorithms, namely, random forest, multilayer perceptron, and support vector machine (SVM), to detect fake news. They used XGBoost as an ensemble learner, achieving an accuracy of 92%.

Thota et al. [28] presented a paper on detecting fake news using natural language processing. They used TF-IDF and Word2Vec with a dense neural network based on the news headline. In another paper on fake news detection, Jwa et al. [14] implemented a model using bidirectional encoder representations from transformers (BERT). The deep contextualizing nature of BERT has yielded strong results, including the ability to determine the relationship between the headline and the body of a news article.

2.1.3 Forgery Detection

As the name suggests, image forgery detection consists of trying to detect malicious information that is conveyed through images. In 2018, Zhang et al. [35] developed a "fauxtography" detector which could detect images which are misleading on social media platforms.

Palod et al. [19] passed pretrained Word2Vec comment embeddings through an LSTM network to generate a "fakeness" vector, and achieved an F-score of 0.82. Shang et al. [25] proposed a model that involved network feature extraction, metadata feature extraction, and linguistic feature extraction to detect clickbait in YouTube videos. The network feature extraction used comments in the videos and extracted semantic features. In the linguistic feature extraction, they relied on document embedding for comments using Doc2Vec, and they also employed a metadata module. In 2019, Reddy et al. [22] implemented a model using word embedding and trained on a support vector machine (SVM). In [7], Dong et al. have proposed a "deep similarity-aware attentive model" that focuses on the relation between the titles that are misleading and the target content. This method was quite different from traditional feature engineering and seemed to work reasonably well. In [24], Setlur considered a semi-supervised confidence network along with a gated attention based network. Based on a small labeled dataset, this method gave promising results.

In many of the above approaches, only the textual information given by the title and the description, along with the metadata features, have been taken into consideration while training a model. An exception is the work in [25], where the authors have also used comments to extract features. It is also worth noting that the embedding layers of Word2Vec, BERT, and Doc2Vec have been used in all of the implementations mentioned above.

In this research, we experiment with multiple embedding layers, including BERT, DistilBERT, and Word2Vec. In previous research, BERT has proven to be effective because of its deep contextualizing nature [14]. A combination of multiple models, known as ensemble learning, has given interesting results in [28], and we also consider ensemble models in the form of random forest classifiers.

2.1.4 Hoax Detection

Articles in which facts are knowingly misrepresented can be viewed as hoaxes. These reports provide deceptive information to readers and present it as legitimate facts. One of such examples can be a fake story about a celebrity death. In [27], the authors have proposed a technique that uses logistic regression for classifying hoaxes. In the model proposed, they have used features based on user interaction and have achieved an accuracy of 99%. Zaman et al. [33] employed a nïve Bayes algorithm which uses the feedback from users as an input to verify if a news is a hoax. Kumar et al. [16] have proposed a method which uses random forest classifier to classify the credibility of the articles on Wikipedia. They achieved an accuracy of 92%. Hoax detection is, though, a less explored area, as compared to the topics discussed above.

2.2 Natural Language Processing

Natural language processing (NLP) is the ability of a machine to process and understand the language of a human. It is used to solve many real-world problems,

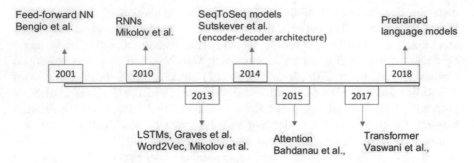

Fig. 2 NLP advancements in recent years [21]

such as machine translation, question answering, and predicting words. Figure 2 shows a timeline of some recent advances in NLP.

In the early 1990s, statistical and probabilistic approaches were employed to train NLP algorithms. However, with the arrival of the Web, the amount of data grew considerably, and such algorithms became inadequate. In 2001, Bengio et al. experimented with feedforward neural networks. Later, recurrent neural networks (RNN) and long short-term memory (LSTM) models were introduced [12]. As of 2012, techniques such as latent semantic indexing (LSI), latent semantic analysis (LSA), and support vector machines (SVM) became popular in the NLP domain. Part of speech (POS) tagging is a commonly used approach.

In 2013, Tomas et. al. introduced Word2Vec, which is used to generate vector representations of words. These embeddings are obtained from the weights of a relatively simple neural network, and the vectors can capture important semantic information, based on the cosine distance between Word2Vec embeddings [23].

Global vector for word representation (GloVe) was introduced in 2014 and is an attempt to combine the benefits of LSA, LSI, and Word2Vec. It is based on the occurrence of a word in the entire corpus. CNNs and LSTMs have become popular for NLP related tasks in recent years, as such models can capture effectively utilize sequential information [12]. LSTM is a highly specialized type of RNN that mitigates the gradient issues that occur with plain vanilla RNNs. Gated recurrent unit (GRU) is a variant of LSTM introduced in 2014 that is lighter, in the sense of having fewer parameters that need to be trained.

Sutskever et al. [26] proposed a sequence-to-sequence learning approach which uses an encoder-decoder architecture. In fact, such encoder-decoder models appear to be the main language modeling frameworks for NLP tasks today. The concept of an attention mechanism was proposed by Bahdanau et al. [3] in 2015 to overcome the limitation of fixed vector length for input sentences in sequence-to-sequence models [31]. Attention provides information about the importance of a part of a sentence during the decision process.

To better deal with the inherent complexity of attention mechanisms, transformers were introduced [18]. Transformer includes multiple stacks of encoder-decoder architecture, where at each step in the processing, the model takes the output of

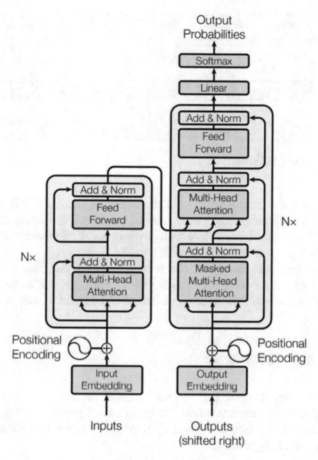

Fig. 3 Architecture of a transformer [18]

the previous step as an input. Figure 3 shows the architecture of a transformer where the decoder is on the right and the encoder is on the left. Initially, the input tokens are converted to embedding vectors. Since this model does not have any RNN units, position indices are stored in a n-dimensional vector space in the form of embeddings. There are three fully connected layers in this particular attention mechanism, namely, the input key K, the value V, and the query Q, which is a matrix of queries. The algorithm defines weights for words based on all the words in K, and it generates a vector representation for all words based on multi-head attention [18]. The other processes include context fragmentation, and multiple parallel attention layers. Some example of deep learning models that make use of transformers include BERT, RoBERTa, mBERT, and DistilBERT.

Bidirectional encoder representations from transformer (BERT) uses a transformer which is based on attention to learn the contextual relation between words. It involves an encoder which reads the input, and decoder which predicts the output. It

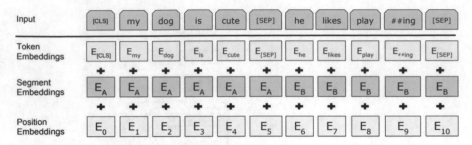

Fig. 4 Input for the BERT model [15]

is called bidirectional because instead of reading input sequentially from a specific direction, the transformer reads the sequence of words in both directions. This helps in learning the context of words based on previous and subsequent words. Figure 4 illustrated the input pattern used in a BERT model.

BERT has four pretrained versions with different layers, hidden nodes, and parameters. Each of these BERT models can be fine-tuned for a specific task by adding additional layers. DistilBERT is a lighter and a faster variant of BERT.

2.3 Learning Techniques

In this section, we discuss the various machine learning techniques that we have employed in this research. Specifically, we have performed experiments for YouTube clickbait detection based on logistic regression, random forest, and MLP, with various embedding mechanisms.

2.3.1 Logistic Regression

Logistic regression is a supervised learning algorithm that is used for categorical data where some parameter—which depend upon the input features and the output—is a categorical prediction. In Logistic regression, a sigmoid function is fitted on the data. The formula for the sigmoid function is

$$\sigma(w^T x + b) = \frac{1}{1 + e^{-(w^T x + b)}} \tag{1}$$

which produces a value in the range of 0–1, and hence it can be interpreted as a probability. The clickbait detection problem can be treated as a type of binomial logistic regression, where the output can be either zero or one [1].

2.3.2 Random Forest

A random forest is based on simple decision trees—a large group of decision trees operate together in an ensemble-like manner. Each tree is trained on a subset of the data and features, a process known as boostrap aggregation, or bagging. In bagging, the data for each tree is randomly selected with replacement [32]. The final prediction of the random forest can be obtained via a simple voting scheme. A random forest mitigates the tendency of individual decision trees to overfit the training data.

The important hyperparameters in a random forest are n-estimators, n-jobs, max-features, and min-sample-leaf. The n-estimators parameter represent the number of trees that are constructed. Typically, adding more trees increases performance at the cost of computation time. The max-features parameter is the number of features required to split at a specific node. The parameter n-jobs is the number of processors that work in parallel.

2.3.3 Multilayer Perceptron

A multilayer perceptron (MLP) is a basic type of feedforward neural network that includes input and output layers, along with at least one hidden layer. An MLP with two hidden layers is illustrated in Fig. 5.

The output layer of an MLP can be used for prediction or classification. Next, we briefly discuss regularization and activation functions; see [9] for additional details on these and related topics.

Neural network models are prone to overfitting. An overfitted model is very effective in classifying the training data but it obtains poor accuracy in predicting the test data—in effect, the model has "memorized" the training data, rather than learning from the training data. One useful technique to prevent overfitting is the use of dropouts, where some number of nodes are ignored during various training steps [9]. This simple approach forces nodes that would otherwise atrophy to become active in the learning process.

An activation function is used to determine the output of node in a neural network. There are multiple types of activation functions, including tanh, sigmoid, ReLU, and leaky ReLU [9]. In this research, we have experimented with ReLU and tanh.

3 Implementation

This section includes details on the implementation used in this research. We discuss the setup used to train and execute the various machine learning models, the experimental design, and so on.

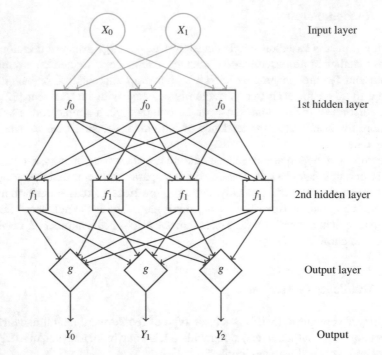

Fig. 5 MLP with two hidden layers

Table 1 Host machine configuration

Component	Details
Model	ASUS ZenBook
Processor	Intel(R) Core(TM) i7-8565U CPU @ 1.80GHz 1.99 GHz
RAM	16.0 GB
System type	64-bit OS
Operating system	Windows 10

3.1 Hardware and Software

In this research, we used multiple Conda virtual environments for each implementation. Conda is an open source package and environment management system which runs on multiple operating systems [5]. The host machine was configured as given in Table 1. All the training and the experiments were run on the host machine.

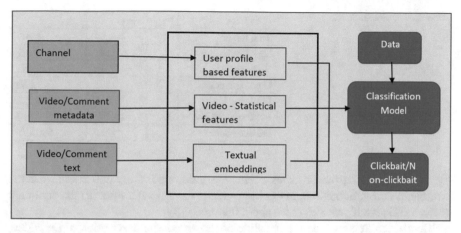

Fig. 6 Overview of clickbait classification model

3.2 Approach

Our clickbait detection experiments are based on a set of labeled videos. The problem is formulated as a binary classification problem where for each video a machine learning algorithm classifies it is clickbait or non-clickbait. The information from multiple sources (e.g., title, description, comments) are combined and fed to the classification model. The performance is evaluated and analyzed by multiple measures, specifically, precision, recall and the F-score.

There are three types of features considered in this research. The first involves features from the profile of the user who posted the video (subscriptions count, views count, and videos count). The second type of feature is based on extracting textual information from the video (title and description). The third component involves statistical features related to the video (like count, dislike count, like-dislike ratio, views, and number of comments). A classification model performs binary classification (clickbait or non-clickbait) based on some combination of these features. An overview is provided in Fig. 6.

3.3 Features

Features that provide information regarding the reputation of the channel and the videos include the number of subscribers of the YouTube channel, the number of likes or upvotes, and the age of the channel. These statistical features represent the response of viewers to the channel. Previous related work claimes that videos that are clickbait tend to have a relatively small number of subscribers and likes [30].

Usually, the number of views for the clickbait and non-clickbait videos are quite similar [34]. Useful information in determining the credibility of a video is given

Table 2 Dataset statistics

Data Item	Min	Mean	Max
Title length	10	54	107
Description length	15	1131	5162
View count	21	5,660,978	2,543,466,463
Comment count	0	522	49,060
Like count	0	49,615	13,542,232
Subscriber count	977	10,200	23,695,417
Dislike count	0	1320	516,171

by the dislike ratio, the favorites count, the video age, the views count, and the comments count. Sometime, in clickbait videos the uploader disables the comment section. This itself provides clues about the video [30].

Textual features include the headline of the video, the description of the video, and the comments by the viewers. YouTubers who upload clickbait usually employ techniques which are deceptive. They use catchy and exaggerated phrases for the title and description of the video. Some common phrases are "viral," "top." "won't believe", "epic", and similar. We tokenize the text and embed it in classification models using various embedding techniques, including Word2Vec, BERT, and DistilBERT

3.4 Dataset

Every month, billions of people visit YouTube and the videos are watched for over a billion hours. A large number of videos are also uploaded by the users. In fact, YouTube is a platform where people can generate revenue by uploading videos and gaining viewership for their videos.

In this research, the evaluation is done on a dataset of 8219 labeled videos, where 4300 are non-clickbait and 3919 are clickbait. The dataset was crawled from the Google YouTube API for the list of video IDs fetched from the Github source [17]. These sources were randomly selected and manually verified by the authors. The statistics for various parameters are shown in Table 2.

3.5 Experiments

In this research, we experimented multiple techniques including multiple language modeling techniques. We used Word2Vec, BERT, and DistilBERT for word embeddings. Architecture for the individual models is also shown. A grid search was used for training and building the models to obtain the best set of parameters. All experiments are based on an 80–20 training-testing split of the data, and 5-fold cross validation was used in each case. In this section, we briefly describe each of our models, and in the next section we give the results for each experiment.

3.5.1 Experiment I: Logistic Regression with Word2Vec

In this experiment, we used a Word2Vec model provided by Gensim [10] to generate the vector representations of words in the dataset. A logistic regression model is trained on these embeddings along with additional features, specifically, comments count, likes count, dislikes count, and subscriptions count for the channel.

3.5.2 Experiment II: Random Forest with Word2Vec

In this experiment, a random forest classifier is trained on the Word2Vec embeddings. We again used the Word2Vec model provided by Gensim. The values tested for n-estimators is 10, 20, 30, 50, and 100. The set of input features are title, description, and metadata features such as comments count, likes count, dislikes count, and subscriptions count for the channel.

3.5.3 Experiment III: MLP with Word2Vec

In this case, we again use the Word2Vec model provided by Gensim. The embedding for title and description is concatenated with the metadata features of the video and is fed to an MLP for classification. The batch size is 10 for 40 epochs. The activation functions used are ReLU and sigmoid. Figure 11 in the Appendix provides the overall architecture of the model.

Note that we use two input embedding layers for textual data (namely, title and description), which are then concatenated together. After this step, the output from the dense layer is flattened and concatenated with the input for the metadata features. Finally, a fully connected layer is used for classification.

3.5.4 Experiment IV: MLP with DropOut, Batch Normalization, and Word2Vec

This experiment is an optimization of the previous experiment. In this model, additional dense layers, along with batch normalization and dropout rate of 0.5, are employed. We have used parametric rectified linear units (PReLU) as the activation function. The batch size is again 10 for 40 epochs. Figure 12 in the Appendix illustrates the overall architecture of the model.

In this model, the output from the embedding layers for the textual data is concatenated, followed by a fully connected dense layer, batch normalization, and activation. This output is finally flattened and concatenated with the metadata features.

3.5.5 Experiment V: MLP with BERT

In this experiment, we have used BERT embedding for title and description of the video. The advantage of using BERT as an embedding model is that it provides context-based representation for each word in a sentence. In contrast, Word2Vec provides representations which are fixed irrespective of where the word is used in the sentence. The pretrained model of BERT that is used in this experiment has 12 layers, 110M parameters, and 768 hidden layers. The BERT tokenizer is used to split the words into tokens and attention masks are used for padding. The mask value of one is for tokens that are not masked, while the value zero means that the token is added by padding and should not be considered for attention. The model uses Adam optimizer, and the batch size is 10 for 5 epochs, and we have used sequence of length 180 for this experiment. Figure 13 in the Appendix shows the model architecture.

Note that the output of the BERT embedding layer is followed by a dense layer, which is then concatenated with the metadata features. After this, a dropout layer followed by a fully connected layer is used for classifying the data.

3.5.6 Experiment VI: MLP with DistilBERT

DistilBERT is a faster, lighter model that is a variant of BERT—it runs 60% faster and has 45% fewer parameters than BERT [6]. For this experiment, we have used a pretrained DistilBERT model. The embeddings for tile and description are fed into a MLP and, later, concatenated with the metadata features of the video and the YouTube channel. The model uses Adam optimizer and the batch size is 10 for 5 epochs. Figure 14 in the Appendix gives details on this model architecture.

Note that in this model, the input from the metadata features is concatenated. Of course, the output layer is a dense layer that is used for classification.

4 Results

Recall that in experiment I, we have use logistic regression with Word2Vec embeddings for the features title and description, along with the metadata features. In this case, we achieve an accuracy of 52% with just title as input, and an accuracy of 70% with all of these features. This model is fast to train and much simple to implement. Figure 7 shows the ROC curve for this logistic regression model.

Experiment II involves using a random forest classifier based on the title, description, likes count, dislikes count, comments count, and subscriptions count.

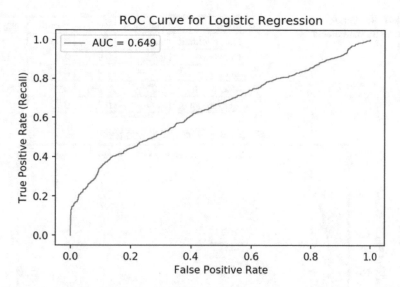

Fig. 7 ROC curve for logistic regression

Table 3 Random forest based on title and like/dislike counts

Class	Precision	Recall	F-score	Support
Non-clickbait	0.81	0.80	0.81	1275
Clickbait	0.80	0.81	0.80	1182
Accuracy	–	–	0.80	2457
Macro avg	0.80	0.80	0.80	2457
Weighted avg	0.80	0.80	0.80	2457

We used Word2Vec embeddings for title and description. We trained this model in multiple sets of inputs. The first set of inputs includes just the title and metadata features. The last set of inputs included all the features. Not surprisingly, we find that the accuracy improves as more features are added.

Table 3 shows precision and accuracy of 80.1% for the model with the first set of input features, that is, title and two metadata features for likes count and dislikes count.

Table 4 shows the report for this experiment when we use the title and all the metadata as features. The accuracy for this experiment is 92.5%. The report shows the precision and recall of the model in classifying clickbait and non-clickbait videos. The model performs slightly better in classifying non-clickbait videos.

Figure 8 shows the ROC curve for the random forest model where the input features included title, description, and all the metadata features, that is, count, dislikes count, comments count, subscriptions count, and views count. The AUC for this model is 0.95 with an accuracy of 94%. This shows that the model performs well and that adding more features increases the accuracy of the model.

In experiment III a simple MLP is used for classification, based on Word2Vec embeddings for title and description that are concatenated with metadata features.

Table 4 Random forest based on title and metadata features

Class	Precision	Recall	F-score	Support
Non-clickbait	0.93	0.93	0.93	1275
Clickbait	0.92	0.92	0.92	1182
Accuracy	–	–	0.93	2457
Macro avg	0.93	0.93	0.93	2457
Weighted avg	0.93	0.93	0.93	2457

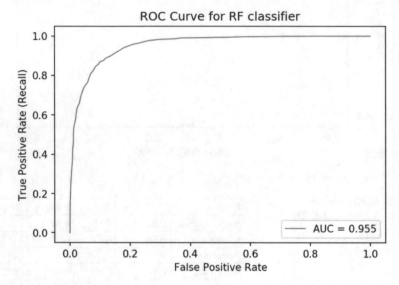

Fig. 8 ROC curve for random forest

In this case, the test accuracy is observed to fluctuate during the training process, but the best average accuracy achieved is better than 91%. Figure 9a shows the accuracy for this experiment over the 30 training epochs.

In experiment IV, a modified MLP is used with batch normalization and PReLU as an activation function. In this case, the accuracy is slightly worse than in experiment III, although the training is more stable, as can be observed in Fig. 9b.

In experiment V, we have used a transfer learning model based on BERT for word embeddings. This experiment with BERT gives an accuracy of 94.5%. In this experiment the length of the input sequence is fixed at 180 characters. Figure 9c shows the plot for accuracy over training epochs for both the train and validation sets. Note that the number of epochs is small due to the extended training time required, as compared to other models considered.

In experiment VI, we have used a lighter variant of BERT model for the word embeddings, namely, DistilBERT. The accuracy achieved in this case is around 92%. This model is significantly faster to train than the BERT, although the accuracy obtained with BERT is slightly better than using DistilBERT. Table 5 shows the precision and recall for experiment VI, while Fig. 9d shows the training and test accuracy over epochs.

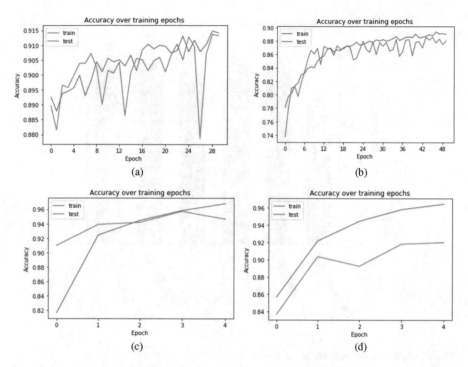

Fig. 9 Accuracy over training epochs for experiments III through VI. (**a**) Experiment III. (**b**) Experiment IV. (**c**) Experiment V. (**d**) Experiment VI

Table 5 Results for experiment VI (MLP with DistilBERT)

Class	Precision	Recall	F-score	Support
Non-clickbait	0.92	0.95	0.93	884
Clickbait	0.93	0.89	0.91	754
Accuracy	–	–	0.92	1638
Macro avg	0.92	0.92	0.92	1638
Weighted avg	0.92	0.92	0.92	1638

In Fig. 10 we summarize the results of our six experiments in terms of accuracy (to two decimal places). Note that in the bar graph in Fig. 10, "LR" is shorthand for linear regression, "RF" is short for random forest, and "MLP+" is used to denote our MLP model that includes dropout and batch normalization. The bars from left-to-right represent experiments I through VI, as discussed in Sects. 3.5.1 through 3.5.6, above.

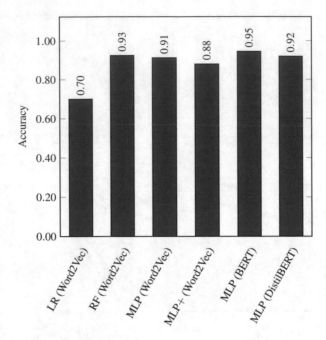

Fig. 10 Accuracy comparison of experiments

5 Conclusion and Future Works

The goal of this research was to utilize state-of-the-art techniques to classify YouTube videos as clickbait or non-clickbait. A YouTube video has multiple characteristics that can serve as useful features for such classification. We leveraged three main types of such features, namely, user profile, video statistics, and textual data. In this research, multiple classification techniques were considered, including logistic regression, random forest, and MLP, and we employed Word2Vec, BERT, and DistilBERT as language models. The best accuracy was achieved using an MLP classifier based on BERT embeddings, but a the more lightweight DistilBERT performed almost as well. We also confirmed that the accuracy of the models could be increased by adding more features. Although there is relatively little work that is directly comparable to the research reported here, our results—as summarized in Fig. 10—improve significantly on the previous work discussed in Sect. 2.1.1.

For future work, more features can be included. For instance, the transcript of the video might contain useful information. For example, the "distance" between the transcripts and the title could provide important insight, as the content of clickbait videos often differs significantly from the title. The network structure of the comments and replies, which represents the semantic features and attributes, can also be considered [25].

In this research, we experimented with BERT, Word2Vec, and DistilBERT for word embeddings. For future work, DocToVec embeddings could also be considered. We used random forest classifier, and other ensemble techniques could be considered, including, such as XGBoost. In addition, a careful analysis of the relative strength of the various features considered in this research would be useful. Finally, we plan to experiment with state-of-the-art attentive language models, such as XLNet, which is claimed to be better than BERT for determining long-term dependencies [29].

Appendix: Model Architectures

See Figs. 11, 12, 13, and 14.

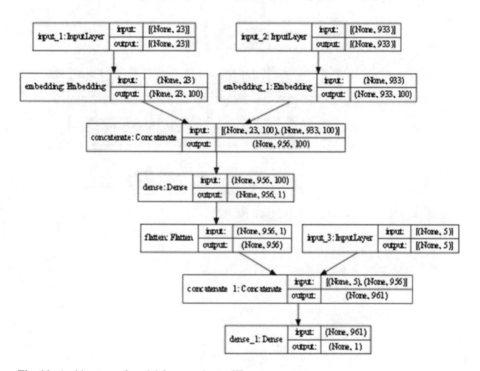

Fig. 11 Architecture of model for experiment III

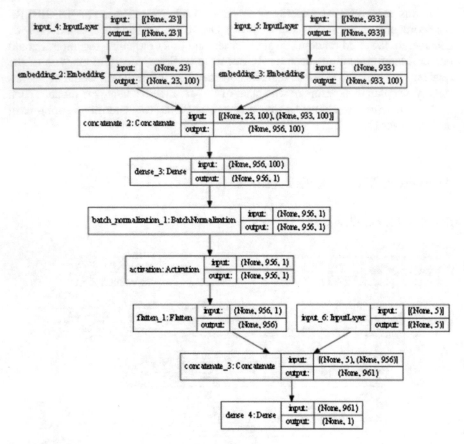

Fig. 12 Architecture of model for experiment IV

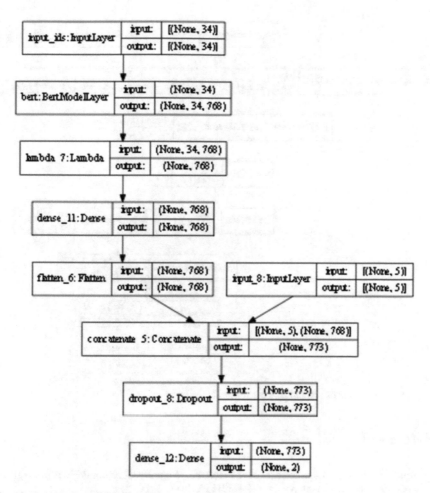

Fig. 13 Architecture of model for experiment V

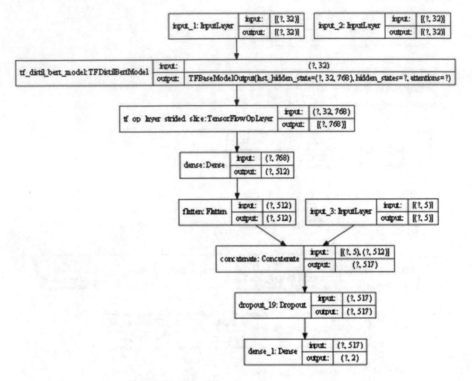

Fig. 14 Architecture of model for experiment VI

References

1. Apoorva Agrawal. Logistic regression simplified. https://medium.com/data-science-group-iitr/logistic-regression-simplified-9b4efe801389, 2017.
2. Iftikhar Ahmad, Muhammad Yousaf, Suhail Yousaf, and Muhammad Ovais Ahmad. Fake news detection using machine learning ensemble methods. *Complexity*, 2020, 2020.
3. Dzmitry Bahdanau, Kyunghyun Cho, and Yoshua Bengio. Neural machine translation by jointly learning to align and translate. *https://arxiv.org/abs/1409.0473*, 2014.
4. Abhijnan Chakraborty, Bhargavi Paranjape, Sourya Kakarla, and Niloy Ganguly. Stop click-bait: Detecting and preventing clickbaits in online news media. In *IEEE/ACM International Conference on Advances in Social Networks Analysis and Mining*, ASONAM, pages 9–16, 2016.
5. Conda. https://docs.conda.io/en/latest/.
6. Distilbert — transformers 4.5.0.dev0 documentation. https://huggingface.co/transformers/model_doc/distilbert.html.
7. Manqing Dong, Lina Yao, Xianzhi Wang, Boualem Benatallah, and Chaoran Huang. Similarity-aware deep attentive model for clickbait detection. In *Pacific-Asia Conference on Knowledge Discovery and Data Mining*, pages 56–69, 2019.
8. Aviad Elyashar, Jorge Bendahan, and Rami Puzis. Detecting clickbait in online social media: You won't believe how we did it. https://arxiv.org/abs/1710.06699, 2017.
9. Franckepeixoto. A simple overview of multilayer perceptron (MLP) deep learning. https://www.analyticsvidhya.com/blog/2020/12/mlp-multilayer-perceptron-simple-overview/, 2020.

10. Gensim — python framework for fast vector space modelling. https://pypi.org/project/gensim/.

11. Maria Glenski, Ellyn Ayton, Dustin Arendt, and Svitlana Volkova. Fishing for clickbaits in social images and texts with linguistically-infused neural network models. https://arxiv.org/abs/1710.06390, 2017.

12. Alex Graves. Generating sequences with recurrent neural networks. https://arxiv.org/abs/1308.0850, 2013.

13. Jason Hennessey. 12 surprising examples of clickbait headlines that work. https://www.searchenginejournal.com/12-surprising-examples-of-clickbait-headlines-that-work/362688, 2020.

14. Heejung Jwa, Dongsuk Oh, Kinam Park, Jang Mook Kang, and Heuiseok Lim. exBAKE: Automatic fake news detection model based on bidirectional encoder representations from transformers (BERT). *Applied Sciences*, 9(19), 2019.

15. Samia Khalid. BERT explained: A complete guide with theory and tutorial. https://towardsml.com/2019/09/17/bert-explained-a-complete-guide-with-theory-and-tutorial/, 2019.

16. Srijan Kumar, Robert West, and Jure Leskovec. Disinformation on the Web: Impact, characteristics, and detection of Wikipedia hoaxes. In *Proceedings of the 25th International Conference on World Wide Web*, pages 591–602, 2016.

17. Saurabh Mathur. GitHub. https://github.com/saurabhmathur96/clickbait-detector.

18. Maxime. What is a transformer? https://medium.com/inside-machine-learning/what-is-a-transformer-d07dd1fbec04, 2020.

19. Priyank Palod, Ayush Patwari, Sudhanshu Bahety, Saurabh Bagchi, and Pawan Goyal. Misleading metadata detection on YouTube. In *European Conference on Information Retrieval*, pages 140–147, 2019.

20. Martin Potthast, Tim Gollub, Matthias Hagen, and Benno Stein. The clickbait challenge 2017: Towards a regression model for clickbait strength. https://arxiv.org/abs/1812.10847.

21. Praboda Chathurangani Rajapaksha and Rajapaksha Waththe Vidanelage. Clickbait detection using multimodel fusion and transfer learning. https://tel.archives-ouvertes.fr/tel-03139880/document, 2020.

22. KV Sankar Reddy, K Sai Nihith, M Sasank Chowdary, and TR Krishna Prasad. An efficient word embedded click-bait classification of YouTube titles using SVM. In *Symposium on Machine Learning and Metaheuristics Algorithms, and Applications*, pages 175–184, 2019.

23. Moiz Saifee. Recent advancements in NLP (1/2). https://medium.com/swlh/recent-advancements-in-nlp-1-2-192ac7eefe3c, 2019.

24. Amrith Rajagopal Setlur. Semi-supervised confidence network aided gated attention based recurrent neural network for clickbait detection. https://arxiv.org/abs/1811.01355, 2018.

25. Lanyu Shang, Daniel Yue Zhang, Michael Wang, Shuyue Lai, and Dong Wang. Towards reliable online clickbait video detection: A content-agnostic approach. https://arxiv.org/abs/1907.07604, 2019.

26. Ilya Sutskever, Oriol Vinyals, and Quoc V Le. Sequence to sequence learning with neural networks. https://arxiv.org/abs/1409.3215, 2014.

27. Eugenio Tacchini, Gabriele Ballarin, Marco L Della Vedova, Stefano Moret, and Luca de Alfaro. Some like it hoax: Automated fake news detection in social networks. https://arxiv.org/abs/1704.07506, 2017.

28. Aswini Thota, Priyanka Tilak, Simrat Ahluwalia, and Nibrat Lohia. Fake news detection: A deep learning approach. *SMU Data Science Review*, 1(3):10, 2018.

29. Phylypo Tum. A survey of the state-of-the-art language models up to early 2020. https://medium.com/@phylypo/a-survey-of-the-state-of-the-art-language-models-up-to-early-2020-aba824302c6, 2020.

30. Deepika Varshney and Dinesh Kumar Vishwakarma. A unified approach for detection of clickbait videos on YouTube using cognitive evidences. *Applied Intelligence*, pages 1–22, 2021.

31. Ashish Vaswani, Noam Shazeer, Niki Parmar, Jakob Uszkoreit, Llion Jones, Aidan N Gomez, Lukasz Kaiser, and Illia Polosukhin. Attention is all you need. https://arxiv.org/abs/1706.03762, 2017.

32. Tony Yiu. Towards data science: Understanding random forest. https://towardsdatascience. com/understanding-random-forest-58381e0602d2, 2019.
33. Badrus Zaman, Army Justitia, Kretawiweka Nuraga Sani, and Endah Purwanti. An Indonesian hoax news detection system using reader feedback and naïve Bayes algorithm. *Cybernetics and Information Technologies*, 20(1):82–94, 2020.
34. Savvas Zannettou, Sotirios Chatzis, Kostantinos Papadamou, and Michael Sirivianos. The good, the bad and the bait: Detecting and characterizing clickbait on YouTube. In *IEEE Security and Privacy Workshops*, SPW, pages 63–69, 2018.
35. Daniel Yue Zhang, Lanyu Shang, Biao Geng, Shuyue Lai, Ke Li, Hongmin Zhu, Md Tanvir Amin, and Dong Wang. Fauxbuster: A content-free fauxtography detector using social media comments. In *IEEE International Conference on Big Data*, Big Data, pages 891–900, 2018.
36. Yiwei Zhou. Clickbait detection in tweets using self-attentive network. https://arxiv.org/abs/ 1710.05364, 2017.

Survivability Using Artificial Intelligence Assisted Cyber Risk Warning

Nikolaos Doukas, Peter Stavroulakis, Vyacheslav Kharchenko, Nikolaos Bardis, Dimitrios Irakleous, Oleg Ivanchenko, and Olga Morozova

Abstract The dependence of everyday human endeavours to information systems of different sorts is continuously increasing, simultaneously as important activities such as work and healthcare are evolving so as to exploit the capabilities of computers and networks. At the same time, malicious cyber activities are becoming ever more often and more destructive as criminals also exploit technological progress. In this context, the necessity for system survivability is becoming more important than expecting that computer system security will avert all possible attacks. Artificial intelligence is a technology that is achieving maturity and contributing in a variety of applications. This chapter presents approaches for applying artificial technology schemes in order to promote survivability by detecting evidence of cyber attacks. This chapter presents three recently proposed schemes that detect such behavior in different contexts. The first scheme aims at the detection of threats within data from emails, programs and network traffic. The second scheme pursues the detection of unexpected system behavior by using a clone of the operational system. The third scheme is focusing on the use of redundant resources, such as those encountered in cloud computing schemes, and on the events following a cyber-attack that is already partially successful and affecting the pooled computing resources. The scheme can be used as a toolkit for preventing the negative effects of computer virus cyber-attacks and ensuring high availability for cloud pooled resources.

N. Doukas · N. Bardis · D. Irakleous
Hellenic Army Academy, Athens, Greece
e-mail: nd@ieee.org; bardis@ieee.org; diracleous@sse.gr

P. Stavroulakis (✉)
Technical University of Crete, Chania, Greece
e-mail: pete_tsi@yahoo.gr

V. Kharchenko · O. Morozova
National Aerospace University "KhAI", Kharkiv, Ukraine
e-mail: v.kharchenko@csn.khai.edu; o.morozova@khai.edu

O. Ivanchenko
National Technical University "Dnipro Polytechnic", Dnipro, Ukraine
e-mail: vmsu12@gmail.com

© The Author(s), under exclusive license to Springer Nature Switzerland AG 2022
M. Stamp et al. (eds.), *Artificial Intelligence for Cybersecurity*, Advances in
Information Security 54, https://doi.org/10.1007/978-3-030-97087-1_12

1   Introduction

Economic, cultural and virtually all sectors of everyday life are over a period of several decades now becoming increasingly dependent on information systems and the electronic services provided via them, while critical fields like medicine and defense are following the same trend. With cyber-crime and other forms of malicious cyber activity continuing to rise, information systems need to display increased survivability i.e., the capacity to continue operating despite attempted or partially successful cyber-attacks. Artificial intelligence-based techniques are therefore being sought that are capable of providing early warnings about intrusions to information systems.

During the last two years, a large proportion of the population saw their ability to work becoming dependent to the existence of internet connectivity and the health of computers. The measures taken to prevent the spreading of COVID19 caused a dramatic increase in the use of computer systems for working remotely while services involving critical sectors such as finance, health, commerce etc. are being offered on-line. The already existing requirement for effective cyber-security has become even more crucial and urgent, whilst new cybersecurity threats are emerging and evolving rapidly. System users and operators require trustworthy and adaptive frameworks to ensure security. The use Artificial Intelligence techniques has also been increasing over the years, with the technology exhibiting maturity. Computer information systems are designed to get more and more intelligent, becoming able to perform difficult decision-making tasks and in less time. In military environments mission areas are studied using simulation and object-oriented architectures [1]. They imitate the human cycle of sensing, reasoning, and acting quite satisfactorily that their significance in process automation [2]. Furthermore, machine learning leverages analytical model building to provide more than the expected performance [3]. It is evident that security is a crucial factor for the uninterruptible operations within this highly complex environment [4, 5]. AI algorithms are therefore being employed as a tool to constantly monitor computer information systems and produce warnings of imminent cyber-threats.

With an ever-increasing number of human activities depending on information systems, the notion of survivability of an information system has been proposed [5–7] that describes the ability of such a system to avert aspiring cyber attackers, avoid total collapse and maintain a reduced service level during a successful attack and promptly recover after the attack has been stopped. Cyber-security efforts aim to promote survivability by focusing on the three R's, namely robustness, response and resilience [8]. Artificial Intelligence (AI) based cyber defend systems follow the same principles and AI techniques have been proposed that pursue increases in survivability by targeting the 3Rs. AI techniques promote robustness by enhancing a system's ability to maintain expected behavior in the event when it is processing unexpected input by developing self-testing and self-healing software [8]. Such input may arise from errors, random events or malicious activity. In the context of response, AI enables a system to defeat an attack without intervention and simultaneously optimize its response strategy and adjust its aggressiveness based on previous successes [8]. For example, systems exist that create their own honeypots for attackers and their own decoys. Finally, resilience is promoted by AI by enhancing the system's ability to detect threats and anomalies and hence increasing their ability to withstand attacks [8].

Given the complexity of current information systems, as well as of cyberattacks, that are inherently of a deceptive nature, constant monitoring is required for survivability and the 3R's [8]. Monitoring is used for detecting deviations of the actual system from expected behavior early, in order to trigger the appropriate response. An AI system watching for such deviations needs profound knowledge of the expected behavior. This can be achieved via the use of a clone copy of the deployed system that acts as a control system, operating in a controlled environment and providing benchmarks for the real systems expected behavior [8]. The sensitivity of this type of monitoring can be adjusted to be compatible to the requirements of the application. A formal model for developing the clone system has been presented [9] that is suitable for operation in conjunction with a decision support system for promoting business goals, including cyber resilience.

This chapter presents three recently proposed schemes that detect such behavior in different contexts.

The first scheme aims at the detection of threats within data. Malicious code fragments are a critical risk to progressively more complex military computer systems. Data from emails, programs and network traffic are collected and analyzed to provide the datasets to model the threats and provide the tools to enhance detection algorithms and evaluate existing protection schemata. In this work, open datasets of threats for training and testing AI detection algorithms are used that have been classified to benign and malicious code based on the features extracted. Natural Language Processing algorithms have been used to train the classifiers using a combination of the methods to provide better results. The overall detection rate achieved is 87.76% in the tests and provides the basis for the usage of this methodology and the integration to existing protection schemata.

The second scheme is targeting the facilitation of detection of unexpected system behavior. One approach to this end, is the use of a clone of the original system that is deployed in a controlled environment and its behavior is considered as a benchmark for the expected behavior of the original system. AI algorithms are trained via adversarial exercises and simulated attacks to recognize divergence between the two systems and produce relevant cyber risk warnings. The presented technique focuses on the implementation of the clone system with emphasis on decision support and the prediction of breakdowns, optimization of service and quality improvement.

The third scheme is focusing on the use of redundant resources, such as those encountered in cloud computing schemes, and on the events following a cyber-attack that is already partially successful and affecting the pooled computing resources. A mathematical model is proposed that considers the impact of the malicious activity on resources and the failures of individual machines. A Semi-Markov approach is used to create a technical subsystem that monitors states in order to solve the problem of analyzing overall availability level, detecting failures and quantifying the impact of the operation of malicious software. The scheme can be used as a toolkit for preventing the negative effects of computer virus cyber-attacks and ensuring high availability for cloud pooled resources.

2 Related Work

An availability model has been presented [10] for an Infrastructure–as–a–Service (IaaS) Cloud with multiple pools of physical machines (PMs). An independent, autonomous mathematical model was proposed that considers abrupt failures of PMs of the pool caused by the impact of deliberate malicious activity, hardware and software failures or other unforeseen interactions with on information resources of the IaaS Cloud. The model is constructed using a stochastic Semi-Markov (SMP) approach. The model employs monitoring of the states of the observed system and its subsystems in order to solve task of determining and analyzing the overall availability level for the IaaS Cloud resource, to the extent that this is affected by the failures and negative impact of the malicious computer viruses and other cyber threats. The study presents the results compared to benchmark steady state availability for the IaaS Cloud, failure rates and repair rates of the PMs that were obtained via observation. For the presented results, overall estimates of availability are obtained, considering the consequences of the activation of two types of malicious computer viruses by using the monolithic SMP model for an IaaS Cloud with three pools of PMs. Therefore, two additional branches of deliberate malicious impacts on PMs resources are required to be implemented by using proposed SMP availability model for an IaaS Cloud.

From the above considerations, it is concluded that the overall effort to promote survivability against cyber-threats, it is necessary to use AI to monitor the content of the emails, data in the databases, scripts, executable code that may contain malicious code. This widespread range of sources of possible cyber-threats should undergo a scanning and cleaning process before being used. Multiple approaches for detection methodologies have been proposed which demonstrates that the problem of data monitoring for potential dangers is a hard problem [8]. The proposed methodologies can be categorized as signature and non-signature based approaches [9]. The contribution of AI techniques is fundamental towards the aim of detecting all types of threats in data. Relevant datasets have been created with annotations if they belong to malicious code [10]. Machine learning algorithms include Supervised, unsupervised, reinforcement methods. The three schemes that were outlined above and are presented in this chapter promote the use of artificial intelligence in order to (i) produce early warning indicators of cyber threats in data, (ii) detect divergent system behavior that could be a sign of ongoing malicious activity within the system and (iii) monitor the availability of pooled cloud resources and the operational state of their individual physical machines during the period that they are suffering from the impact of computer viruses. In all these cases, the final goal is the assurance of increased survivability for systems.

3 Security Infringement Detection

In this section a technique for detecting security infringements in information systems is presented, that is based on a combination of learning techniques. More specifically, the techniques considered are Linear classifiers, Naïve Bayes classifiers, Decision trees with Random Forest technique and Convolutional neural networks employing deep learning.

PE Format

From a practical point of view, part of the analysis is based on the examination of Portable Executable (PE) files, a common type of files in the Windows Operating System. They include .exe, .dll, and .sys files. All these files include a PE header, which is a set of instructions to the Windows OS about the analysis of the code that follows. The fields of PE header are usually used as features for the detection of malicious software [11]. Programming libraries in Python can be used to extract the values of PE header.

Many fields in PE files do not follow a strict organization. There exist redundant fields and spaces which can be replaced by malicious code.

3.1 Static Analysis of Code

When static analysis is used, the sample code is tested without being executed. The obtained information may be the PE of the file [12–14] or even more specialized like YARA signatures [15]. In this section, several features will be presented that can be extracted from the executable files via statistical analysis. These features are used in the experiments performed on the dataset to train and the classification algorithms.

It has been proposed in literature [10] that emails be classified using machine learning classifier in a cloud computing system. The security requirements of defense information systems and cloud computing infrastructure have been analyzed and benchmarks of the necessary performance have been determined. Information system users and security software are aware that malware is more likely to be embedded in files of certain types such as executable, shell script etc. Hence attackers are adjusting by masking the true nature of malicious code by hiding it inside normally harmless files or files of unknown type [13]. It is therefore necessary to develop techniques capable of determining the type of file given byte sequences correspond to, without depending on the standard identification criteria normally used by operating systems e.g. file extensions, file headers etc. This part of the study was focused on scanning files in order to definitively determine the type of the file by examining and recognizing the nature of its contents. A database of files was created for this purpose by scraping various internet sources such as GitHub and malware repositories, providing current samples of both benign and malicious files [13]. A base-64 encoding was used for binary data, when this was not already present. Script

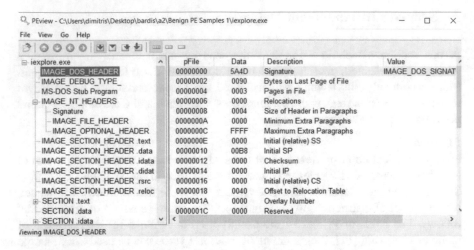

Fig. 1 Exploring the PE header of a benign executable using a hex editor

files, e.g. Javascript, were processed in plain text format. The obtained files are thoroughly filtered, adapted and selected for the testing environment.

In Fig. 1 the loading process of an executable file is visualized, and the code image is examined.

Various parts of the file can hence be examined, including the special field e_magic in DOS header which contains the MZ character series that corresponds to 0x4D 0x5A sequence, and the special header field PE defined as the IMAGE_NT_HEADERS structure. For the purposes of determining the file type, the fact that for static analysis the file is not loaded in memory is advantageous, since in this way the significant risk of malicious code concealment is avoided.

The malicious data is obtained from public databases which include real data and is available for comparison. This PE dataset contains 425 different samples, 378 benign and 47 malicious [13, 14].

Natural Language Processing (NLP) provides a wide range of techniques in text selection, analysis, and estimation. Since the content of the test and training databases is encoded as character sequences, that may be considered as text, the problem of classifying the file content as benign or malicious may be approached as an NLP problem [13]. It is noted that the data has been already annotated for the benign and malicious code fragments. Thus, supervised learning is used to design the classifiers.

Similarity between files can be measured using the hash functions used in cryptographic applications. The requirement in this case it to calculate quantitative similarity scores for file comparison, via similarity hashing algorithms. It hence becomes possible to detect modifications involving copying, insertions, deletions and tampering of the content. These scores can then by used a distance measure for clustering algorithms. Similarity hashing is more suitable in the current context, since it provides information about the nature and the extent of the intervention to the data, rather than a decision on whether two files are identical [13].

3.2 Methodology

The code sequences of the database are processed by means of static analysis of their respective files that are used for training and applying the following classification techniques.

- Linear classifier with Gaussian calculation of the parameters
- Naïve Bayes classifier
- Decision trees with Random Forest technique to solve regression and classification problem
- Convolutional neural networks employing deep learning for malware detection

The overall process is to create a dictionary of all the test cases. All data are created as arrays of words using suitable delimiters. Duplicate entries are removed resulting a dictionary of unique words. Data is transformed into vectors where the entries are used as input to the algorithms.

The header is analyzed by parsing into items or tokens that may be bytes, strings, or any other combination. A sequence of N such items is called N-Gram and this model has been used in many quantitative studies like computational linguistics, speech recognition, bioinformatics, etc. [16, 17]. The N-Gram model has been successfully used to create Markov chains for statistical prediction and text generation from a corpus [13]. In the context of malware detection, N-Gram counts and assortments are used as a basis of the local statistical analysis of the core data to provide a malicious code identification tool. In this study, data was analyzed in 4-Grams of characters and the 10 most frequent ones in each file were used as features. Once this process is completed, a global set of the most useful N-Grams is determined, since for any database of significant size, it is infeasible to consider the large number of all possible N-Grams. For this purpose, the N-Grams that should be chosen are the ones that present the highest discriminating capabilities i.e. ones that are more often observed in a specific type of data, based on the particular dataset. The resulting sequences are generally quite long scoring highly complicated search space. The above feature selection is utilized to enhance performance and reduce the dimensionality of the problem.

The filtering approach deals with the selection of the samples and the assignment of labels. The dataset is separated into learning and test subsets, created as the N-Grams from data and the feature vectors and normalized. The splitting is such that the proportion of each of the two types of data is equal in the global, the training and the test sets.

The classification algorithms receive as input both the data from the PE headers and the features extracted from the N-Gram analysis. Each item is mapped to a numerical vector x using hashing vectorizer, combined with a Term-Frequency

Times Inverse Document Frequency (TFIDF) transformer [18], the text data is converted into numeric form [18–20]. These phases correspond to the considerations and the processing already described in the current Section.

For this study, instead of optimizing a single classifier, a composite scheme utilizing several classifiers was used. A majority decision rule was then used in order to produce the final classification. The distance between points can be calculated with standard distances, like Euclidean, Manhattan or Chebyshev [21]. The Euclidean distance was used in the experiments in this study. A brief description of the classifiers used will be given in the following paragraphs, with detailed emphasis only on aspects that are different from their standard form described in literature. Further details about these schemes are widely available e.g., [21–24].

For a given classification scheme, decision thresholds and similar decision parameters obtained by training may require post adjustment in order to tune the different false positive/false negative detection probabilities required in a particular application.

Binary Classification Method

For a given item yielding the observations feature vector x, a discrimination function f is given as

$$f(w, x) \in \mathbb{R}$$

where w is a set of parameters selected so as to achieve the best separation of the data. In linear models for classification have the general form

$$f(w, x) = w^T x + w_0$$

which are inadequate in most situations since the classes are not linearly separable. In the multiple classes case, a k-rank discrimination function is used.

Naïve Bayes Algorithm

Data belong to two classes, i.e., benign and malicious. We have

$$P(\text{Benign}|k) = \frac{P(\text{Benign})\,P(k|\text{Benign})}{P(k)}$$

and

$$P(\text{Malicious}|k) = \frac{P(\text{Malicious})\,P(k|\text{Malicious})}{P(k)}$$

where k is the vector of N features.

Assuming that the features in k are mutually independent, the algorithm calculates the probabilities for new each sample case and compares the two probability values. The larger value is the winner. In the case where the two values are equal the algorithm cannot provide an answer.

Decision Trees

Using supervised learning a decision tree can be constructed to according to an if-logic. The data are classified using well defined questions like a calculated quantity is over a given value. Pro-pruning procedures can be used like minimum tree depth, the maximum leaf nodes, and the minimum samples for each leaf, are used to minimize the size of the evaluated tree. In general, decision trees offer low generalization.

A particular consideration in the case of the Decision Tree was the imbalance caused in the classification mechanism by the fact that benign files were more amply available than malicious ones. The reasons for this imbalance can be readily comprehended by considering the fact that users tend to promptly delete any file they perceive as a threat, e.g. based on the information from an antivirus program. The imbalance can be corrected using several schemes [21]. One scheme involves using class weights for each class that are inversely proportional to the frequency of this class in the data. Another approach is to restore balance by randomly repeating data samples of the least populated class within the iterations of the training epochs. Alternatively, samples from the most populated class can be discarded, for the purpose of equalizing the effect. All these methods were shown to improve the classification performance.

Random Forest

After the calculation of the N-Gram, the names of the units and their number that belong to the PE hear of each file, while the data that cannot be analyzed are omitted. Using a hashing vectorizer the text data are converted into numerical values. A random forest classifier [22] is used for the training, test, and validation data.

Convolutional Neural Network

The Convolutional Neural Network CNN is constructed following the steps [23, 24].

- Declaration of the required programming libraries
- Create a list of files and locations
- The bytes of the file are stored to an array
- Data are divided to train, validate and test
- Optimizer rate is defined
- The structure of the neural network is created
- The model is constructed

For a given number of epochs and batch size, the data are used to train the CNN.

The Aggregated Model

Once trained all models an aggregation of all results takes place. Each method has its own strong aspects, and a soft voting procedure takes place to get the final classification decision. In Table 1 a proposed aggregation classifier is presented.

Table 1 Outline of the composite classification scheme used

Steps	Details
Data preparation	PE headers are being processed
Feature extraction	Dictionaries and hashing
Training base classifiers	Naïve Bayes
	Random Forest
	Convolution Neural Network
Feature identification	Is the pattern recorded in the database as known case
Soft voting engine	Validate the result, find a weighted average, insert in the quarantine, update the database

Fig. 2 Confusion matrix for binary classification

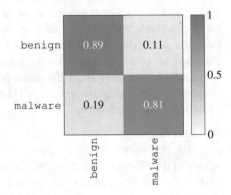

3.3 Results

Applying the combined detection algorithm to the provided dataset the confusion matrix in Fig. 2 is derived. True positive and true negative detection rates are both over 80%. This performance is considered satisfactory as a base for the malicious activity warning system application. With this starting point, the sensitivity of the scheme may be adaptively adjusted for the probability of a false alarm, in order to provide the required compromise between the frequency of alerts and the level of security required. Application of the scheme on the entire dataset produces a correct classification ratio of 87.76%. The efficiency of algorithm varies according to the size of training data.

For the particular case of the Deep Learning Malware Detector algorithm Table 2, the changes in performance with an increasing number of epochs is summarized. The loss function is the objective function that is used to rank and compare the candidate solutions. Epochs are full training cycles used it the training iteration. The results demonstrate that large increases in the number of epochs are not necessary in order to achieve the best possible accuracy of decisions.

Table 2 Deep learning testing accuracies

Epochs	Loss	Accuracy
1	0.3860	0.8604
5	0.3534	0.8906
10	0.3431	0.8813
60	0.3054	0.8996
200	0.4362	0.8709

3.4 Evaluation

The problem of using AI techniques in order to analyze the contents of files for the purpose of determining if potentially malicious code is present was analyzed and formulated as a NLP problem. Existing NLP software tools were used in order to design an end-to-end scheme for processing stored data and producing warning of the potential existence of malware, using machine leaning. The preprocessing require in order to extract the necessary features from the raw data was determined based on particular considerations arising from the nature of the problem. Datasets and data dictionaries were obtained from the performed training of the schemes that can be reused in order to repeat the detection task in other contexts. The datasets may be enriched with additional examples when these are available, in order to improve the classification performance. The framework designed is of particular importance for military applications due to the ability for tuning its sensitivity.

The application of the scheme in specific contexts may be easily tested and benchmarked before deployment to the production environment. This scheme simultaneously illustrates, both in theoretical and practical terms, the feasibility and benefits of using AI classifiers in order to produce indicators of cyber threats and malicious cyber activity. The scheme is essentially used to develop a model for the potential malicious activity. Using this model, it significantly improves security compared to customized rules and datasets. The data and the software tools and libraries used are open source. They can hence take advantage of improvements proposed by the community and continue building knowledge on the appearance of new threats. The composite classification approach used provided, combined with the NLP formulation of the problem provided superior results compared to those expected form applying the individual methods. Numerical data of the benchmark performance of this technique were produced that can be used in future enhancements. The scheme is suitable for demanding applications, such as military environments.

4 Digital Twin Cyber Resilience Decision Support

In the introduction, the notion of how using a duplicate system operating in a controlled environment for the prompt and early detection of cyber-attacks was presented. A formal model supporting the specification, implementation and deployment of digital twin (DT) systems using AI and Internet of Things (IoT) concepts was recently presented [9]. The formal model contributes to the design of DTs capable of detecting diversions from the expected behavior and hence supporting decisions and giving early warnings of cyber threats.

A DT is an exact copy of the actual system under modeling, that exhibits identical dynamic behavior in the environment, but under controlled conditions. The purpose of this system is to provide benchmark or reference behavior for the expected behavior of the actual system so as to enable automated anomaly detection. Ideally, the duplicate system has the same physical structure as the actual one. The overall system concept is illustrated in Fig. 3. However, since the actual system model may include non-replicable entities such as people or behaviors, some of the duplicate subsystems may have to be virtualized. This technology, together with simulation primitives, exhibits a wide variety of prospective applications for the purposes of predicting breakdowns, anomalies and cyber-attacks, optimizing service plans, and optimizing performance. The technology has been demonstrated to cooperate with IoT, AI and Virtual and Augmented Reality subsystems [9]. Thus, DT is a purely digital replica system that exhibits the same behavior as the real-world object, process or system in a controlled environment. DTs can also be considered at subsystem level to formulate components for creating twins of larger systems.

The DT concept arises from the Industry 4.0 movement and has been developed in order to facilitate the following [9].

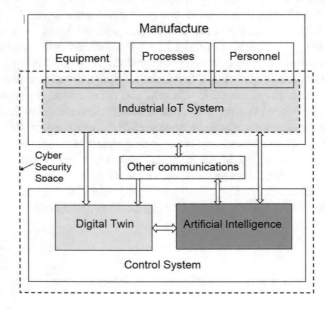

Fig. 3 Overall system concept

- The detection of divergence in behavior due to physical faults, adversarial intervention or erroneous input,
- Planning by means of the prediction of outcomes
- Global system optimization and decision making.

The concept of DT is related to previous concepts of computer aided design and to the notions of online customer profiles, but current DTs involve four significant differences:

- The model reliability with an emphasis on how they support specific performance aims;
- Communication with the real world, for monitoring and control in real time;
- The use of advanced big data analytics and artificial intelligence to open innovative deduction perspectives;
- The capability of evaluating *what-if* scenarios and conducting realistic exercises.

DT systems are combined with AI subsystems in order to promote survivability thereby improving performance and reducing downtime. Machine learning algorithms for manufacturing are shaped and tuned to the specific challenges of systems—such as reducing losses, improving process stability, limiting downtime and detecting anomalies [9].

Applications of AI in currently active operational systems include the predictive maintenance of production computer systems and machines, the use of image processing technologies for the automatic sorting of items such as consumer products, like batteries or food, or user communication using text-based dialog systems e.g., in chatbots. AI supported DT systems are critically dependent on the availability of databases containing real-world or high-quality, artificial performance data. Furthermore, a prerequisite for the successful deployment of AI enabled DT systems is advanced digital maturity of the organization. In physical terms, this implies e.g., the installation of suitable digital sensor systems in many applications along the field of operations. In terms of management know-how, the development of AI applications requires the availability of knowledge of data analysis and/or computer science. However, since this knowledge is not or only to a limited extent available to many organizations, external services are a solution [9]. Modular designs are feasible, in the context of which specific isolated cognitive AI-based services, such as image or face recognition or the conversion of speech into text can be delegated to external entities, like cloud processing resources, in the case where the size of the organization does not permit in-house support for all functionalities.

Additionally, due to the continuously increasing potentials of AI supported applications, the information processing environment paradigms are evolving. Applications involving pattern recognition tend for example to become more autonomous and cost efficient [9]. Simultaneously, AI systems bring improvements in the system's ability to forecast user and environment dependent parameters such as customer actions and wishes, adversarial activities and external interventions. As a result, innovative operational models arise that focus on adapting to the deployment environment, including user behavior. As the AI's acquaintance with the user, the equipment and the environment progresses, deviations from the expected norm become easier to be identified as potential cyber-attacks.

Further amplification of the benefits of the application of AI systems, data stores of limited scope need to be integrated with others in broad range data platforms that promote cross-application collaboration and the creation of digital ecosystems. This necessity is caused by the fact that AI is critically dependent on the volume of training data, including synthetic ones generated from digital models. The training data must include information regarding all possible states of the entities of interest. The quality, the representativeness and the robustness of the training data significantly affect the performance of the system and its overall effectiveness.

Even though AI does not demand the existence of analytical knowledge about the problem to be solved, such as the divergence from the expected behavior caused by the actions of a cyber intruder, data describing all occurrences that may be encountered during normal operations.

Possible benefits from the application of AI include the following [9].

i. Collection of statistics necessary for highlighting disruptions from normal behavior occurring over the entire dataset.
ii. Identification of critical situations encountered at operation time that are not provided for with the predictive analytics;
iii. Processing of large volumes of data produced at operation time and identifying and assessing inconsistencies.

Current research activities on Artificial Intelligence in operations data analysis focuses on the following topics [9].

i. Hierarchical and distributed neural networks-based system with combined relearning
ii. Big Data analytics for multi drone fleets-based monitoring adverse occurrences, such as accidents in remote locations
iii. Deep learning of neural networks for image recognition in space monitoring and manufacturing
iv. Machine vision of autonomous systems
v. Expert systems for logistics based on fuzzy logic
vi. Text recognition using deep learning neural networks
vii. Application of AI for development and implementation of IoT for industry domains.

4.1 Landscape Model Development

The concept of the Landscape has been proposed [9] that is an instrument for representing and analyzing the state of technological development in an entity. Entities may be systems of different sizes, from an information system to a country. It is significant regarding the Industry 4.0 movement which includes modern technologies such as AI, Digital Twins, IIoT etc. The landscape usually consists of a disorganized set of technologies, actors involved, development and implementation. Such a collection cannot serve as a model and be used in any analysis using formal techniques [9]. It has been proposed to describe the landscape in terms of a formal model as

$$LS = \{Reg, Tech, Ent, t, M_{tt}, M_{et}\}$$

where Reg is a region or location or set of regions, $\{Reg_i\}, i = 1, \ldots, n$, $Tech$ is a technology or set of technologies $\{Tech_j\}, j = 1, \ldots, m$, Ent is a technology or set of entities or enterprises $\{Ent_k\}, k = 1, \ldots, p$, t is the time or time-slot of interest, M_{TT} is a mapping $M : T \rightarrow T$ connecting technologies, and M_{ET} is a mapping $M : E \rightarrow T$ describing technologies included in entities.

The model can be described by

- different technologies, locations etc.
- metrics obtained from n, m, p and the cardinalities of the sets $Reg, Tech, Ent$

The set-theoretical model of the landscape can be also represented as a connected graph with weighted nodes and links. More specifically,

- the set of nodes is defined by $Tech$
- the weight of each node is given by the number of entities including the technology
- the weight of each connection is given by the number of unique technologies included by the two entities.

For the case of enterprises, this model allows the evaluation of the most developed technologies and the description of the activities of each enterprise.

A variant of the set-theoretical model of the landscape has been presented [25] that is applicable for the case of Digital Twins. The model is given by

$$DT = \{PE, VM, Ss, DD, CN\}$$

where PE are physical entities, VM are virtual models, Ss are services, DT are data and CN are connections [25]. In this context, the physical entities need to be represented as virtual models in the digital twin, in order to reproduce their behavior. They can be units, systems or systems of systems.

The virtual models are representations of the PEs that maintain their physical and operational properties and present the same behavior for the same events. Additionally, they follow the same rules or logical abilities, such as reasoning, evaluation and decision making.

The data comes at different times, from different sources, are multidimensional and heterogeneous. Some of the data may be actual observations, some may be artificially generated and finally, some may be the product of knowledge of the functionality of the system. Data from multiple sources may be fused according to the needs of the application.

The services cover services offered and services received by the DT. The offered services can be simulation, verification, monitoring, optimization etc., while received services include data services, knowledge banks, computing services etc.

The DTs are connected with their real duplicates to perform complex operations and analyses. Each DT contains six connections

- physical entities to virtual model
- physical entities to data
- physical entities to services
- virtual models to data
- virtual models to services
- services to data

Digital twins designed using such models have been developed for production lines, training personnel, business process optimization, smart cities, construction, healthcare, shipping etc. In the context of cyber resilience, it is proposed that DTs are used for parallel and dynamic monitoring. This concept [8] employs continuous monitoring of the operational system under observation. Due to the diverse and unpredictable nature of possible cyber attacks, the proposed approach for detecting such events is by detecting divergence of the observed system behavior from expected system behavior. This detection needs to be early and prompt in order for the possible attack to be adequately deterred. It is hence proposed that the DT is operated in parallel to the operational system and an AI monitoring system is used to compare the performance of the two systems. The AI system receives input from the data connection and obtains monitoring and telemetry information. Additional training for recognizing divergent behavior is provided to the AI system by executing simulated attacks on the DT. The AI system hence produces security alerts regarding detected divergent behavior. The sensitivity of the alerts is configurable via suitable thresholds of the severity of the attack. Current research involves the additional development of the set-theoretical model so as to further formalize the design and implementation of the DT and its software and hardware components.

5 Semi-Markov Cloud Availability Model

The capability of assessment of the level of survivability achieved following the consequence of cyber-attacks before and after the introduction of the AI survivability promoting schemes is an indispensable tool for the successful development of such schemes.

Cloud Infrastructure–as–a–Service (IaaS) is an extensively used and appreciated cloud computing model with applications to a diverse variety of tasks in different operational environments, manufacturing installations, as well as in the scientific domain. The successful deployment of IaaS Cloud implementations critically depends on the existence of robust solutions to the problem of maintaining availability and guaranteeing cybersecurity for the cloud infrastructure components [10]. Therefore, the challenge of ensuring the availability level of the IaaS Cloud in an environment of diverse cybersecurity threats becomes a particularly significant component of the cybersecurity effort at national level. In order to address this problem, cloud service providers and users of cloud services require techniques capable of determining the effective cybersecurity level for IaaS Cloud, taking into account reliability characteristics of physical machines (PMs) in the process. Such resources typically include different types of servers based on virtual and real physical computer systems components.

Given that cybersecurity and reliability for cloud infrastructure components and availability of the IaaS Cloud are all elements of the overall system survivability, it has been proposed [10] to employ a monolithic Semi-Markov (SMP) model for the purpose of quantifying the overall availability level of the cloud infrastructure. As the global monitoring parameter, the steady state availability was used in order to derive state information [10].

In the research presented in [10] the monolithic SMP model was used for obtaining overall estimates of availability in the context where the effects of two types of malicious computer viruses were required to be studied for an IaaS Cloud instance encompassing three pools of PMs. For this purpose, the proposed SMP availability model for an IaaS Cloud [10] was used in order to implement two branches of deliberate malicious impacts on PMs resources.

Virtually all cloud service providers and users are currently appreciating the necessity to apply significant amounts of effort in order to maintain availability and promote cybersecurity of the IaaS Cloud. A variety of modeling approaches have been employed as instruments for the development of a toolkit necessary for preventing the adverse impact of deliberate malicious activity and ensuring increased availability level of the cloud services [10]. These approaches include models based on [10]

- stochastic non-state-space
- state-space,
- continuous-time Markov chains and
- discrete-time Markov chains (DTMCs).

Other schemes presented in literature employ non-Markovian approaches in order to solve identical tasks in preference to Markov models [10]. Alternative SMP models have also been proposed [10]. The research presented in [10] involves the development of two types of SMP models. The first proposed SMP model is one that can be solved through usage of Embedded Markov Chains for Cloud Systems [10]. The second proposed SMP model is also a model based on embedded DTMCs. The SMP based approach is then employed for the modeling and the determination of the steady-state availability of an IaaS Cloud with three pools of PMs. The modelling includes consideration of sudden failures and deliberate malicious impacts [10].

The model is not bound to a particular cloud architecture, but considers a generic simplified structure for the implementation of the PMs pools belonging to a single IaaS Cloud. According to this structure, an IaaS Cloud consists of hot, warm and cold pools of PMs. Hot pool PMs are powered on and operational. Warm pool PMs are also powered on, but are not active. Finally, cold pool PMs are turned off. Additionally, it assumes that all PMs in a pool belong to the same kind: a hot pool contains only fully operational PMs, a warm pool also includes normal working machines, but these PMs are not ready and a cold pool contains only turn off PMs. A specialized Technical and Information Monitoring System (TIMS) is used to monitor the sequence of states the system is following. Furthermore, TIMS is responsible for performing repair, remove or replacement operations for failed PMs. The entire range of deliberate malicious effects on software and hardware components of IaaS Cloud is also detected by TIMS [10]. The example considered involves the spreading of the WanaCry and Petya ransomware within the system.

The aim of the model is not to provide comprehensive screening and absolute protection from ransomware attacks; the operation of model is based on considering the stochastic process of the spreading of the impact of the malware on the cloud infrastructure. The ultimate purpose for using the Semi-Markov model is to obtain availability estimates of the timeframe for users to observe the effects of the ransomware to the system [10].

The model design is based on the assumption that there exists a relation between the ransomware attack and the reduction of the overall availability level and performance of the IaaS Cloud. Suppose that attack develops in accordance with familiar scenario, namely: first phase, when virus penetrates to physical machine and tries to impact information resource allocated by the PM (WannaCry pattern); second phase, when virus spreads by using of cryptography and ransomware techniques (Petya pattern). In Fig. 4, the finite graph is illustrated of the SMP availability model considering deliberate double insidious malicious impacts on information resources of IaaS Cloud. According to the model description given earlier on, the model also considers that the IaaS Cloud consists of three identical PMs, which are deployed as hot, warm and cold physical machine, respectively [10].

The proposed model additionally contains the TIMS component which, together with additional devices, are responsible for monitoring the system and detecting unauthorized intrusions. Vendors and users of cloud computing platforms are generally incorporating rigorous and effective monitoring systems, that comprise the Monitoring plane [10]. The Monitoring plane provides the functionality necessary to detect multiple instances of unauthorized penetrations employing different points of access. The model proposed can be used as an additional analytical toolkit to develop of anomaly detection technique based on considering different adverse effects, such as sudden failures of PMs and separate deliberate hacker attacks. The SMP availability model considering the adverse effects of deliberate malicious activities on information resources of IaaS Cloud consists of 20 states. Two branches for the activation and evolution of the viruses within the system are modeled. First branch is branch of activation of the WannaCry virus and second branch is the branch of the dispersion of the Petya virus. Table 3 separates the state transitions occurring in Fig. 4 for the case of each of the malwares that cause them in the context of the SMP availability model. The model is considering the impact of malicious activities internal to the information resources of the IaaS Cloud consisting of three

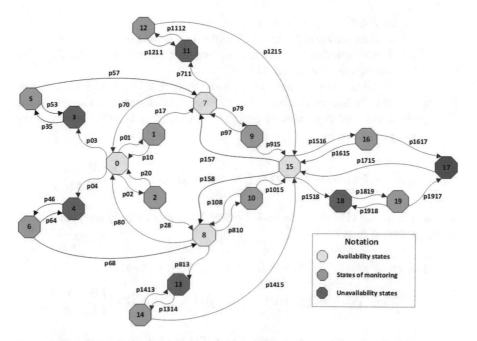

Fig. 4 Finite graph of the SMP availability model for the IaaS Cloud

Table 3 System state transitions caused by the two types of malware [10]

WannaCry virus	Petya ransomware
$0 \xrightarrow{P_{0,3}} 3 \xrightarrow{P_{3,5}} 5$	$0 \xrightarrow{P_{0,4}} 4 \xrightarrow{P_{6,4}} 6 \xrightarrow{P_{6,8}} 8$
$5 \xrightarrow{P_{5,7}} 7$	$7 \xrightarrow{P_{7,11}} 11 \xrightarrow{P_{11,12}} 12 \xrightarrow{P_{12,15}} 15$
	$8 \xrightarrow{P_{8,13}} 13 \xrightarrow{P_{13,14}} 14 \xrightarrow{P_{14,15}} 15$
	$15 \xrightarrow{P_{15,18}} 18 \xrightarrow{P_{18,19}} 19 \xrightarrow{P_{19,17}} 17$

PMs, as well as the impact of external malicious actions. Furthermore, according to previous experience [10] the monitoring system may be used, which this is a really effective means in order to achieve timely detection, but not prevention, of attacks on cloud assets and resources. In Fig. 4, the following conventions are observed for the presentation of the states of the second model: available states are in yellow color, unavailable states are in red color, control and monitoring states for TIMS system are green color.

As illustrated in Fig. 4, if three PMs fail the IaaS Cloud becomes unavailable. Consider the occasion where the system state is unavailable for the IaaS Cloud. Following that, the IaaS Cloud becomes available when the model has been in the states s_0, s_7, s_8, s_{15}. In state s_0, the IaaS Cloud is operational, because all three PMs are available. The opposite situation can only arise when system enters states $s_3, s_4, s_{11}, s_{13}, s_{18}$. These states may be described as states of viruses' attacks, when the system is unavailable due to hidden failures of PMs. The subset of states for TIMS involves all remaining states, namely $s_1, s_2, s_5, s_6, s_9, s_{12}, s_{14}, s_{16}, s_{19}$.

Indeed, the states $s_1, s_2, s_5, s_6, s_9, s_{12}, s_{14}, s_{16}, s_{19}$ are states, when system maintainers will have the ability to exploit the findings of TIMS in order to determine and solve control or monitoring tasks, including measures to implement the defensive features, regarding the viruses' activity.

In order to solve task pertaining to the SMP modeling for cloud infrastructure, it is proposed [10] that the control of the technical and information states the hot PMs perform over the deterministic period of time T, and transitions from states s_i to states s_j are given by

$$Q_{0,1}(t) = Q_{012}(t) = Q_{7,9}(t) = Q_{8,10}(t) = Q_{15,16}(t) = \begin{cases} 0 & \text{if } t < T \\ 1 & \text{if } t > T \end{cases}$$

It is also proposed [10] that transitions for the TIMS from state s_i to s_j state occur during period of time τ_c as

$$Q_{1,0}(t) = Q_{2,0}(t) = Q_{5,3}(t) = Q_{6,4}(t) = Q_{9,7}(t) = Q_{10,8}(t)$$

$$= Q_{12,11}(t) = Q_{14,13}(t) = Q_{16,15}(t) = Q_{19,18}(t) = \begin{cases} 0 & \text{if } t < \tau_c \\ 1 & \text{if } t > \tau_c \end{cases}$$

The transitions for branch of activation of the WannaCry virus can be written based on hypoexponential distribution as [10]

$$Q_{10(t)} = Q_{10(t)} = \begin{cases} 1 - \alpha e^{-\lambda_1 t} + \beta e^{-\lambda_2 t} & \text{if } t < T \\ 0 & \text{otherwise} \end{cases}$$

where $a = \frac{\lambda_2}{\lambda_2 - \lambda_1}, \beta = \frac{\lambda_1}{\lambda_2 - \lambda_1}$.

Then, the transitions for branch of development of the Petya virus can be written based on hyperexponential distribution as [10]

$$Q_{0,4}(t) = Q_{6,8}(t) = Q_{7,11}(t) = Q_{8,13}(t)$$

$$= Q_{12,15}(t) = Q_{14,15}(t) = Q_{15,18}(t)$$

$$= Q_{19,17}(t) = \begin{cases} \rho(1 - e^{-\lambda_3 t}) + \omega(1 - e^{-\lambda_4 t}) & \text{if } t < T \\ 0 & \text{otherwise} \end{cases}$$

where $\rho \in [0, 1], \omega = 1 - \rho$.

For other states, the exponential and Erlang-k, $(k = 2)$ distributions are used in order to describe all times to sudden failures and recoveries of PMs respectively. The cumulative distribution functions for these states are stipulated in Table 4 [10].

Next, if by using the steady-state probability vector, all previous equations and total probability relation $\sum_{i=0}^{19} \pi_i = 1$, the required result is obtained as $A = \pi_0 + \pi_7 + \pi_8 + \pi_{15}$, where $\pi_0, \pi_7, \pi_8, \pi_{15}$ are steady states for states s_0, s_7, s_8, s_{15}.

Table 4 Transitions for failures and recoveries of PMs for the IaaS cloud

Transitions for PMs	CDFs for time transitions
1 → 7	
2 → 8	
9 → 15	$\exp(\lambda_s)$
10 → 15	
16 → 17	
7 → 0	
8 → 0	$\text{Erlang}(2, \mu_1)$
15 → 7	
15 → 8	$\text{Erlang}(2, \mu_2)$
17 → 15	

Simulation results presented in [10] demonstrate the capability of the SMP to model the expected behavior of the system, where the modelling prediction of the availability matches the observed values and reducing the spreading rate of the viruses, the availability increases.

The modeling results have several theoretical and practical implications [10]. Theoretical perspectives involve the development of Semi-Markov availability models with special states. This type of models may be solved using embedded DTMCs. Practical perspectives relate to the availability assessments of IaaS Cloud and possibility to optimize the architecture and diversification of specific services to be provided. AI techniques are an ideal technique to be employed for this optimization.

Future research could be dedicated to specifying numerical values of parameters for modeling availability assessments of IaaS Cloud with three pools of physical and virtual machines using AI for continuous tracking of such parameters.

6   Future Work

As it was explained in the introduction section of this chapter, cybersecurity is a problem concerning not solely the technical community, but society in general due to its dependence on information systems. AI techniques have been identified as a feasible means of processing large volumes of data for the purpose of identifying threats and divergent behavior. The technology of AI assisted cybersecurity has not yet reached the required level of maturity [26]. There exist several issues that need to be further studied. The success of any AI algorithm is highly dependent on the quantity and quality of the data used for its training.

Current techniques for AI based cybersecurity are primarily based on data originating internally from the organization that they concern. Internal data are naturally closer and more fitted to the organization's internal structure and may allow a quicker learning curve about detecting previous attacks and existing threats. However, the ability to exploit external data offers the prospective for better resilience and hence survivability, by considering the broader trends and developments in the Cyberspace. To this extent, data from GitHub was used in first scheme presented in this chapter, in order to support the recognition of benign files. The fact that a new dataset can be regularly rebuilt in an automatic way, gives the scheme the ability to adapt to emerging threats. Further data sources need to be exploited such SourceForge, search engines, feeds of threat intelligence data by industry and hacker forums that would provide more insight into software like exploit development kits, trends in threat design etc. Additionally, exploitation of data from public or private cyber threat reporting repositories and stores should be considered.

Adversarial Machine Learning (ML) is a growing class of techniques that aim to deceive algorithms by generating data that can pass as rel data. Malicious users are using such techniques for a variety of purposes, including generating AI driven system attacks [26]. Adversarial ML defense focuses on threat modeling, attack simulation, countermeasures, detection and evasion [26]. These schemes attain learning with small datasets and can hence quickly adapt to evolving environments, similarly to human actors. The second scheme presented in this chapter contributes in this direction by providing the infrastructure for realistic attack simulation detection and countermeasure exercises. Due to the dynamic nature of cyberattacks, it is proposed in literature that AI algorithms should not be allowed to take the relevant decisions, but rather support human operators in deciding. The second scheme presented in this paper should be used to become a fundamental part of the code of an AI based cybersecurity decision support system. To the same end, the third scheme presented, involving the SMP used for modeling system availability is suitable for providing such decision support systems with insight into the survivability prospects of cloud systems and the health of each one of their pooled resources.

7 Conclusions

Three techniques were presented, capable of producing indicators of malicious activity within information processing systems that could be associated with the presence of cyber-attacks. The first scheme concerned the detection of threats within data. Executable code, e-mail messages and network packets were processed via AI algorithms in order to model the threats and achieve effective detection of computer viruses and other dangerous content. Open databases and Natural Language processing were employed in order to train different types of classifiers and optimize the results. The second scheme was related to the detection of unexpected system behavior that can be associated to an ongoing cyber-attack. Such behavior can be identified by operating a clone system of the system under observation and using its behavior as a benchmark for the expected behavior for the original system. A technique has been proposed that enables the design and implementation of the clone systems in order to facilitate decision support. AI algorithms can be trained in order to detect diversions of the observed from the expected behavior and produce timely warnings. The final scheme focused on the cloud processing paradigm and was related to the examination of its availability and the detection failures of its constituent subsystems. The analysis was based on a Semi Markov model that enables monitoring of system states, analyzing availability and measuring the impact of the activation of computer viruses within the system. These results contribute to the prevention of the adverse effects of computer viruses and the assurance of high availability of computer systems.

References

1. M. A. Nacar B. Kasım, A. B. Çavdar and E. Çayırcı. Modeling and simulation as a service for joint military space operations simulation. *The Journal of Defense Modeling and Simulation*, 18(1):29–38, 2019.
2. M. J. North and C. M. Macal. *Agent-Based Modeling and Computer Languages*, pages 865–889. Springer Link, 2020.
3. J. van Oijen G. Poppinga M. Hou J. Roessingh, A. Toubman and L. Luotsinen. Machine learning techniques for autonomous agents in military simulations–multum in parvo. In *2017 IEEE International Conference on Systems, Man, and Cyber-netics (SMC)*, pages 3445–3450, October 2017.
4. Y. G. Kim J. Koo and S. H. Lee. Security requirements for cloud-based C4I security architecture. In *2019 International Conference on Platform Technology and Service (PlatCon)*, pages 1–4, January 2019.
5. N. Doukas O. P. Markovskyi P. Stavroulakis M. Kolisnyk V. Kharchenk and N. G. Bardis. *Reliability, Fault Tolerance and Other Critical Components for Survivability in Information Warfare*, volume 990, pages 346–370. Springer, Cham, 2017.
6. Peter Stavroulakis. *Reliability, survivability and quality of large-scale telecommunication systems: case study: Olympic games*. John Wiley and Sons, 2004.
7. Peter Stavroulakis Doukas, Nikolaos and Nikolaos Bardis. *Review of Artificial Intelligence Cyber Threat Assessment Tech-niques for Increased System Survivability.*, pages 207–222. Springer, Cham, 2021.
8. Taddeo M. McCutcheon T. and Floridi L. Trusting artificial intelligence in cybersecurity is a double-edged sword. *Nature Machine Intelligence*, 1(12):557–560, 2019.
9. Vyacheslav Kharchenko, Oleg Illiashenko, Olga Morozova, and Sergii Sokolov. Combination of digital twin and artificial intelligence in manufacturing using industrial IoT. In *2020 IEEE 11th international conference on dependable systems, services and technologies (DESSERT)*, pages 196–201. IEEE, 2020.
10. Oleg Ivanchenko, Vyacheslav Kharchenko, Borys Moroz, Leonid Kabak, and Kyrylo Smoktii. Semi-Markov availability model considering deliberate malicious impacts on an infrastructure-as-a-service cloud. In *2018 14th International Conference on Advanced Trends in Radio-electronics, Telecommunications and Computer Engineering (TCSET)*, pages 570–573. IEEE, 2018.
11. Diomidis Spinellis. Reliable identification of bounded-length viruses is np-complete. *IEEE Transactions on Information Theory*, 49(1):280–284, 2003.
12. Yogesh Bharat Parmar. *Windows Portable Executor Malware detection using Deep learning approaches*. PhD thesis, Dublin, National College of Ireland, 2020.

13. Emmanuel Tsukerman. *Machine Learning for Cybersecurity Cookbook: Over 80 recipes on how to implement machine learning algorithms for building security systems using Python.* Packt Publishing Ltd, 2019.
14. S Lee, K Lee, et al. Packed PE file detection for mal ware forensics. *Computer Science and Its Applications*, 2009.
15. David N Palacio, Daniel McCrystal, Kevin Moran, Carlos Bernal-Cárdenas, Denys Poshyvanyk, and Chris Shenefiel. Learning to identify security-related issues using convolutional neural networks. In *2019 IEEE International Conference on Software Maintenance and Evolution (ICSME)*, pages 140–144. IEEE, 2019.
16. Tina Rezaei, Farnoush Manavi, and Ali Hamzeh. A PE header-based method for malware detection using clustering and deep embedding techniques. *Journal of Information Security and Applications*, 60:102876, 2021.
17. Nitin Naik, Paul Jenkins, Roger Cooke, Jonathan Gillett, and Yaochu Jin. Evaluating automatically generated YARA rules and enhancing their effectiveness. In *2020 IEEE Symposium Series on Computational Intelligence (SSCI)*, pages 1146–1153. IEEE, 2020.
18. Shahzad Qaiser and Ramsha Ali. Text mining: use of TF-IDF to examine the relevance of words to documents. *International Journal of Computer Applications*, 181(1):25–29, 2018.
19. Nexus. Freeware hex editor. https://mh-nexus.de/en/hxd/. Accessed: 2021-09-30.
20. Mohd Zaki Mas' ud, Shahrin Sahib, Mohd Faizal Abdollah, Siti Rahayu Selamat, and Choo Yun Huoy. A comparative study on feature selection method for n-gram mobile malware detection. *Int. J. Netw. Secur.*, 19(5):727–733, 2017.
21. Abdullah Elen and Emre Avuçlu. Standardized variable distances: A distance-based machine learning method. *Applied Soft Computing*, 98:106855, 2021.
22. Zeinab Khorshidpour, Sattar Hashemi, and Ali Hamzeh. Evaluation of random forest classifier in security domain. *Applied Intelligence*, 47(2):558–569, 2017.
23. Zhiwei Gu, Shah Nazir, Cheng Hong, and Sulaiman Khan. Convolution neural network-based higher accurate intrusion identification system for the network security and communication. *Security and Communication Networks*, 2020, 2020.
24. Iraj Elyasi Komari, Mykola Fedorenko, Vyacheslav Kharchenko, Yevhenia Yehorova, Nikolaos Bardis, and Liudmyla Lutai. The neural modules network with collective relearning for the recognition of diseases: Fault-tolerant structures and reliability assessment. *Neural Networks*, 1:3, 2020.
25. Qinglin Qi, Fei Tao, Tianliang Hu, Nabil Anwer, Ang Liu, Yongli Wei, Lihui Wang, and AYC Nee. Enabling technologies and tools for digital twin. *Journal of Manufacturing Systems*, 2019.
26. Sagar Samtani, Murat Kantarcioglu, and Hsinchun Chen. Trailblazing the artificial intelligence for cybersecurity discipline: a multi-disciplinary research roadmap, 2020.

Machine Learning and Deep Learning for Fixed-Text Keystroke Dynamics

Han-Chih Chang, Jianwei Li, Ching-Seh Wu, and Mark Stamp

Abstract Keystroke dynamics can be used to analyze the way that users type by measuring various aspects of keyboard input. Previous work has demonstrated the feasibility of user authentication and identification utilizing keystroke dynamics. In this research, we consider a wide variety of machine learning and deep learning techniques based on fixed-text keystroke-derived features, we optimize the resulting models, and we compare our results to those obtained in related research. We find that models based on extreme gradient boosting (XGBoost) and multi-layer perceptrons (MLP) perform well in our experiments. Our best models outperform previous comparable research.

1 Introduction

Today, popular forms of biometric authentication include fingerprints and facial recognition. However, such biometric techniques do not resolve all authentication issues. For example, studies show that the elderly are reluctant to use facial recognition and fingerprint recognition for authentication on mobile phones, while young people prefer to type instead of using other ways to authenticate [20]. Therefore, some passive biometric have recently emerged. In this research, we consider biometric based on keystroke dynamics. Such techniques are applicable to the authentication problem, and can also potentially play a role in intrusion detection.

Keystroke dynamics are derived from typing behavior. This approach typically relies on features such as the duration of keyboard events, the duration of the "bounce," the time difference between each character, and so on [38]. Such data can be collected through monitoring keyboard input and recording, for example, the time intervals between each keystroke. However, it is worth noting that a biometric based

H.-C. Chang · J. Li · C.-S. Wu · M. Stamp (✉)
San Jose State University, San Jose, CA, USA
e-mail: han-chih.chang@sjsu.edu; jianwei.li@sjsu.edu; ching-seh.wu@sjsu.edu;
mark.stamp@sjsu.edu

© The Author(s), under exclusive license to Springer Nature Switzerland AG 2022
M. Stamp et al. (eds.), *Artificial Intelligence for Cybersecurity*, Advances in
Information Security 54, https://doi.org/10.1007/978-3-030-97087-1_13

on keystroke dynamics is unlikely to be powerful enough to serve as a standalone authentication technique, and hence keystroke dynamics generally must be used in conjunction with other types of authentication, such as passwords [30]. In its related role as an IDS, keystroke dynamics may be competitive with other approaches [38].

Compared with popular biometric technologies such as fingerprints and iris scans, keystroke dynamics has some advantages. First, in terms of hardware, keystroke features can be gathered through a simple API interface, with the collected data then passed to a model for evaluation. Hence, no additional hardware deployment is involved, which reduces the cost. Second, as alluded to above, keystroke information can be obtained in a more passive and natural manner, which eases the collection burden on users. Third, keystroke dynamics can be used in an ongoing, real-time IDS mode to judge whether current behavior is consistent with a specific user's previous behavior. In contrast, in a typical username and password authentication scenario, such passive monitoring is not an option. Therefore, keystroke dynamics can serve to enhance security beyond the authentication phase.

Of course, there are also some disadvantages to using keystroke dynamics for authentication. One issue is that if a user has an injured hand or is simply distracted or overly emotional, their typing patterns may not be consistent with the patterns used for training. Furthermore, another disadvantage is that typing patterns may vary based on different keyboards, or even due to new applications or software updates, which indicates that models must be updated regularly. Although such concerns are legitimate, it is clear that these issues can be mitigated, and hence the utilization of keystroke dynamics is likely to increase in the near future.

In this research, we analyze various keystroke dynamics data and train machine learning and deep learning models to distinguish between users. Features include individual key presses and flight time, among others. Note that for the sake of user privacy, we do not store sequences of actual keystrokes, and hence the text itself is not used for modeling purposes.

We consider a wide variety of learning techniques, including k-nearest neighbors (k-NN), random forests, support vector machines (SVM), convolutional neural networks (CNN), recurrent neural networks (RNN), long short-term memory (LSTM) networks, extreme gradient boosting (XGBoost), and multilayer perceptrons (MLP).

Much of the previous research in this field is based on multiclass models trained on relatively small amounts of data per user. There are several inherent problems with such an approach. For example, if a new user is added, or the typing content (e.g., password) is changed, the model needs to be retrained. Furthermore, until recently, most work in this field considered only traditional statistical machine learning methods, with limited use of modern deep learning techniques. In contrast, we focus on modern machine learning and deep learning techniques, and we are able to improve on previous related work.

The remainder of this paper is organized as follows. Section 2 discusses relevant background topics, including introducing the learning techniques considered. We provide a survey of previous work in Sect. 3, while Sect. 4 describes the dataset used in our experiments. Our experimental results are presented in Sect. 5. Lastly, Sect. 6

summarizes our main results and we include a discussion of possible directions for future work.

2   Background

In this section, we discuss keystroke dynamics in general and we consider previous work in this field. In the next section, we introduce the dataset and the various machine learning models that we use in this research.

2.1   Keystroke Dynamics

According to [27], "keystroke dynamics is not what you type, but how you type." Most previous work on typing biometrics can be divided into either classification based on a fixed-text or authentication based on free-text [30]. For fixed text, the text used to model the typing behavior of a user and to authenticate the user is the same. This approach is usually applied to short text sequences, such as passwords. Classification can be based on various timing features related to the characters typed [10]. Moreover, by combining a password along with a username, such a system can be further strengthened [26]. A comprehensive discussion related to the fixed-text data problem can be found in [30].

As for the free text case, the text used to model typing behavior of a user and to authenticate the user is not necessarily the same. This approach is usually applied to long text sequences, and can be viewed as a continuous form of authentication or as an intrusion detection system (IDS). Again, in this paper we only consider the fixed-text problem.

Previously, many different distance-based methods have been applied to keystroke dynamics. More recently, machine learning techniques have been considered, including support vector machines (SVM), recurrent neural networks (RNN), and so on [38]. The learning techniques evaluated in this paper are introduced below.

2.2   Learning Techniques

For our experiments, we have considered a wide variety of learning techniques. We introduce these learning techniques in this section.

2.2.1 Random Forest

A random forest [14] is a supervised, decision tree-based machine learning method that is often highly effective for classification and regression tasks. This technique consists of a large number of individual decision trees, where each decision tree is based on a subset of the available features, and a subset of the training samples. The subsets used for each decision tree are selected with replacement. A majority vote or averaging of the component decision trees is used to determine the random forest classification.

2.2.2 Support Vector Machine

Support vector machines (SVM) [30] are a powerful class of supervised machine learning techniques. The key idea of an SVM is to construct a hyperplane, so that the data can be divided into categories [34]. The so-called "kernel trick" enables us to efficiently deal with nonlinear transformations of the feature data. As with random forests, SVMs often perform well in practice.

2.2.3 K-Nearest Neighbors

The k-nearest neighbors (k-NN) algorithm [24] is an intuitively simple technique, whereby we classify a sample based on the k nearest samples in the training set. In spite of its simplicity, k-NN often performs well, although overfitting is a concern, especially for small values of k. Both k-NN and random forest are neighborhood-based algorithms, although the neighborhood structure determined by each is significantly different.

2.2.4 T-SNE

The method of t-distributed stochastic neighbor embedding (t-SNE) is a non-linear dimensionality reduction technique that was originally proposed in [35]. It is typically used for data visualization, to reduce the dimensionality of the feature space, and for clustering. In contrast to the more well-known principal component analysis (PCA), t-SNE is better able to capture non-linear relationships in the data.

2.2.5 XGBoost

XGBoost, the name of which is derived from extreme gradient boosting, is a popular technique that has played an important role in a large number of Kaggle competitions. In comparison to the simpler AdaBoost technique, XGBoost has advantages in terms of dealing with outliers and misclassifications.

Data augmentation consists of generating synthetic data based on an existing dataset. Such "fake" data can be used to make up for a lack of data for a given problem. Data augmentation has often proved valuable in practice. We consider data augmentation in our XGBoost experiments.

2.2.6 LSTM and Bi-LSTM

Long short-term memory (LSTM) is a highly specialized recurrent neural network (RNN) architecture that is able to better deal with the vanishing and exploding gradient issues that plague plain "vanilla" RNNs [34]. Consequently, LSTMs generally perform much better over longer sequences as compared to vanilla RNNs.

A bi-directional LSTM (bi-LSTM) combines two LSTMs, one computed in the forward direction and another computed in the backward direction. Bi-LSTMs are well-suited to sequence labeling tasks and have proven to be strong at modeling contextual information in natural language processing (NLP) tasks.

In our LSTM and bi-LSTM experiments, we consider two different encoding methods. In addition to the standard raw feature encoding, we also experiment with one-hot encoding. Assuming that a feature can take on m possible values, a feature value of k has a one-hot representation consisting of a binary vector of length m with a 1 in the kth position and 0 elsewhere. When training, one-hot encoding has a natural interpretation as a vector of probabilities, and hence it is well suited to training involving a softmax output layer, for example.

We also consider attention mechanisms. The idea of an attention mechanism is intuitively simple—we want to force the model to focus on some specific aspect of the training data. Attention is somewhat related to regularization, in the sense that we reduce the potential for over-reliance on some parts of the training data, which can lead to various pathologies, including overfitting.

2.2.7 Convolutional Neural Network

Convolutional neural networks (CNN) are designed to deal effectively and efficiently with local structure. CNNs have proven their worth in the realm of image analysis. Most CNN architectures include convolutional layers, pooling layers, and a fully-connected output layer.

2.2.8 Multi-Layer Perceptron

The structure of a generic multi-layer perceptron (MLP) includes an input layer, one or more hidden layers, and an output layer. Each node, or neuron, in a hidden layer includes a nonlinear activation function, which is the key to the ability of an MLP to deal with challenging data. To mitigate overfitting, we employ dropouts for regularization in our MLP experiments [23].

3   Previous Work

In this section, we first consider distance-based methods. Then we discuss more recent work that relies on various machine learning techniques.

The concept of keystroke dynamics first appeared in the 1970s and was focused on fixed-text data [12]. In subsequent years, Bayesian classifiers based on the mean and variance in time intervals between two or three consecutive key presses were applied to the problem [28]. The result in [28] claim a classification accuracy of 92% on a dataset with 63 users.

Typical of early work in this field are nearest neighbor classifiers based on various distance measures. Initially, Euclidean distance or, equivalently, the L_2 norm was used. In contrast to the L_2 norm, the L_1 norm (i.e., Manhattan distance) makes it easier to determine the contributions made by individual components, and it is more robust to the effect of outliers. In [24], it is shown that among all distance-based techniques, the best performance is obtained from a nearest neighbor classifier that uses a scaled Manhattan distance.

Neither the L_1 nor the L_2 norm deal effectively with statistical properties, and hence statistical-based distance measures have also been considered. For example, Mahalanobis distance has been widely used in keystroke dynamics research [4].

Recently, research in keystroke dynamics has been heavily focused on machine learning techniques. Such research includes k-nearest neighbors (k-NN) [37], K-means clustering [17], random forests [25], fuzzy logic [15], Gaussian mixture models [16], and many other approaches. In the remainder of this section, we discuss some relevant examples of machine learning based research focused on fixed-text keystroke dynamics.

In [36], support vector machines (SVM) are used to extract features from the data that are then used for classification. Another popular machine learning technique has been used in keystroke dynamics is hidden Markov models (HMM). An HMM includes a Markov process that is "hidden" in the sense that it can only be indirectly observed [33]. In [7], an HMM is used to learn the time intervals in keystroke dynamics.

A number of neural network architectures have also been applied in keystroke dynamics in recent years [5, 22]. Deep learning techniques have also been successfully applied to classification and have achieved better performance, as compared to previous techniques, such as those considered in [29]. Deep networks usually require a relatively long time to train, and hence Adam optimization and leaky rectified linear unit (leaky relu) activation functions are often used to speed up the learning process [23].

In [2], a genetic algorithm known as neuro evolution of augmenting topologies (NEAT) is considered. This algorithm achieves a high accuracy on a custom dataset.

In [8], keystroke dynamics authentication based on fuzzy logic is considered, and an accuracy of 98% is achieved. This model evolves in the sense that it can update keystroke templates when a user login is successful. The research in [21] uses extreme gradient boosting (XGBoost), random forest, multilayer perceptron (MLP), and other machine learning methods to perform multiclass classification on the Carnegie Mellon University (CMU) dataset, which is the same dataset considered in this paper. In [21], a highest accuracy of 93.79% is achieved using XGBoost. However, these authors do not discuss hyperparameter tuning, and thus it may be possible to improve on their results.

As the name suggests, the equal error rate (EER) is the point where the false acceptance rate (FAR) and false rejection rate (FRR), at which point the sum of the FRR and FAR is minimized. The value of the EER is serves as an indicator of the performance of a system, enabling the direct comparison of different biometrics— the lower the value of EER, the better the performance of the system. The EER is easily obtained from an ROC curve.

The authors of [6] propose using convolutional neural networks (CNN) for authentication based on keystroke dynamics. Their model architecture is very similar to that in [18], with the main ideas deriving from a sentence classification task. They feed time-based feature vectors into the model directly instead of reshaping the vectors into matrices. They also explore the influence of different kernel sizes, different numbers of kernels, and different numbers of neurons in the fully connected layer. Their model is evaluated on an open fixed-text keystroke dataset, and their best equal error rates (EER) are 2.3 and 6.5% with and without data augmentation, respectively.

Time-based features and pressure-based features are considered in [1]. By combining the information of these two kinds of features, the authors achieve good performance. In addition, they deal with typos—when a typo is recognized, the duration of keystroke time between the wrong key and back-space key is ignored, as is the duration between the back-space key and the correct key.

Another study considers deep belief networks (DBN) to extract hidden features, which are then used to tune a pre-trained neural network [9]. The authors of [9] claim that deep learning techniques significantly outperform other algorithms on the CMU fixed-text dataset.

The CMU keystroke dataset is a well known public fixed-text dataset and has been extensively studied. The use of a common dataset enables research to be directly compared. In [24], the authors introduce this dataset and achieve a baseline result with an EER of 9.6%. There are now many studies that use this same dataset and outperform this baseline result. For example, in [6], the authors obtain an EER of 2.3%, based on a CNN with data augmentation, while in [23], an EER of 3% is attained using a multi-layers perceptron (MLP).

As an aside, we note that other keystroke features might be of interest. For example, keystroke acoustics for user authentication are considered in [32]. In this research, a dataset containing 50 users results in an EER of 11%, which shows that acoustical information can be informative. However, an advantage of keystroke dynamics is that such information is easily collected directly from any standard keyboard.

Table 1 Keystroke features in CMU dataset

Notation	Number	Summary	Description
H	11	Hold time	The length of time that a key is pressed
DD	10	Down-down	The length of time from one key press to the next key press
UD	10	Up-down	The length of time from one key being released until the next key is pressed
Total	31	–	—

4 Dataset

The Carnegie-Mellon University (CMU) fixed-text dataset is used for all experiments considered in this paper. The CMU dataset commonly serves to benchmark techniques in keystroke dynamics research [3, 6, 11, 13, 21, 23, 31]. This dataset includes 51 users' keystroke dynamics information, where each user typed the password ".tie5Roanl" a total of 400 times, consisting of 50 repetitions over each of 8 sessions. Between sessions, a user had to wait at least one day, so that the day-to-day variation of each subject's typing was captured [24]. Furthermore this password was chosen to be representative of a strong 10-character password, as it contains a special symbol, a number, lowercase letters, and a capital letter. Each time this password is typed, 31 time-based features were collected, as listed in Table 1. Note that the Enter key is pressed after typing the 10-character password. Hence, there are 11 keystrokes, consisting of 10 consecutive pairs.

Individual keystrokes in a sequence can be viewed as words in a sentence, in the sense that we can tie the UD-time and DD-time from two adjacent keystrokes with the duration of the previous keystroke. Following this approach, for each keystroke, we obtain a vector consisting of three features, which we interpret as an 11×3 matrix. Thus, our feature "vectors" consist of a sequence of these matrices. We refer to this matrix as the "fixed keystroke dynamics sequence," which we abbreviate as fixed-KDS.

5 Experiments and Results

This section contains the results of our fixed-text experiments on the CMU dataset. We provide some analysis and discussion of our results.

As mentioned above, in the CMU dataset, the data is arranged as a table with 31 columns, representing the collected information for one timing of the password. For example, one column is H.period which is the hold time for the "." key. The hold time is the length of time when the key was depressed. Another example is the column DD.period.t, is the time interval between when the "." key was pressed until the "t" key was pressed. The overall table is $20,400 \times 31$, where each

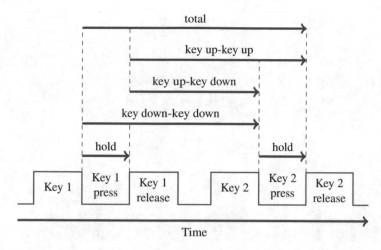

Fig. 1 Keystroke dynamics features

row corresponds to the timing information for a single repetition of the password by a single subject. Figure 1 illustrates the timing relationship between consecutive keystrokes.

5.1 Data Exploration

There are 31 timing features in the CMU dataset, which can be divided into three groups which we denote as DD, UD and H. Here, we analyze the data to determine whether there is any significant difference among these three groups. For this data exploration, we have randomly selected six of the 51 subjects for analysis.

In Fig. 2a, each line graph represents the 400 input feature vectors corresponding to a given subject. From this figure, we observe that most of the feature vectors are fairly consistent in that they follow a similar pattern for a given subject. This indicates that subjects tend to be relatively consistent with respect to this particular feature group. This observation can be seen as a positive indicator of the potential to successfully classify the subjects. However, when the six subjects' average cases are compared in Fig. 2b, the results show that the subjects have somewhat similar typing patterns.

The analogous results for the key-up key-down features are shown in Fig. 3. We observe that this data is similar to key-down key-down data in Fig. 2.

In Fig. 4a, we compare the six subjects based on the hold-time feature, and here the differences are more pronounced. In particular, the average cases in Fig. 4b reveal more substantial differences. These results indicate that the hold duration should be a strong feature for distinguishing users.

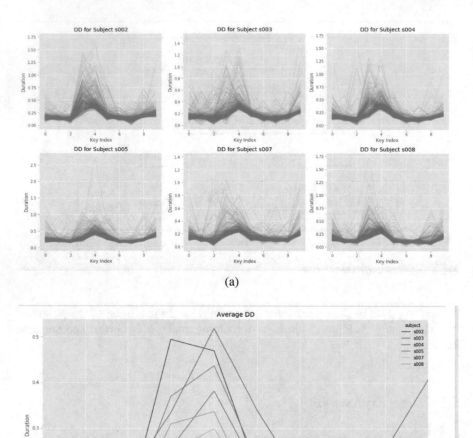

(a)

(b)

Fig. 2 Key-down key-down for six subjects (400 keystrokes). (**a**) Individual. (**b**) Average

To further explore the data, we apply t-SNE as a clustering technique to gain insight into how the data is distributed. In this case, we consider a subset consisting of the first seven subjects, using all 400 records for each of these subjects. The result in Fig. 5 show that the subjects can be clustered into different groups. This is again promising, as it indicates that we should have success in distinguishing users.

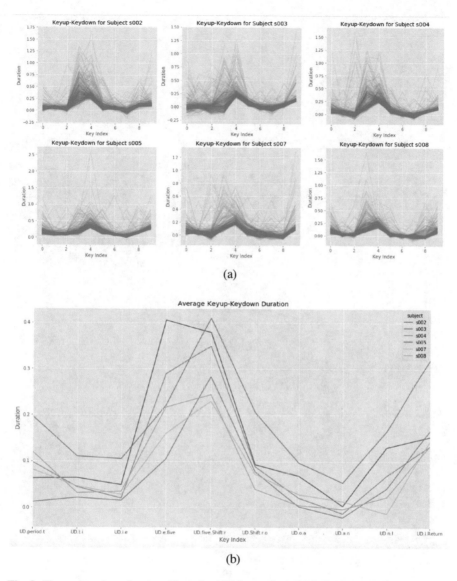

Fig. 3 Key-up key-down for six subjects for (400 keystrokes). (**a**) Individual. (**b**) Average

5.2   Classification Results

In this section, we give our classification results. Here, we experiment with k-NN, random forest, SVM, XGBoost, MLP, CNN, RNN, and LSTM.

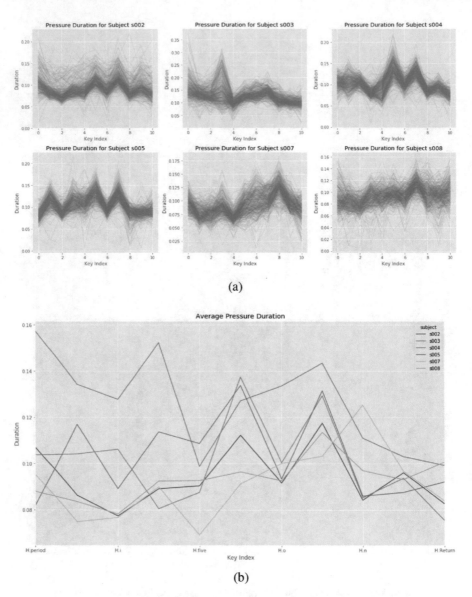

Fig. 4 Hold time for six subjects (400 iterations). (**a**) Individual. (**b**) Average

5.2.1 *K*-Nearest Neighbor Experiments

We optimize with respect to three parameters of the *k*-NN algorithm, namely, the number of neighbors, the weight function used for prediction, and the distance category. As in all of our parameter tuning experiments, we employ a Bayes model to generate a suit of parameters with the highest probability being the best result.

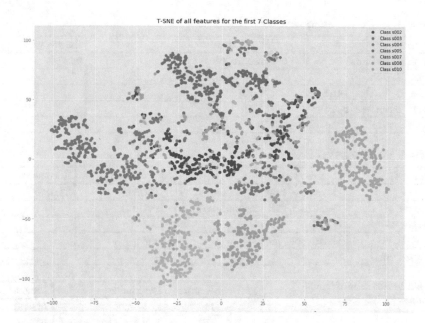

Fig. 5 T-SNE of features of seven subjects

Table 2 Results for k-NN

Parameter	Search space	Accuracy
n_neighbors	[**5**,50]	82.27%
weight	Uniform, **distance**	
p	[**1**, 2, 3]	
Experiments	50	

Table 2 shows the search space for each parameter and the best accuracy achieved. Boldface entries in Table 2 are used to indicate the optimal parameter values.

5.2.2 Random Forest Experiments

We optimize four parameters of the random forest algorithm, namely, the number of decision trees, the maximum depth of each decision tree, the minimal number of samples in a leaf node, and the minimum number of samples required to split. Again, we make use of different combinations of values of these parameters to build a Bayes model, which generates a set of parameters that will, with high probability, yield the best result. Table 3 shows the range considered for each of these parameters, the optimal values that we found (in boldface), and the best result obtained.

Table 3 Results for random forest

Parameter	Search space	Accuracy
n_estimaters	[100, **1000**]	93.55%
max_depth	None, 5, 10, 15, 20, 30, **35**, 40	
min_samples_leaf	[**1**, 10]	
min_samples_split	[**2**, 5]	
Experiments	50	

Table 4 Results for SVM

Parameter	Search space	Best value	Accuracy
C	Real(1e-6, 1e+6, log-uniform)	920,319	88.02%
Gamma	Real(1e-6, 1e+1, log-uniform)	0.61620	
Degree	[1, 8]	8	
Kernel	linear, poly, rbf	rbf	
Experiments	50	–	

Table 5 Accuracy of four features for XGBoost

Feature	Description	Accuracy
H	Hold time	76.91%
DD	Key-down key-down	76.39%
UD	Key-up key-down	81.10%
All	H, DD and UD	95.15%

Table 6 Selected parameters for XGBoost

Parameters	Value
learning-rate	0.21
n-estimators	1000
max-depth	2
min-child-weight	1.4

5.2.3 Support Vector Machine Experiments

Here, we consider four parameters of an SVM, namely, the value of the regularization parameter, the kernel function, and the two coefficients of the kernel function. Again, a Bayes model is built to search the optimal values of these parameters. The search space for each parameter, the optimal values, and the best accuracy are given in Table 4.

5.2.4 XBGoost Experiments

Next, we classify the samples using XGBoost. Here, we consider each of the three feature groups (DD, UD, and H) individually, as well as the combination of all three. The multi-classification results for the 51 subjects are shown in Table 5 and the model parameters used to achieve these results are given in Table 6.

Table 7 Results for XGBoost

Description	Data size	Accuracy
No augmentation	16,320	95.42%
Augmentation	48,960	96.39%

Table 8 Results for MLP

Model	Parameters				
	Input-channel	Output-channel	Num-layers	Learning-rate	Accuracy
MLP	31	100	3	0.001	95.96%

Based on these results, we conduct further experiments with XGBoost. Given the fairly limited size of the training data, we apply a simple data augmentation strategy—we randomly perturb each timing feature based on a range of $(-0.02, 0.02)$. In this experiment, we set the augmentation ratio to two, meaning that the amount of augmented date is two times the amount of original data. We find that this data augmentation provides a slightly improvement in the accuracy, as shown in Table 7.

5.2.5 Multilayer Perceptron Experiments

Our generic MLP consists of four fully connected layers, in which the number of neurons are 512, 256, 144, and 51, respectively. The output of the last layer is fed into a softmax function to calculate the corresponding probability for each class. A rectified linear unit (relu) activation function and a batch normalization layer are used in the first and second dense layers. We use the cross entropy loss function for this model—additional parameters are listed in Table 8. This MLP model yields an impressive accuracy ot 95.96%.

5.2.6 Convolutional Neural Network Experiments

The input for our CNN model is the fixed-KDS data structure, which we discussed in Sect. 4. The architecture of our CNN is based on that of the so-called textCNN in [18], which is used to process sequential text data. The key idea is to apply multiple rectangular kernels, instead of more typical square kernels. Specifically, the width of all kernels is the same as the embedding size for each word, so the output for each convolution is a one-dimension vector. Then multiple max-pooling layers are used to process these vectors to yield one feature for each kernel. Finally, these generated features are concatenated into a one-dimension vector, and multiple fully-connected layers are used to produce the class prediction. Our CNN model is illustrated in Fig. 6.

In our keystroke dynamics model, we view each keystroke event as a "word" and each keystroke sequence as a "sentence." In this way, six different convolution

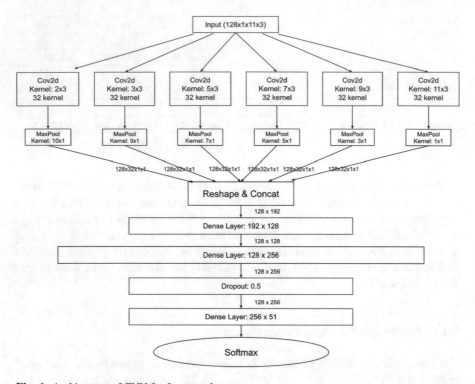

Fig. 6 Architecture of CNN for free-text datasets

kernels are applied to this sequential data, and continuous max-pooling layers extract the most important feature from each kernel. Then the concatenated vector is fed into three dense layers, with a softmax function is used to generate the probability for each class. In addition, a dropout layer is added after the penultimate layer. The cross entropy loss function is used. For these CNN experiments, the best result we obtain is an accuracy of 92.57%.

5.2.7 Recurrent Neural Network Experiments

The architecture of our RNN-based neural network is shown in Fig. 7. The input data for this model is the fixed-KDS, as discussed in Sect. 4. The idea behind this model comes from the field of sentiment analysis. Since keystroke data is inherently sequential, we applying a two-layers bi-directional RNN. In this experiment, the cross entropy loss function is used, and the best result we obtain is an accuracy of 93.45%.

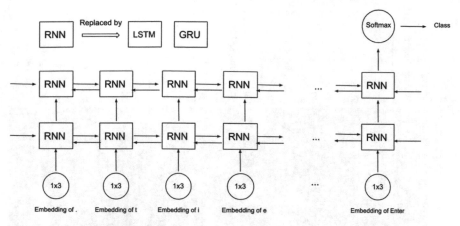

Fig. 7 Architecture of bi-RNN

Table 9 Results for LSTM and bi-LSTM with one-hot encoding

Model	Parameters				Accuracy
	Input-size	Hidden-size	Num-layers	Learning-rate	
LSTM	31	5	1	0.3	91.28%
Bi-LSTM	31	5	1	0.3	90.02%

5.2.8 LSTM Experiments

Next, we apply both LSTM and bi-LSTM with one-hot encoding. In these experiments, one-hot encoding is applied on both the subject and the timing features, which then serve as the feature vectors for the LSTM and bi-LSTM. The accuracies we obtain for LSTM and bi-LSTM are shown in Table 9. Although these results are reasonably strong, they are not competitive with our XGBoost experiments.

We further consider a bi-LSTM with attention, primarily as a way of analyzing feature importance. The attention matrix in the form of a heatmap, appears in Fig. 8. This matrix consists of the weights determined by the attention layer. In this matrix, the x-axis represents the 31 features, while the y-axis is based on 20 consecutive training samples. We observe that after several epochs, the attention seems to have a tendency to converge to specific features—at the end of the training, we find the most significant features are DD.period.t, DD.e.five, UD.Shift.r.o, DD.n.l, and DD.l.Return.

5.3 Summary and Discussion

We summarize our experimental results for the CMU fixed-text dataset in Fig. 9. The result shows that among all models we have considered, XGBoost with

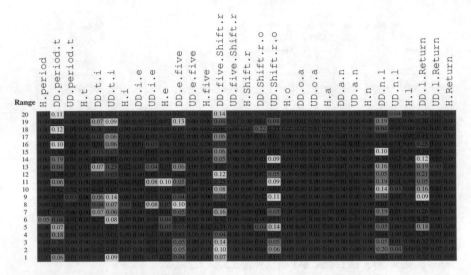

Fig. 8 Attention matrix heatmap

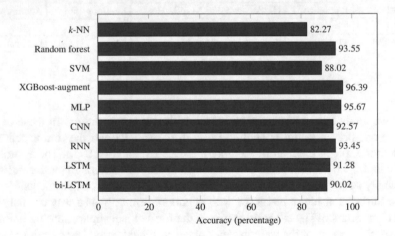

Fig. 9 Summary of our results

data augmentation, denoted XGBoost-augment, achieves the highest accuracy at 96.39%.

While XGBoost with data augmentation achieves the best results, MLP does nearly as well. When comparing the training times for these two models, we find that XGBoost with data augmentation take 18 minutes to train, while MLP requires about half an hour. Both of these are very reasonable training times, but if great efficiency during training is required, then XGBoost with data augmentation may be the better choice.

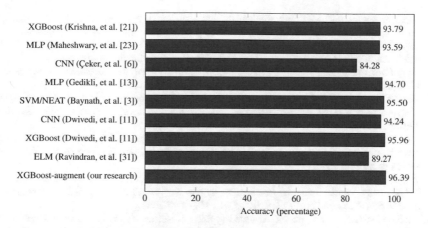

Fig. 10 Comparison to previous work

In Fig. 10 we provide a comparison of our best result to previous work. We see that our best accuracy of 96.39% offers a modest improvement over previous work in this field.

6 Conclusion and Future Work

In this paper, we tested and analyzed a wide variety of machine learning techniques for biometric authentication based on fixed-text typing characteristics. We found that XGBoost with data augmentation performed best, with MLP performing nearly as well. Our results improved upon previous research involving the same dataset.

There are many avenues available for future work. For example, we model optimization and model fusion would be interesting. For model optimization, we could consider techniques from contrastive learning and self-supervised techniques to see whether these approaches can improve our model.

As another example of possible future work, the robustness of various techniques can be evaluated using an algorithm known as POPQORN [19]. The idea behind POPQORN is to observe the effect of outside disturbances to the model and thereby measure its robustness.

References

1. Faisal Alshanketi, Issa Traore, and Ahmed Awad Ahmed. Improving performance and usability in mobile keystroke dynamic biometric authentication. In *2016 IEEE Security and Privacy Workshops*, SPW, pages 66–73, 2016.
2. Purvashi Baynath, K. M. Sunjiv Soyjaudah, and Maleika Heenaye-Mamode Khan. Machine learning algorithm on keystroke dynamics pattern. In *2018 IEEE Conference on Systems, Process and Control*, ICSPC, pages 11–16, 2018.

3. Purvashi Baynath, K. M. Sunjiv Soyjaudah, and Maleika Heenaye-Mamode Khan. Machine learning algorithm on keystroke dynamics fused pattern in biometrics. In *2019 Conference on Next Generation Computing Applications*, NextComp, pages 1–6, 2019.
4. Saleh Ali Bleha, Charles Slivinsky, and B. Hussien. Computer-access security systems using keystroke dynamics. *IEEE Transactions on Pattern Analysis and Machine Intelligence*, 12(12):1217–1222, 1990.
5. Marcus Brown and Samuel Joe Rogers. User identification via keystroke characteristics of typed names using neural networks. *International Journal of Man-Machine Studies*, 39(6):999–1014, 1993.
6. Hayreddin Çeker and Shambhu Upadhyaya. Sensitivity analysis in keystroke dynamics using convolutional neural networks. In *2017 IEEE Workshop on Information Forensics and Security*, WIFS, pages 1–6, 2017.
7. Wendy Chen and Weide Chang. Applying hidden Markov models to keystroke pattern analysis for password verification. In *Proceedings of the 2004IEEE International Conference on Information Reuse and Integration*, IRI, pages 467–474, 2004.
8. Lucian Constantin. *PC World*: AI-based typing biometrics might be authentication's next big thing. https://www.pcworld.com/article/3162010/ai-based-typing-biometrics-might-be-authentications-next-big-thing.html, 2017.
9. Yunbin Deng and Yu Zhong. Keystroke dynamics user authentication based on Gaussian mixture model and deep belief nets. https://www.hindawi.com/journals/isrn/2013/565183/, 2013.
10. Paul S. Dowland and Steven M. Furnell. A long-term trial of keystroke profiling using digraph, trigraph and keyword latencies. In Yves Deswarte, Frédéric Cuppens, Sushil Jajodia, and Lingyu Wang, editors, *Security and Protection in Information Processing Systems*, pages 275–289. Springer, 2004.
11. Chaitanya Dwivedi, Divyanshu Kalra, Divesh Naidu, and Swati Aggarwal. Keystroke dynamics based biometric authentication: A hybrid classifier approach. In *2018 IEEE Symposium Series on Computational Intelligence*, SSCI, pages 266–273, 2018.
12. George E. Forsen, Mark R. Nelson, and Raymond J. Staron (Jr.). Personal attributes authentication techniques. https://books.google.com.tw/books?id=tbs4OAAACAAJ, 1977.
13. Ahmet Gedikli and Mehmet Efe. A simple authentication method with multilayer feedforward neural network using keystroke dynamics. https://web.cs.hacettepe.edu.tr/~onderefe/PDF/2019medprai2.pdf, 2020.
14. Ahmet Melih Gedikli and Mehmet Önder Efe. A simple authentication method with multilayer feedforward neural network using keystroke dynamics. In Chawki Djeddi, Akhtar Jamil, and Imran Siddiqi, editors, *Pattern Recognition and Artificial Intelligence*, pages 9–23. Springer, 2020.
15. S. Haider, A. Abbas, and A. K. Zaidi. A multi-technique approach for user identification through keystroke dynamics. In *2000 IEEE International Conference on Systems, Man and Cybernetics*, SMC 2000, pages 1336–1341, 2000.
16. Danoush Hosseinzadeh and Sridhar Krishnan. Gaussian mixture modeling of keystroke patterns for biometric applications. *IEEE Transactions on Systems, Man, and Cybernetics, Part C (Applications and Reviews)*, 38(6):816–826, 2008.
17. Pilsung Kang, Seong-seob Hwang, and Sungzoon Cho. Continual retraining of keystroke dynamics based authenticator. In Seong-Whan Lee and Stan Z. Li, editors, *Advances in Biometrics*, pages 1203–1211. Springer, 2007.
18. Yoon Kim. Convolutional neural networks for sentence classification. https://arxiv.org/abs/1408.5882, 2014.
19. Ching-Yun Ko, Zhaoyang Lyu, Tsui-Wei Weng, Luca Daniel, Ngai Wong, and Dahua Lin. POPQORN: Quantifying robustness of recurrent neural networks. In *Proceedings of the 36th International Conference on Machine Learning*, pages 3468–3477, 2019.
20. Marc Alexander Kowtko. Biometric authentication for older adults. In *IEEE Long Island Systems, Applications and Technology Conference 2014*, LISAT, pages 1–6, 2014.

21. Gutha Jaya Krishna, Harshal Jaiswal, P. Sai Ravi Teja, and Vadlamani Ravi. Keystroke based user identification with XGBoost. In *2019 IEEE Region 10 Conference*, TENCON 2019, pages 1369–1374, 2019.
22. Daw-Tung Lin. Computer-access authentication with neural network based keystroke identity verification. In *Proceedings of International Conference on Neural Networks*, ICNN'97, pages 174–178, 1997.
23. Saket Maheshwary, Soumyajit Ganguly, and Vikram Pudi. Deep Secure: A fast and simple neural network based approach for user authentication and identification via keystroke dynamics. http://iwaise.it.nuigalway.ie/wp-content/uploads/2017/02/DeepSecure.pdf, 2017.
24. Roy A. Maxion and Kevin S. Killourhy. Comparing anomaly-detection algorithms for keystroke dynamics. In *2009 IEEE/IFIP International Conference on Dependable Systems & Networks*, DSN, pages 125–134, 2009.
25. Roy A. Maxion and Kevin S. Killourhy. Keystroke biometrics with number-pad input. In *2010 IEEE/IFIP International Conference on Dependable Systems & Networks*, DSN, pages 201–210, 2010.
26. Soumik Mondal and Patrick Bours. Combining keystroke and mouse dynamics for continuous user authentication and identification. In *2016 IEEE International Conference on Identity, Security and Behavior Analysis*, ISBA, pages 1–8, 2016.
27. Fabian Monrose and Aviel Rubin. Authentication via keystroke dynamics. In *Proceedings of the 4th ACM Conference on Computer and Communications Security*, pages 48–56, 1997.
28. Fabian Monrose and Aviel D. Rubin. Keystroke dynamics as a biometric for authentication. *Future Generation Computer Systems*, 16(4):351–359, 2000.
29. Yohan Mulionoa, Hanry Hamb, and Dion Darmawan. Keystroke dynamic classification using machine learning for password authorization. *Procedia Computer Science*, 135:564–569, 2018.
30. Nataasha Raul, Radha Shankarmani, and Padmaja Joshi. A comprehensive review of keystroke dynamics-based authentication mechanism. In *International Conference on Innovative Computing and Communications*, pages 149–162, 2020.
31. Sriram Ravindran, Chandan Gautam, and Aruna Tiwari. Keystroke user recognition through extreme learning machine and evolving cluster method. In *2015 IEEE International Conference on Computational Intelligence and Computing Research*, ICCIC, pages 1–5, 2015.
32. Joseph Roth, Xiaoming Liu, Arun Ross, and Dimitris Metaxas. Investigating the discriminative power of keystroke sound. *IEEE Transactions on Information Forensics and Security*, 10(2):333–345, 2015.
33. Mark Stamp. A revealing introduction to hidden Markov models. https://www.cs.sjsu.edu/~stamp/RUA/HMM.pdf, 2004.
34. Mark Stamp. *Introduction to Machine Learning with Applications in Information Security*. Chapman and Hall/CRC, Boca Raton, 2017.
35. Laurens van der Maaten and Geoffrey Hinton. Visualizing data using t-SNE. *Journal of Machine Learning Research*, 9(86):2579–2605, 2008.
36. Enzhe Yu and Sungzoon Cho. GA-SVM wrapper approach for feature subset selection in keystroke dynamics identity verification. In *Proceedings of the International Joint Conference on Neural Networks*, pages 2253–2257, 2003.
37. Robert S. Zack, Charles C. Tappert, and Sung-Hyuk Cha. Performance of a long-text-input keystroke biometric authentication system using an improved k-nearest-neighbor classification method. In *2010 Fourth IEEE International Conference on Biometrics: Theory, Applications and Systems*, BTAS, pages 1–6, 2010.
38. Yu Zhong and Yunbin Deng. A survey on keystroke dynamics biometrics: Approaches, advances, and evaluations. https://sciencegatepub.com/books/gcsr/gcsr_vol2/GCSR_Vol2_Ch1.pdf, 2015.

Machine Learning-Based Analysis of Free-Text Keystroke Dynamics

Han-Chih Chang, Jianwei Li, and Mark Stamp

Abstract The development of active and passive biometric authentication and identification technology plays an increasingly important role in cybersecurity. Keystroke dynamics can be used to analyze the way that a user types based on various keyboard input. Previous work has shown that user authentication and classification can be achieved based on keystroke dynamics. In this research, we consider the problem of user classification based on keystroke dynamics features collected from free-text. We implement and analyze a novel a deep learning model that combines a convolutional neural network (CNN) and a gated recurrent unit (GRU). We optimize the resulting model and consider several relevant related problems. Our model is competitive with the best results obtained in previous comparable research.

1 Introduction

Recently, passive biometric authentication and identification techniques have received considerable attention. In this research, we consider such a passive biometric based on keystroke dynamics. The resulting technique is applicable to the authentication problem, and can also potentially play a role in intrusion detection.

Keystroke dynamics are based on typing behavior, such as the duration of keyboard events, the duration of key presses, the time difference between key presses, and so on. Such data can be collected from a standard keyboard by monitoring input and recording the time intervals between each keystroke. Keystroke dynamics may not be strong enough too be used as a standalone authentication system and hence it is typically combined with another type of authentication, such as a password. In its role as an IDS, keystroke dynamics may be competitive with other techniques.

H.-C. Chang · J. Li · M. Stamp (✉)
San Jose State University, San Jose, CA, USA
e-mail: han-chih.chang@sjsu.edu; jianwei.li@sjsu.edu; mark.stamp@sjsu.edu

© The Author(s), under exclusive license to Springer Nature Switzerland AG 2022
M. Stamp et al. (eds.), *Artificial Intelligence for Cybersecurity*, Advances in
Information Security 54, https://doi.org/10.1007/978-3-030-97087-1_14

Compared with most other biometric technologies (e.g., fingerprint), keystroke dynamics has advantages. For example, no special hardware is required to collect keystroke features, and keystroke data can be obtained in a passive manner, which reduces the burden on users. In addition, keystroke dynamics can be used as part of a real-time intrusion detection system (IDS)—in contrast to a typical username and password authentication system, where ongoing monitoring is not a realistic option. In addition to playing a role in authentication, keystroke dynamics can serve to enhance security after a user has been authenticated.

Disadvantages of keystroke dynamics based authentication might arise if a user has an injured hand, a user is distracted, or the hardware (e.g., keyboard) has changed. We believe that that these—and related—disadvantages can be overcome, and we expect that the use of keystroke dynamics in security applications will increase in the future.

In this paper, we analyze free-text keystroke dynamics data and train deep learning models to distinguish between users. The features we consider are related to keystroke timing, and are all obtained from standard keyboards. Furthermore, we do not utilize the characters that are typed, and hence user privacy is maintained.

We focus primarily on a novel model architecture that combines a convolutional neural network (CNN) and a gated recurrent unit (GRU). We experiment with non-timing features and we employ concepts from Siamese networks [10], and we experiment with pre-trained models and attention.

Here, we only consider free-text data [2]. As our free-text dataset, we use [33], which contains keystroke data from 148 participants. The primary goal of this research is to achieve a high accuracy from a biometric system that uses features derived from keystroke dynamics. Among other desirable properties, we would ideally like our models to be scalable and robust.

We experiment with a far wider variety of models and parameters than in any previous work. Among many other aspects of the keystroke dynamics problem that we consider, we determine feature importance for some of our complex models using "explainability" concepts. For our best model, we obtain extremely high accuracy, and an equal error rate that is comparable to previous work.

The remaining sections of this paper are organized as follows. Section 2 discusses relevant background topics, with the focus on a survey of previous work. Section 3 describes the dataset used in our experiments and outlines the machine learning techniques that we employ. Our free-text experimental results are presented in Sect. 4, while Sect. 5 summarizes our main results, and include suggestions for future work.

2   Background

In this section, we discuss keystroke dynamics and we provide a selective survey of previous work in this field. We also introduce the machine learning models that we employ in this research.

2.1   Keystroke Dynamics

According to [25], "keystroke dynamics is not what you type, but how you type." Most previous work on typing biometrics can be divided into either classification that relies on fixed-text or authentication based on free-text data [28]. For fixed-text, the text used to model the typing behavior of a user and to authenticate the user is the same (e.g., a password). This approach is usually applied to short text sequences. Classification is generally based on timing features related to the character typed [7]. An thorough discussion of the fixed-text problem can be found in [28].

For the free-text case, the text used to model typing behavior of a user and the text used to authenticate the same user is, in general, not the same. Free-text usually implies long text sequences, and can be viewed as a continuous form of authentication or as part of an IDS.

In the past, distance based methods were popular for the analysis of keystroke dynamics. More recently, machine learning techniques have come to the fore, including support vector machines (SVM), recurrent neural networks (RNN), hidden Markov models (HMM), k-nearest neighbors (k-NN), multilayer perceptrons (MLP), and so on [36].

Next, we discuss relevant previous work. Then, in the subsequent section, we the dataset and learning techniques used in our experiments.

2.2   Previous Work

In this section, we first consider distance-based methods. Then we discuss more recent work that is based on a wide variety of (mostly) modern machine learning techniques.

2.2.1   Distance Based Research

The concept of keystroke dynamics can be traced back to the 1970s, at which time the analysis was focused on fixed-text data [8]. Subsequently, Bayesian classifiers based on the mean and variance in time intervals between two or three consecutive key presses were applied to keystroke dynamics data [26]. For example, in [26] the authors claim an accuracy of 92% over a dataset consisting of 63 users.

Typical of relatively early work in this field are nearest neighbor classifiers based on distance measures. Euclidean distance or, equivalently, the L_2 norm was often used. More success was found using the L_1 norm (i.e., Manhattan distance), which makes it easier to single out the contribution made by individual components, In addition, the L_1 norm is more robust with respect to outliers. In [21], the best result obtained from a distance-based technique uses a nearest neighbor classifier based on a scaled Manhattan distance. Subsequently, statistical-based distance measures—

such as Mahalanobis distance—were used with success in keystroke dynamics research [3].

The equal error rate (EER) is a point on the ROC curve where the sum of the false accept rate (FAR) and false rejection rate (FRR) is minimized. The EER is a commonly used measure of success for biometric systems—the lower the value of EER, the better the performance of a system. For example, in [21] the authors achieve an equal error rate (EER) of 0.096.

2.2.2 Machine Learning Based Research

Recently, research in keystroke dynamics has been dominated by machine learning techniques. Such techniques include K-nearest neighbors (k-NN) [35], K-means clustering [16], random forests [22], fuzzy logic [11], Gaussian mixture models [14], and many, many more.

In comparison to the fixed-text problem, the number of research studies involving free-text data is much smaller. In [34] it is claimed that the amount of research done with fixed-text was eight times as much as that for free-text, as of 2013.

Free-text presents several challenges as compared to fixed-text. For example, in free-text, the number of keys typed can differ greatly. There may also be word-specific dependencies in free-text [31] that would not be relevant in fixed-text.

As an aside, we note that other keystroke features might be of interest. For example, keystroke acoustics are considered in [30], where a dataset containing 50 users yields an EER of 11%. This results shows that acoustic-based typing information can be useful for authentication. However, an advantage of keystroke dynamics based on timing features is that such information can be easily collected from any standard keyboard.

3 Implementation

In this section, we first introduce the keystroke dynamics dataset considered in this research. Then we discuss the various learning techniques that we have applied to this free-text dataset.

3.1 Dataset

For our free-text data, we choose to use the so-called Buffalo dataset [33], which was collected by researchers at SUNY Buffalo. This dataset contains long fixed-text and free-text keystroke data from 157 subjects, with each subject using the keyboard over three sessions [33]. For the fixed-text, users were requested to type Steve Jobs' Stanford commencement speech, which was split into three pieces. In free-text, users are requested to answer two survey style questions and one scene description. The time duration within each session was about 50 minutes, with about 5700 keystrokes, on average, and hence over 17,000 for the three sessions combined. Furthermore, there was a 28 day time interval between sessions, and four different types of keyboards were used. In this paper, we only consider the Buffalo free-text keystroke data.

Note that the Buffalo free-text dataset is divided into two subsets, referred to as a "baseline" subset and a "rotation" subset. In the baseline subset, there are 75 users using the same type of keyboard across all three sessions. For the rotation subset, there are 74 users using three different types of keyboard across their three sessions. The data collected on each keystroke consists of the name of the key, the key event (key-down or key-up), and a timestamp (measured in milliseconds).

3.2 Techniques Considered

In our free-text experiments, we also employ machine learning models that are somewhat more complex and experimental in nature, as compared to those typically used in comparable research. Next, we briefly discuss both of the models we consider.

3.2.1 BERT

Bidirectional encoder representations from transformers (BERT) is a language model developed by Google [6] that is designed to serve as an encoder in an encoder-decoder model. The BERT encoder converts words into vector representations that a corresponding decoder can then use to generate the output. BERT training is divided into two major steps which are known as pre-training and fine-tuning. In the pre-training phase, Google has used a large amount of text data to train the model in an unsupervised manner. In fine-tuning, labeled training data is used to adapt the model to a specific problem. In our experiments, we use the pre-trained model and the fine-tuning is performed based on the words typed by a user.

3.2.2 CNN-GRU Model

We propose a novel hybrid CNN-GRU model that is designed to learn from a sequence of individual keystroke features. This model was inspired by related work in [20].

The core idea of using a GRU is that it can take advantage of sequential information in a user's typing behavior. Since a GRU is a type of recurrent neural network, it has the ability to learn the current characteristics of the input based on previous characteristics. In addition, we use a CNN before the GRU, with the aim of providing enhanced features to the GRU. In effect, this CNN step can be viewed as a form of feature engineering. In our CNN, the length of the convolutional kernel corresponds to the number of sequences that are covered. The convolution operation has the ability to produce a "higher-level" keystroke signature. Subsequently, these signatures serve as the input to the GRU. After training the GRU, a user's keystroke behavior pattern is obtained.

We also implement dropout within the proposed model. The idea of dropout was introduced in [32] as a regularization technique for deep neural network. As the name indicates, for dropouts we randomly drop nodes (neurons) along with their connections from the neural network during training. This has the effect of training each mini-batch over a different network. Dropouts serve to prevent overfitting by forcing nodes that would likely otherwise atrophy to be active in the learning process.

For this proposed model, we use the `BCEWithLogitalLoss` activation function, as opposed to the more typical sigmoid. According to [27], `BCEWithLogitalLoss` is more stable than sigmoid or `BCELoss`.

4 Free-Text Experiments

This section provides our experiments and results for the machine learning techniques discussed above when applied to the Buffalo free-text dataset. We also provide some analysis of our experiments.

The Buffalo free-text dataset contains three different sessions. Thus, we perform threefold cross validation with each session serving as a fold. We use the notation s01-train-s2-test to mean that we use sessions 0 and 1 for training, with session 2 reserved for testing—the notation for the other folds is analogous.

4.1 Text-Based Classification

First, we attempt to classify based on the text typed by users. Ultimately, we want to classify users based on their keystroke dynamics, but by first focusing solely on the characters typed, we can see how much information is contained in users' differing responses, as opposed to their typing characteristics.

For our text-based experiment, we use BERT for multi-classification of the 148 users in the free-text part of the Buffalo dataset, based on words typed. This experiment, which we refer to as BERT-word, was complicated by various typos that had to be corrected.

The result of this experiment is extremely poor, and testing loss does not decrease significantly. One problem may be that the dataset is too small, as the training data for each user consists of only about 15 lines of words, with 2 lines for testing. Note

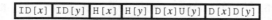

Fig. 1 Keystroke timing feature vector

that each line contains about 18 words. In a typical NLP application, we would have hundreds of times more data for training and testing.

The bottom line is that we are not able to classify users based on the actual text that was typed with an accuracy greater than guessing. Although additional experiments might be helpful, it appears that there is little useful information contained in the raw text. We now turn our attention to keystroke dynamics based models.

4.2 Keystroke Dynamics Models

Here, we apply and analyze our novel CNN-GRU architecture, which we outlined in Sect. 3.2.2, above. This model is based only on keystroke dynamics, as opposed to the actual text typed by a user.

The goal here is to build a model that could be used as part of an ongoing intrusion detection system (IDS). That is, the model would be used to periodically verify the identity of a user in real time.

For our dataset, we employ the baseline subset of the Buffalo free-text, in which each of 75 users typed across three sessions. After determining the basic parameters of our model, we consider a wide variety of modifications.

4.2.1 Features

We consider three types of features in our experiments, namely, timing features, rate features (discussed below), and "fusion." For the fusion case, we simply combine the timing and rate features.

First, we consider timing features only. In this case, we transform the free-text keystroke data into a fixed-length keystroke sequence, then further convert the sequence into a keystroke vector. The format of the keystroke vector is presented in Fig. 1, where x and y are consecutive keystrokes. Note that x and y are simply numeric values representing the position in the keystroke sequence—the key that was pressed in not recorded. That is, we do not use what was typed as a feature, just how it was typed. This assures that a user's typing is not revealed by our analysis.

The notation H $[x]$ is the hold duration of the key, U $[x]$ is the key-up time (the timestamp when key is released), and D $[x]$ is the key-down time (the timestamp of the key was pressed). Therefore, D $[x]$ U $[x]$ is the time duration that a key was pressed until it was released, and D $[x]$ D $[y]$ is the time duration between two consecutive keys being pressed. In all cases, x and y indicate any key being pressed.

After obtaining the keystroke vectors, we normalize the timing features. As discussed in [37], this normalization results in features with mean 0 and variance 1.

Next, we consider "rate" features. Traditionally, keystroke dynamics is based on timing, as discussed above. However, other typing habits can also serve as indicators of typing behavior. For example, features that relate to the use of the left and right hands may help the model to distinguish between users.

In addition to conventional timing features, we further consider seven features consisting of the rate at which various keys are pressed. Specifically, we consider rate at which each of the delete, left shift, right shift, left caps, right caps, control key, and the combination of left and right arrow keys are pressed. These features will surely be useful for distinguishing a user's handedness, but they may also be useful as more general indications of typing style.

As a first experiment, we consider different numbers of key-strokes for analysis. Specifically, we consider conventional key-length 100 and rate features with key-length 500. For each user, we append the rate features after the timing features. We have tested models with three different kinds of features, namely, timing only, rate only, and combined. The accuracy of the timing feature only is 84.72%, while the accuracy improves to 90.82% after combining with the rate features.

For some users, the rate features make a large difference, but for others they actually make the results worse. For example, user 1 with timing feature alone has an accuracy is only about 60%. But if we combine with the rate feature, accuracy increases to 95%. In this case, the mixed features seem to be effective. However, for user 7, the timing feature alone yield an accuracy of 82%, but if we include the rate features, the accuracy drops to 50% for the fusion case, while considering rate alone, the accuracy is 87.5%. In some cases it is better to use timing alone while in some other cases, it is better to combine both, and yet other cases it is best to just use rate features. But in most cases, the accuracy is increased when using the combined features. The result of first seven users is shown in Fig. 2.

Finally, we consider feature "fusion," that is, we combine timing and rate features, using the same key length for both. We have also experimented with independent key-lengths, but this does not improve on the results presented here.

Next, we combine the rate features with the timing feature using the idea from Siamese network [10]. A Siamese network has two inputs, which are fed into two neural networks that, respectively, map the inputs into a new space. A distance-related loss function is used to train the network parameters, so that the trained network can measure the similarity of the two inputs. For the rate features, we apply a linear transformation using the `torch.nn.Linear()` module from `PyTorch`, while conventional features use a multi-kernel CNN-GRU. We then concatenate the two features together and pass them through a fully connected layer to obtain the desired output.

The results of our fusion experiments are summarized in Table 1. We observe that when we apply longer key lengths, the result tend to improve. However, the reason we stop at key length 250 is that some users typed much less than was typically the case. Regardless, these experiments show that the this fusion approach is effective, and achieves very high accuracy for this authentication problem.

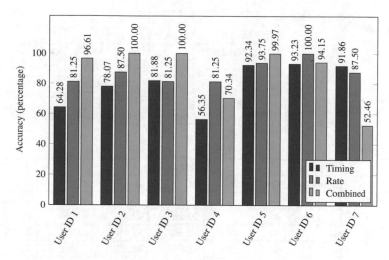

Fig. 2 Accuracy for different users

Table 1 Feature fusion for different key lengths

Model	Parameters		Test session			
	Kernel	Key-length	s0	s1	s2	Average
CNN-GRU	2,2,2	50	89.2%	89.7%	89.6%	89.5%
	2,2,2	150	92.5%	93.1%	93.7%	93.1%
	2,2,2	200	93.4%	93.6%	94.3%	93.8%
	2,2,2	250	94.6%	94.1%	93.8%	94.2%

Table 2 Best result for parameter tuning of kernel

Model	Parameters			Accuracy
	Kernel	Out-channel	RNN-size	
CNN-GRU	2,2,2	32	8	92.10%

4.2.2 Parameter Tuning

Next, we perform parameter tuning on our CNN-GRU model. The hyperparameters that we vary include the learning rate, kernel size of the CNN, and keystroke length, among others. Since our model is implemented in Cuda [12], we are able to efficiently test many parameter values.

As discussed above, our original model uses a single kernel CNN-GRU model and achieves an average accuracy of 84.7%. We consider a multi-kernel CNN, where the kernel is a list so that different combinations can be tested. When we include the rate features, we obtain a best average accuracy of 92.1% using the model parameters in Table 2.

In a multi-kernel model, when the kernel sizes differ, features are observed on different scales—when the kernel size is large, the receptive field is bigger and more of the input is observed. The tradeoff is that larger kernel sizes tend to result

Table 3 Parameter tuning CNN kernel

Model	Kernel	Test session			Average
		s0	s1	s2	
CNN-GRU	2	85.0%	85.6%	83.5%	84.7%
	2,4,6	89.2%	90.1%	90.7%	90.0%
	2,4,6,8	87.9%	88.1%	87.9%	88.0%
	2,2,2	91.7%	92.4%	92.3%	92.1%
	4,4,4	89.6%	90.4%	90.0%	90.3%

Table 4 Parameter tuning CNN out channel

Model	Parameters		Test session			Average
	Kernel	Out-channel	s0	s1	s2	
CNN-GRU	2	16	90.9%	90.4%	90.7%	90.6%
	2	32	92.4%	92.1%	91.4%	92.0%
	2	48	92.6%	92.4%	91.8%	92.3%
	2	64	92.6%	92.2%	91.8%	92.2%
	2	96	91.6%	92.3%	92.3%	92.1%
	2	128	92.1%	92.5%	91.6%	92.1%

Table 5 Parameter tuning CNN convolution

Model	Parameters		Test session			Average
	Conv.	Learning rate	s0	s1	s2	
CNN-GRU	3	0.001	91.1%	90.6%	89.9%	90.5%
	3	0.01	91.6%	92.3%	91.6%	91.8%
	6	0.01	92.1%	92.1%	91.8%	91.8%
	9	0.01	91.5%	92.1%	90.8%	91.5%

in overfitting. Furthermore, when we combine different kernels together, padding issues arise. The purpose of padding is to make the size of the feature map consistent with the size of the original image, with padding determined by the size of the filter and the size of the stride. Padding enables us to use all of the actual data.

Our multi-kernel experiment results are given in Table 3. None of these results improve on our best previous accuracy of 92.3%.

Another parameter of interest in our CNN is the "out channel" which is the number of channels produced by the convolution. The results obtained when experimenting with this parameter are given in Table 4. Here, we obtain a marginal improvement on our previous best accuracy.

We also conduct experiments varying the depths of convolutional layers. These experimental results for three to nine layer modes are summarized in Table 5. Somewhat surprisingly, these result indicate that higher layer models do not seem to improve over our 3-layer model. In the realm of future work, it would be interesting to experiment with other deep networks, such as ResNet, DenseNet, and SENet.

Table 6 Analysis of attention layer

Model	Parameters		Test session			
	Kernel	Attention	s0	s1	s2	Average
CNN-GRU	2	—	90.16%	88.37%	88.54%	89.02%
	2	✓	89.91%	89.17%	88.92%	89.00%
	2,2,2	—	91.70%	92.40%	92.30%	92.10%
	2,2,2	✓	92.00%	92.40%	92.50%	92.30%

Table 7 Attention and rate features (baseline subset)

Model	Hyperparameter	Value	Test session			
			s0	s1	s2	Average
CNN-GRU	CNN kernel	2	95.4%	94.5%	94.0%	94.6%
	CNN out	48				
	RNN size	8				
	Learning rate	0.001				
	Weight-decay	0.00001				
	Step-scheduler	70				
	Key-length	250				
	Epochs	80				
	Attention	Yes				
	Rate features	7				

Next, we experiment with an attention mechanism in both the single kernel and multi-kernel cases. When using conventional features only, we obtain a marginal improvement, as summarized in Table 6.

After experimenting with additional parameter tuning, we find the best model uses the parameters in Table 7. These parameters will be used in all subsequent experiments.

4.2.3 Fine Tuning

In this section, we consider a model for multi-classification of all 75 users (as discussed above) and then use this model in a pre-trained mode to construct a binary classification model. We refer to this two-step process as "fine tuning."

Note that here we consider binary classification of each user, and hence each user will have their own model. We then consider the average case for each of the resulting 75 models to obtain our accuracy results. In the multi-classification stage, we construct a single model among 75 users that we then use as a pre-trained model to construct each of the binary classifiers via "fine-tuning."

We have experimented with a wide variety of parameters at the multiclass stage. We obtain a best result of 76.96% using the parameters shown in Table 8.

Table 8 Best parameters for multi-classification (baseline subset)

Model	Hyperparameter	Value	Test session		
			s0	s1	s2
CNN-GRU	CNN kernel-size	3	77.74%	77.38%	76.96%
	CNN out-channel	192			
	RNN size	32			
	Learning rate	0.01			
	Weight-decay	1e−5			
	Step-scheduler	[80,350,390]			
	Key-length	250			
	Epochs	400			

Table 9 Fine-tune results (baseline subset with freeze)

Model	Parameters		Test session			
	Freeze	Learning rate	s0	s1	s2	Average
Fine-tune	—	0.01	96.7%	97.4%	97.1%	97.1%
	—	0.001	97.3%	97.3%	96.9%	97.2%
	—	0.0001	94.5%	95.2%	94.6%	94.7%
	✓	0.01	94.4%	94.7%	93.6%	94.2%
	✓	0.001	94.0%	93.9%	94.0%	94.0%
	✓	0.0001	84.5%	84.5%	85.1%	84.7%

Table 10 Test and validation ratios

User ID	Test			Validation		
	Pos	Neg	Ratio	Pos	Neg	Ratio
1	310	12,797	0.02422	2643	115,319	0.0229
2	182	12,925	0.01408	1673	116,289	0.0144
3	248	12,859	0.01929	2109	115,853	0.0182
4	173	12,934	0.01338	1697	116,265	0.0146
5	184	12,923	0.01424	1607	116,355	0.0138

After obtaining our best pre-trained model, we further apply it in a binary classification model. In the best case, we obtain an accuracy of 97.2%, as summarized in Table 9.

It can be seen from the result in Table 10 that the ratio of positive samples to negative samples in the validation and test sets is roughly equal, which means that we have data with similar distributions for evaluation and testing. In Table 10, we consider user IDs from 1 to 5.

In addition, both the training and validation sets use random sampling, while the test set runs on all samples. Therefore, in the test set, the number of samples for label 0 (not users) is much more than the number of samples for label 1 (users). The results shown in Table 11 indicate that a model that can better discriminate samples with label 0 will be stronger.

Table 11 Test and validation

	Test session			
Data	s0	s1	s2	Average
Val	89.67%	87.98%	87.70%	88.45%
Test	94.74%	89.22%	90.05%	91.34%

Table 12 Label-swap

	Test session			
Metric	s0	s1	s2	Average
Accuracy	94.74%	89.22%	90.05%	91.34%
Precision	99.60%	99.58%	99.49%	99.57%
Recall	95.05%	89.45%	90.40%	91.63%
F1 score	97.25%	94.04%	94.46%	95.25%

We further analyze the precision, recall, F1 score, and perform parameter adjustments for the labels 0 and 1. Note that originally, label 1 is used as the positive label (user). However, since the data is so imbalanced, we swap the positive and negative label to observe the result on the multi-kernel (2-2-2) model. These label switching results are shown in Table 12. We observe that this model has a strong ability to detect intruders—the precision in the best case is virtually 100%. This experiment indicates that our model would work well as an IDS.

4.2.4 GRU with Word Embedding

In addition to keystroke features, we experiment with some text-based features in out GRU model. This does raise security concerns, since we must record what a user actually types, as opposed to simply using keystroke dynamics. But, we want to determine whether this additional level of detail can result in an improved model.

We use the nn.Embedding for word vectors in PyTorch, which provides a mapping between words and their corresponding vectors. The embedding weights can be trained, either by random initialization or by pre-trained word vector initialization. This technique can be used to determine the positional relationship of two keys on the keyboard. For example, keys that are positioned next to each other can be classified as adjacent, and their vector should be similar. Furthermore, other relationships can be determine, such as keys that are pressed with the left hand, as compared to those that are pressed with the right hand.

We implement this word embedding in our GRU model. Note that no CNN model is used in this experiment. We compare the experimental results for different dimensions of word embedding vectors with and without attention. The result of these experiments are summarized in Table 13. Note that embedding weights are trained by random initialization. Unfortunately, this model suffers from overfitting, as can be seen from the graphs in Fig. 3.

In an attempt to deal with this overfitting issue, we use Word2Vec [23, 24] to generate vector embeddings. Specifically, we train on a sentence in two directions

Table 13 Word embedding for GRU

| Model | Parameters | | Test session | | | |
	Embedding size	Attention	s0	s1	s2	Average
GRU	3	—	85.53%	83.68%	83.54%	84.25%
	3	✓	85.10%	84.52%	82.67%	84.10%
	5	—	83.22%	81.98%	80.76%	81.99%
	5	✓	83.41%	81.95%	80.64%	82.00%
	10	—	81.31%	79.29%	78.57%	79.72%
	10	✓	80.96%	79.27%	77.99%	79.41%

Uid: 1 Best_acc 0.7164909638554217

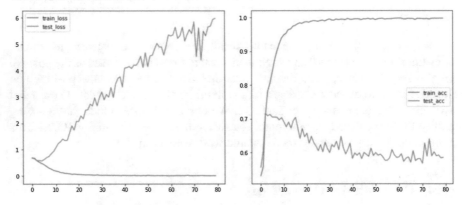

Fig. 3 Word embedding with GRU showing overfitting

Table 14 Word embedding and pre-trained vectors

| Model | Parameters | | Test session | | | |
	Embedding size	Pre-trained	s0	s1	s2	Average
GRU	3	—	85.53%	83.68%	83.54%	84.25%
	3	✓	84.99%	83.47%	82.53%	83.66%

with random text length from 6 to 12 characters. We find that keys that are close together result in a higher score. For example, A and S, which are adjacent on a standard QWERTY keyboard, have a cosine similarity of 0.9918, while A and P, which are on opposite ends of the keyboard, have a cosine similarity of 0.7891.

The experimental results obtained of comparing random initialization and initialization with pre-trained vectors using a GRU model are shown in Table 14. Again, these results still clearly result in overfitting. When embedding is used, the model can reach a training accuracy of about 0.998, but the loss during testing increases.

When we apply word embedding in the CNN-GRU model, the overfitting issue is resolved—these results are given in Table 15. However, the result of word embedding do not outperform our previous experiment. This result is significant, since it shows that for our free-text dataset, there is nothing to be gained by using

Table 15 Word embedding for CNN-GRU

| Model | Parameters | | Test session | | | |
	Kernel	Embedding	s0	s1	s2	Average
Word-CNN-GRU	2	2	91.28%	93.07%	88.56%	90.07%
CNN-GRU	2,2,2	None	94.74%	89.22%	90.05%	91.34%

Table 16 CNN-transformer results

| Model | Test session | | | |
	s0	s1	s2	Average
CNN-encoder	89.89%	86.22%	84.97%	87.03%

the actual text typed by a user, as compared to simply using keystroke dynamics. Since using the text would raise serious privacy concerns, it is beneficial that we do not have to use such data to obtain optimal results.

4.2.5 CNN-Transformer

Next, we experiment with a transformer technique on our CNN-GRU model, which we refer to as the CNN-Transformer model. Specifically, we apply positional encoding before the encoder layer—the result of this experiment is shown in Table 16. This result is significantly worse than our previous best model, so we do not pursue this approach further.

4.2.6 CNN-GRU-Cross-Entropy-Loss

Cross entropy can be used to determine how close the actual output is to the expected output. This loss function combines the two functions of `nn.LogSoftmax()` and `nn.NLLLoss()`. This function is considered useful when dealing with an imbalanced training set, and we have such a dataset.

In this experiment, we change the activation function from

```
BCEWithLogitsLoss
```

to

```
CrossEntropyLoss.
```

Furthermore, we calculate the output during training and testing with `softmax` and `argmax`, respectively. The result for this experiment is given in Table 17. We see that this is a strong model, indicating that imbalance may be an issue that we should address. Nevertheless, this experiment does not improve on our best results.

Table 17 CNN-GRU-cross-entropy results

Model	Test session			
	s0	s1	s2	Average
CNN-GRU-cross-entropy	98.3%	96.7%	95.2%	96.7%

Table 18 Multi-classification on rotation subset

Model	Parameters				
	Kernel	Out-channel	RNN-size	Epochs	Accuracy
CNN-GRU	2	96	8	120,240	58.22%
	16	192	64	40,80	49.16%

Table 19 Fine-tuning on rotation subset (without freeze)

Model	Learning rate	Test session			
		s0	s1	s2	Average
CNN-GRU	0.001	86.9%	83.7%	91.1%	87.2%
	0.01	89.8%	86.7%	93.2%	89.9%

4.2.7 Rotation Subset

To this point, our best model uses the fine-tuning technique discussed in Sect. 4.2.3. Based on this model, we also consider the so-called rotation subsets of the Buffalo free-text dataset, in which 73 users use different keyboards in different sessions.

In this experiment, we build a multi-classification model on the rotation subset and obtain the result shown in Table 18. Compared to our previous multi-classification results, the classification here is much worse. This is to be expected, since each session of the data is obtained from a different keyboard.

Next, we use a multi-classification model as a pre-trained model to build binary classifiers, analogous to the fine-tuned models discussed above. The fine-tune results based on the rotation subset is shown in Table 19.

From these results, we observe that the use of different keyboards will likely create serious difficulties for modeling based on keystroke dynamics. Thus, we conclude that different models will be needed for the same user when using different keyboards.

4.2.8 Robustness

We also want to consider the robustness of our models. There are various definitions of robustness, but in general, we want to quantify the effect of a changing environment on a model. There is no standard way to measure robustness for keystroke dynamics. Here, we use a technique known as synthetic minority oversampling technique (SMOTE) to generate synthetic data that has similar characteristics to

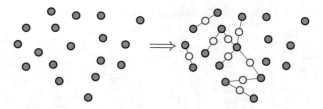

Fig. 4 SMOTE illustrated

Table 20 SMOTE results for CNN-GRU

| Model | Parameters | | | Test session | | | |
	Kernel	SMOTE ratio	Result	s0	s1	s2	Average
CNN-GRU	2,2,2	—	Validation	89.57%	87.56%	87.66%	88.26%
			Test	94.74%	89.22%	90.05%	91.34%
	8	0.1	Validation	79.76%	77.91%	77.93%	78.35%
			Test	98.05%	97.39%	96.50%	97.31%
	8	0.5	Validation	73.47%	71.02%	71.32%	71.94%
			Test	98.49%	98.32%	98.07%	98.26%

the training data. We then measure robustness in the sense of how well our models perform on this SMOTE-generated data.

SMOTE is typically applied as a data augmentation technique to an imbalanced dataset. The idea behind SMOTE is to generate similar samples to the training data using a straightforward interpolation approach [9]. The concept of SMOTE is illustrated in Fig. 4, where the hollow circles on the right-hand side represent augmented data points that are obtained by interpolating between actual data points.

In this set of experiments, we use 6×250 dimensional array as the feature vector, and use the `imbalanced-learn` package in Python package to generate SMOTE data points. First, we use SMOTE to increase the positive samples and apply a "smoothing ratio" of 0.1 which means that we increase in the number of positive samples to 0.1 times the number of negative samples. We also experiment with a smoothing ratio of 0.5. The training and validation results of these experiment are summarized in Table 20.

The performance on the SMOTE data shows that after augmenting, the accuracy decreases on the validation set, while the performance on the test set has improved. The is typically a sign of overfitting, and we speculate that SMOTE has, in effect, introduced noise that has resulted in this overfitting. As further evidence, when adding a higher proportion of SMOTE data, the performance on the validation set reduces further, while performance on the test set improves.

We also perform experiments for different ratios of under sampling. That is, we reduce the number of negative samples to a specified proportion of the positive samples. The results of these experiments are given in Table 21. As expected, these results show minimal change, as compared to the corresponding base models in Table 20.

Table 21 SMOTE results for CNN-GRU with undersampling

Model	Parameters		Result	Test session			
	Kernel	SMOTE ratio		s0	s1	s2	Average
CNN-GRU	2,2,2	—	Validation	89.57%	87.56%	87.66%	88.26%
			Test	94.74%	89.22%	90.05%	91.34%
	8	0.1 under	Validation	82.82%	81.40%	80.80%	81.67%
			Test	93.61%	89.90%	90.47%	91.33%
	8	0.5 under	Validation	82.70%	80.64%	80.49%	81.28%
			Test	93.38%	91.11%	89.96%	91.48%
	8	1.0 under	Validation	82.60%	81.09%	80.77%	81.49%
			Test	93.03%	91.17%	90.01%	91.40%

Table 22 SMOTE ratio 0.5 with label-switching

Metric	Test session			
	s0	s1	s2	Average
Accuracy	98.49%	98.22%	98.08%	98.26%
Precision	98.81%	98.79%	98.81%	98.80%
Recall	99.67%	99.42%	99.24%	99.44%
F1 score	99.24%	99.10%	99.02%	99.12%

We further analyze the precision, recall and F1 score of our models. Note that in these experiments we also perform label switching. As discussed above, label switching may provide a better indication of the utility of a model in the IDS case. The results of these experiments are given in Table 22.

Comparing the result for the "label-switch" case in Table 12, we conclude that the higher the SMOTE ratio, the higher the recall, and the lower the precision. This implies that the model is more capable of capturing data with label 0 (which, due to label switching, represents the positive case), but the accuracy of the model's judging label as 0 is also lower.

4.2.9 Explainability

Next, we briefly consider the "explainability" of our model. That is, we would like to gain some insight into how the model is actually making decisions. Most machine learning techniques are relatively opaque, in the sense that it is difficult to understand the decision-making process. This is especially true of neural network based techniques, and since our model combines multiple techniques, it is bound to be even harder it interpret directly.

Here, we consider local interpretable model-agnostic explanations (LIME) [29] to try to gain insight into the role of the various features in our model. LIME perturbs the input and compares the corresponding outputs. If a small change in an input feature causes the classification to switch, then we can judge the importance of a specific feature to the model's decision-making process.

Fig. 5 LIME illustrated

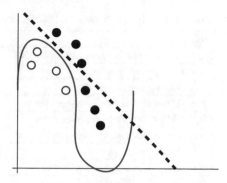

The concept behind LIME is illustrated in Fig. 5, where the solid black circles and hollow red circles represent two categories, and the blue curve is the decision boundary between the classes. LIME uses a simplified linear model—represented in the figure by the dashed line—to predict the classes. Locally, this linear model will likely be very accurate, although it would typically fail badly globally. By using a simple linear model, we can, for example, more easily determine the most relevant features.

In our LIME experiment, we select user 46 and the "s12-train-s0-test" case. For this experiment, the best validation accuracy we obtain is 96.88% and the best test accuracy is 99.31%.

For the sake of brevity, we omit the details of our LIME experiments but, in summary, we find that LIME indicates that our model focuses more on holding time and difference time. In general, it appears that a good model focuses less on key-id, which indicates that it may not be a good feature to distinguish users. For additional details on our LIME experiments, see [5].

4.2.10   Equal Error Rate

The equal error rate (EER) is an objective standard to measure classifiers. In a biometric system, the false accept rate (FAR) is the rate at which a user can authenticate as someone else, whereas the false reject rate (FRR) is the rate at which a user cannot authenticate as themselves. When these rates are equal, the value is called equal error rate. The lower the EER value, the higher the accuracy of the biometrics system. In practice, we would likely not set the system parameters so that the FAR is equal to the FRR. For example, in a financial application that requires high security, the FAR should be very low, which necessitates a somewhat higher FRR. Nevertheless, the EER allows us to easily compare different biometric systems.

Here, we use a sigmoid function for output, so that we can obtain the prediction result in the form of a probability. Then we use the prediction and ground-truth to calculate the confusion matrix. After obtaining labels, we construct an ROC curve to obtain the FPR and TPR and, finally, we determine the EER from the ROC curve.

Table 23 EER for various models (baseline subset)

Model	s01-train-s2-test
Pre-trained word embedding with CNN-GRU	0.1091
CNN-Transformer-encoder	0.1257
CNN-GRU-cross-entropy-loss	0.1502
CNN-GRU-without-sampler-at-best-val	0.0609
CNN-GRU-without-sampler-at-best-eer	0.0611
CNN-GRU-without-sampler-at-non-best-val	0.0594
CNN-GRU-with-sampler	0.0826
CNN-GRU-without-sampler-fine-tune	0.0412 (0.7187-multi)
CNN-GRU-without-sampler-fine-tune	0.0386 (0.7599-multi)

Table 24 EER for various models (all subsets)

Model	s01-train-s2-test
CNN-GRU-attention-without-sampler	0.1389
CNN-GRU-attention-non-without-sampler	0.1239
CNN-GRU-fine-tune	0.1029

Table 25 Best model for EER (all subsets)

Model	Test session			
	s0	s1	s2	Average
CNN-GRU-without-sampler-fine-tune	0.0690	0.0841	0.0557	0.0696

The EER results with s01 for training and s2 for testing for various models discussed above are provided in Table 23. The best EER we obtain is 0.0386, and when we use this model over all sessions we obtain an average EER of 0.0394.

After determining the best result on the baseline subset, we then apply this model to all 148 users. The resulting EER for all subset for different models is shown in Table 24. However, we see that the result is poor, which is apparently due to the rotation subset, where different keyboards are used across sessions, and the fact that some users do not type much data in the baseline subset. Thus, it is more realistic to set a threshold for the key length with different users—in this case, we obtain a best EER result in Table 25. From the last two lines in Table 25, we observe that the model that has the higher multi-classification accuracy results in a lower EER for the fine-tuned case.

4.2.11 Knowledge Distilling

Knowledge distilling is somewhat analogous to explainability. In knowledge distillation we "compress" a model, in the sense that we try to replace a complex model with a much simpler one that achieves comparable results. The goal is to extract the essence of a complex model within a much simpler form.

Table 26 EER results for knowledge distilling

Model	s01-train-s2-test
Student-1-Teacher-1	0.1333
Student-1.5-Teacher-0.5	0.1409
Student-1.99-Teacher-0.01	0.1762
Student-0.99-Teacher-0.01	0.1355
Student-0.5-Teacher-1.5	0.0864
Student-0.5-Teacher-1.99	0.1097
Student-0.01-Teacher-1.99	0.0925

Table 27 Results for tuning `pos-weight` (baseline subset)

Metric	Fine-tune	pos-weight			
		0.1	2	10	50
ERR	0.0395	0.0428	0.0413	0.0392	0.0813
Accuracy	99.33%	99.41%	99.21%	99.03%	96.81%
Precision	78.46%	84.57%	74.82%	67.51%	50.87%
Recall	76.88%	69.45%	78.92%	83.28%	82.77%
F1 score	74.63%	72.92%	73.40%	71.64%	56.64%
AUC	98.84%	98.74%	98.87%	98.88%	98.28%

This method knowledge distilling was first proposed in [4]. Then in [13] a "teacher" and "student" model was proposed from the concept of mentoring. The output of the teacher network is used as a soft label to train a student network. For our model, the EER results obtained based on teacher-student knowledge distilling with different parameters are shown in Table 26.

Note that in this experiment, we simply use a multi-classification model as the teacher and the corresponding binary classification as the student model. With this type of approach, we hope to see that the teacher can improve the results given by the student model, as indicated by a low EER. However, in our experiments the EER is not competitive with our fine tuning model. Hence, we cannot draw any strong conclusions from this experiment, but this is worth pursuing as future work.

4.2.12 Weighted Loss

In the CNN-GRU fine-tune model, we select BCEWithLogitsLoss as our criterion to calculate the loss. However, we did not specify the parameter of pos-weight which is the weight of positive samples in the original experiment. In this experiment, we use the fine-tune model as a backbone and try to tune the positive weights and compare with our previous results. The results of these experiments are given in Table 27.

Note that the results in Table 27 are average among the three different test sessions. From the result in Table 27, we observe that when the value of pos-weight is 0.1, the precision is best, while the recall rate decreases, as compared to the

Table 28 Result for ensembles (baseline subset 1)

		Model		
Metrics	Ensemble	Fine-tune	Softmax	Transformers
ERR	0.082	0.0395	0.1644	0.1449
Accuracy	99.28%	99.33%	96.80%	87.03%
Precision	76.71%	78.46%	41.69%	19.10%
Recall	78.34%	76.88%	71.30%	76.26%
F1 score	74.29%	74.63%	47.46%	23.21%
AUC	95.18%	98.84%	84.23%	93.37%

Table 29 Various model combinations

Model	Combination
$\mathscr{A}$	Fine-tune, pos-w 0.1, pos-w 2, pos-w 10, pos-w 50
$\mathscr{B}$	Fine-tune, pos-w 0.1, pos-w 2, pos-w 10
$\mathscr{C}$	Fine-tune, pos-w 0.1, pos-w 10

Table 30 Results for ensembles (baseline subset 2)

		Model		
Metrics	Fine-tune	$\mathscr{A}$	$\mathscr{B}$	$\mathscr{C}$
ERR	0.0395	0.0285	0.0312	0.0322
Accuracy	99.33%	99.41%	99.69%	99.37%
Precision	78.46%	81.31%	82.98%	82.32%
Recall	76.88%	80.92%	78.79%	77.99%
F1 score	74.63%	78.67%	78.31%	77.75%
AUC	98.84%	99.28%	99.24%	99.18%

original fine-tune model. Moreover, when the value of `pos-weight` is higher, the precision decreases while the recall increases, which nicely illustrates the inherent trade-off between these measures. Based on these results, we can adjust the value of `pos-weight` depending on different application scenarios.

4.2.13 Ensemble Models

In this section, we consider three ensemble models which include various combinations of fine-tuning, softmax, and transformer. The results for these cases are given in Table 28. Note that the results in Table 28 are each an average among three different test sessions. Since the results for the softmax and transformer models are much worse than the fine-tune model, the resulting ensembles only improve slightly in terms of precision and recall, and only for some users.

We further ensemble the models with different `pos-weight` values, with the fine-tune model as the backbone. The parameters of the three ensembles considered, which we denote as models $\mathscr{A}$, $\mathscr{B}$, and $\mathscr{C}$, are specified in Table 29. Note that the results in Table 30 represent the average among the three different test sessions. We observe that the EER, precision, and recall rates all improve with these ensemble techniques.

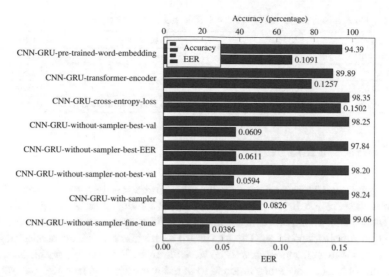

Fig. 6 Accuracies and EER of models (baseline subset)

4.2.14 Discussion

Figure 6 summarizes the results of our free-text experiments, both in terms of EER and accuracy. Note that the best accuracy and the best EER were both achieved with the CNN-GRU-without sampler-fine-tune model.

In our experiment, we have applied a sampler to train so as to deal with the situation of imbalanced data. As for feature engineering, we transform the data into a vector which includes the label of the key and the timing features. Then, we compute the mean and variance and center the data, which enable us to achieve better performance.

We perform parameter tuning on the models and obtain a great improvement in accuracy. The result shows that longer keystroke sequence and larger out-channel size generally result in higher accuracy, while more convolutions do not improve our model. After parameter tuning, we apply an attention layer on the outputs of the GRU and find that some users' accuracy slightly improves. Moreover, we expand the keystroke dynamics features to include rate features. Although the accuracy result of do not greatly improve, the EER does achieve better performance.

We also use a multi-classification model as a pre-trained model for binary classification. The results indicate that such a pre-trained multi-classification model can achieve higher accuracy and a lower EER for the corresponding binary classification model.

Finally, we compare our free-text experiments to previous work. From the results in Table 31, we see that our best model is competitive with the best EER previously obtained, while the accuracy of several of our models exceed 99%.

Table 31 Comparison to previous work for Buffalo free-text dataset

Research	Models	Accuracy	EER
Lu et al. [20]	CNN-RNN	—	0.0236
Ayotte et al. [1]	ITAD metrics	—	0.0530
Huang et al. [15]	SVM	—	0.0493
Our research	Fine-tune CNN-GRU	Greater than 99%	0.0690

5 Conclusion and Future Work

In this paper, we developed and analyzed machine learning techniques for biometric authentication based on free-text data. We focused on a novel CNN-GRU architecture, and we experimented with an attention layer, rate features, pre-trained models, ensembles, and so on. The maximum multiclass classification accuracy that we achieved with our model was 99.31%, with an EER of 0.069. As far as we are aware, this is the best accuracy attained to date for the Buffalo free-text dataset, and our EER is competitive with the best results previously obtained. In addition, for our CNN-GRU model, we considered "explainability" and knowledge distillation, among many other relevant topics.

The high accuracy and low EER for our free-text results indicate that passive authentication and intrusion detection may be practical, based on keystroke dynamics. That is, in addition to an initial authentication at login time, a user can be periodically re-authenticated by passively monitoring typing behavior. In this way, intrusions can be detected in real-time, with a minimal burden placed on users and administrators.

There are many avenues available for future work. For example, we plan to perform extensive model optimization and model fusion. For model optimization, we will consider techniques from contrastive learning [18] and self-supervised learning [19] to see whether these approaches can improve our model.

As another example of possible future work, we plan to evaluate the robustness of our technique using an algorithm known as POPQORN [17]. The idea behind this technique is to observe the effect of outside disturbances to the model and thereby determine its robustness.

Of course, more and better data is always useful, and this is especially true in free-text analysis. Having more realistic and longer-term free-text datasets over a larger number of users would add credence to the results obtained in any research.

References

1. Blaine Ayotte, Mahesh Banavar, Daqing Hou, and Stephanie Schuckers. Fast free-text authentication via instance-based keystroke dynamics. *IEEE Transactions on Biometrics, Behavior, and Identity Science*, 2(4):377–387, 2020.
2. Mario Luca Bernardi, Marta Cimitile, Fabio Martinelli, and Francesco Mercaldo. Keystroke analysis for user identification using deep neural networks. In *2019 International Joint Conference on Neural Networks*, IJCNN, pages 1–8, 2019.
3. Saleh Ali Bleha, Charles Slivinsky, and B. Hussien. Computer-access security systems using keystroke dynamics. *IEEE Transactions on Pattern Analysis and Machine Intelligence*, 12(12):1217–1222, 1990.
4. Cristian Buciluă, Rich Caruana, and Alexandru Niculescu-Mizil. Model compression. In *Proceedings of the 12th ACM SIGKDD International Conference on Knowledge Discovery and Data Mining*, KDD '06, pages 535–541, 2006.
5. Han-Chih Chang. Keystroke dynamics based on machine learning. Master's thesis, San Jose State University, Department of Computer Science, 2020.
6. Jacob Devlin, Ming-Wei Chang, Kenton Lee, and Kristina Toutanova. BERT: pre-training of deep bidirectional transformers for language understanding. http://arxiv.org/abs/1810.04805, 2018.
7. Paul S. Dowland and Steven M. Furnell. A long-term trial of keystroke profiling using digraph, trigraph and keyword latencies. In Yves Deswarte, Frédéric Cuppens, Sushil Jajodia, and Lingyu Wang, editors, *Security and Protection in Information Processing Systems*, pages 275–289. Springer, 2004.
8. George E. Forsen, Mark R. Nelson, and Raymond J. Staron (Jr.). Personal attributes authentication techniques. https://books.google.com.tw/books?id=tbs4OAAACAAJ, 1977.
9. Stephanie Ger and Diego Klabjan. Autoencoders and generative adversarial networks for imbalanced sequence classification. https://arxiv.org/abs/1901.02514, 2020.
10. Romain Giot and Anderson Rocha. Siamese networks for static keystroke dynamics authentication. In *2019 IEEE International Workshop on Information Forensics and Security*, WIFS, pages 1–6, 2019.
11. S. Haider, A. Abbas, and A. K. Zaidi. A multi-technique approach for user identification through keystroke dynamics. In *2000 IEEE International Conference on Systems, Man and Cybernetics*, SMC 2000, pages 1336–1341, 2000.
12. Martin Heller. *InfoWorld*: What is CUDA? Parallel programming for GPUs. https://www.infoworld.com/article/3299703/what-is-cuda-parallel-programming-for-gpus.html, 2018.
13. Geoffrey Hinton, Oriol Vinyals, and Jeff Dean. Distilling the knowledge in a neural network. https://arxiv.org/abs/1503.02531, 2015.
14. Danoush Hosseinzadeh and Sridhar Krishnan. Gaussian mixture modeling of keystroke patterns for biometric applications. *IEEE Transactions on Systems, Man, and Cybernetics, Part C (Applications and Reviews)*, 38(6):816–826, 2008.
15. Jiaju Huang, Daqing Hou, Stephanie Schuckers, Timothy Law, and Adam Sherwin. Benchmarking keystroke authentication algorithms. In *2017 IEEE Workshop on Information Forensics and Security*, WIFS, pages 1–6, 2017.
16. Pilsung Kang, Seong-seob Hwang, and Sungzoon Cho. Continual retraining of keystroke dynamics based authenticator. In Seong-Whan Lee and Stan Z. Li, editors, *Advances in Biometrics*, pages 1203–1211. Springer, 2007.
17. Ching-Yun Ko, Zhaoyang Lyu, Tsui-Wei Weng, Luca Daniel, Ngai Wong, and Dahua Lin. POPQORN: Quantifying robustness of recurrent neural networks. In *Proceedings of the 36th International Conference on Machine Learning*, pages 3468–3477, 2019.
18. Phuc H. Le-Khac, Graham Healy, and Alan F. Smeaton. Contrastive representation learning: A framework and review. https://arxiv.org/abs/2010.05113, 2020.
19. Xiao Liu, Fanjin Zhang, Zhenyu Hou, Zhaoyu Wang, Li Mian, Jing Zhang, and Jie Tang. Self-supervised learning: Generative or contrastive. https://arxiv.org/abs/2006.08218, 2021.

20. Xiaofeng Lu, Shengfei Zhang, and Shengwei Yi. Free-text keystroke continuous authentication using CNN and RNN. *Journal of Tsinghua University (Science and Technology)*, 58(12):1072–1078, 2018.
21. Roy A. Maxion and Kevin S. Killourhy. Comparing anomaly-detection algorithms for keystroke dynamics. In *2009 IEEE/IFIP International Conference on Dependable Systems & Networks*, DSN, pages 125–134, 2009.
22. Roy A. Maxion and Kevin S. Killourhy. Keystroke biometrics with number-pad input. In *2010 IEEE/IFIP International Conference on Dependable Systems & Networks*, DSN, pages 201–210, 2010.
23. Tomas Mikolov, Kai Chen, Greg Corrado, and Jeffrey Dean. https://arxiv.org/abs/1301.3781, 2013.
24. Tomas Mikolov, Ilya Sutskever, Kai Chen, Greg Corrado, and Jeffrey Dean. Distributed representations of words and phrases and their compositionality. https://arxiv.org/abs/1310.4546, 2013.
25. Fabian Monrose and Aviel Rubin. Authentication via keystroke dynamics. In *Proceedings of the 4th ACM Conference on Computer and Communications Security*, pages 48–56, 1997.
26. Fabian Monrose and Aviel D. Rubin. Keystroke dynamics as a biometric for authentication. *Future Generation Computer Systems*, 16(4):351–359, 2000.
27. Adam Paszke, Sam Gross, Francisco Massa, Adam Lerer, James Bradbury, Gregory Chanan, Trevor Killeen, Zeming Lin, Natalia Gimelshein, Luca Antiga, Alban Desmaison, Andreas Kopf, Edward Yang, Zachary DeVito, Martin Raison, Alykhan Tejani, Sasank Chilamkurthy, Benoit Steiner, Lu Fang, Junjie Bai, and Soumith Chintala. PyTorch: An imperative style, high-performance deep learning library. In H. Wallach, H. Larochelle, A. Beygelzimer, F. d' Alché-Buc, E. Fox, and R. Garnett, editors, *Advances in Neural Information Processing Systems 32*, pages 8024–8035. Curran Associates, Inc., 2019.
28. Nataasha Raul, Radha Shankarmani, and Padmaja Joshi. A comprehensive review of keystroke dynamics-based authentication mechanism. In *International Conference on Innovative Computing and Communications*, pages 149–162, 2020.
29. Marco Tulio Ribeiro, Sameer Singh, and Carlos Guestrin. Why should I trust you?: Explaining the predictions of any classifier. In *Proceedings of the 22nd ACM International Conference on Knowledge Discovery and Data Mining*, SIGKDD, pages 1135–1144, 2016.
30. Joseph Roth, Xiaoming Liu, Arun Ross, and Dimitris Metaxas. Investigating the discriminative power of keystroke sound. *IEEE Transactions on Information Forensics and Security*, 10(2):333–345, 2015.
31. T. Sim and R. Janakiraman. Are digraphs good for free-text keystroke dynamics? In *2007 IEEE Conference on Computer Vision and Pattern Recognition*, pages 1–6, 2007.
32. Nitish Srivastava, Geoffrey Hinton, Alex Krizhevsky, Ilya Sutskever, and Ruslan Salakhutdinov. Dropout: A simple way to prevent neural networks from overfitting. *Journal of Machine Learning Research*, 15:1929–1958, 2014.
33. Yan Sun, Hayreddin Çeker, and Shambhu Upadhyaya. Shared keystroke dataset for continuous authentication. In *2016 IEEE International Workshop on Information Forensics and Security*, WIFS, pages 1–6, 2016.
34. Pin Shen Teh, Andrew Teoh, and Shigang Yue. A survey of keystroke dynamics biometrics. *The Scientific World Journal*, 2013, 2013. Article ID 408280.
35. Robert S. Zack, Charles C. Tappert, and Sung-Hyuk Cha. Performance of a long-text-input keystroke biometric authentication system using an improved k-nearest-neighbor classification method. In *2010 Fourth IEEE International Conference on Biometrics: Theory, Applications and Systems*, BTAS, pages 1–6, 2010.
36. Yu Zhong and Yunbin Deng. A survey on keystroke dynamics biometrics: Approaches, advances, and evaluations. https://sciencegatepub.com/books/gcsr/gcsr_vol2/GCSR_Vol2_Ch1.pdf, 2015.
37. Yu Zhong, Yunbin Deng, and Anil K. Jain. Keystroke dynamics for user authentication. In *2012 IEEE Computer Society Conference on Computer Vision and Pattern Recognition Workshops*, pages 117–123, 2012.

Free-Text Keystroke Dynamics for User Authentication

Jianwei Li, Han-Chih Chang, and Mark Stamp

Abstract In this research, we consider the problem of verifying user identity based on keystroke dynamics obtained from free-text. We employ a novel feature engineering method that generates image-like transition matrices. For this image-like feature, a convolution neural network (CNN) with cutout achieves the best results. A hybrid model consisting of a CNN and a recurrent neural network (RNN) is also shown to outperform previous research in this field.

1 Introduction

User authentication is a critically important task in cybersecurity. Password based authentication is widely used, as are various biometrics. Examples of popular biometrics include fingerprint, facial recognition, and iris scan. However, all of these authentication methods suffer from some problems. For example, passwords can often be guessed and are sometimes stolen, and most biometric systems require special hardware [14, 17, 23]. Moreover, research has shown, for example, that the accuracy of face and fingerprint recognition on the elderly is lower than for young people [13]. Thus, an authentication method that can resolve some of these issues is desirable.

Intuitively, it would seem to be difficult to mimic someone's typing behavior to a high degree of precision. Thus, patterns hidden in typing behavior in the form of keystroke dynamics might serve as a strong biometric. One advantage of a keystroke dynamics based authentication scheme is that it requires no specialized hardware. In addition, such a scheme can provide a non-intrusive means of continuous or ongoing authentication, which can be viewed as a form or intrusion detection. Coursera, an online learning website, currently employs typing characteristics as part of its login system [15].

J. Li · H.-C. Chang · M. Stamp (✉)
San Jose State University, San Jose, CA, USA
e-mail: jianwei.li@sjsu.edu; han-chih.chang@sjsu.edu; mark.stamp@sjsu.edu

© The Author(s), under exclusive license to Springer Nature Switzerland AG 2022
M. Stamp et al. (eds.), *Artificial Intelligence for Cybersecurity*, Advances in Information Security 54, https://doi.org/10.1007/978-3-030-97087-1_15

Research into keystroke dynamics began about 20 years ago [16]. However, early results in this field were not impressive. Most of the existing research in keystroke dynamics has focused on fixed-text typing behavior, which is viewed as one-time authentication [2, 5, 12, 14, 23]. Compared with fixed-text keystroke dynamics, the free-text case presents some additional challenges. First, the number of useful features may differ among input sequences. Second, the optimal length of a keystroke sequence for analysis is a factor that must be considered—a longer sequence is slower to process and might include more noise, while a shorter sequence may lack sufficient distinguishing characteristics. Moreover, for free-text keystroke sequences, it is more challenging to extract an effective pattern, thus the robustness of any solution is a concern.

In this paper, we consider the free-text keystroke dynamics-based authentication problem. For this problem, we propose and analyze a unique feature engineering technique. Specifically, we organize features into an image-like transition matrix with multiple channels, where each row and column represents a key on the keyboard, with the depth corresponding to different categories of features. Then a convolutional neural network (CNN) model with cutout regularization is trained on this engineered feature. To better capture the sequential nature of the problem, we also consider a hybrid model using our CNN approach in combination with a gated recurrent unit (GRU) network. We evaluate these two models on open free-text keystroke datasets and compare the results with previous work. We carefully consider the effect of different lengths of keystroke sequences and other parameters on the performance of our models.

The contribution of this paper include the following:

- A new feature engineering method that organizes features as an image-like matrix for free-text keystroke dynamics-based authentication.
- An analysis of cutout regularization as a step in the image analysis process.
- A careful analysis of various hyperparameters, including the length of keystroke sequence in our models.

The remainder of this paper is organized as follows. Section 1 introduces the basic concept of keystroke dynamics-based authentication, and we outline our general approach to the problem. In Sect. 2, we discuss background topics, including the learning techniques employed and the datasets we have used. Section 2 also provides a discussion of relevant previous work. Section 3 describes the features that we use and, in particular, we discuss the feature engineering strategy that we employ to prepare the input data for our continuous classification models. Then, in Sect. 4, we elaborate on the architectures of the various models considered in this paper, and we discuss the hyperparameter tuning process. Section 5 includes our experiments and analysis of the results. Finally, Sect. 6 provides a conclusion and points to possible directions for future work.

Table 1 Use cases for keystroke dynamics-based systems

Text	Scenario	Precision	Recall	Input length
Fixed	One-time authentication	High	High	Short
Free	Intrusive detection	Low	High	Long
Either	Identification	High	Low	Either

2 Background

Authentication is the process that allows a machine to verify the identity of a user. By the nature of the problem, authentication is a classification task. Keystroke dynamics is one of many techniques that have been considered for authentication. One advantage of keystroke dynamics is that such an approach requires no special hardware.

Precision and recall are two metrics used to evaluate classification models. Precision is the fraction of true positive instances among those classified as positive, while recall is the fraction of true positive instances that are correctly classified as such. Table 1 lists some examples of use cases, along with the general degree of precision and recall that typically must be attained in a useful system. Depending on the scenario, too many false positives (i.e., low precision) can render an IDS impractical, but an IDS must detect intrusions (high recall) or it has clearly failed to perform adequately. On the other hand, in the identification problem, we must be confident that our identification is correct (high precision), even if we fail to identify subjects in a number of cases (low recall).

We note in passing that even if the precision and recall are both high, practical usage scenarios for keystroke dynamics based systems may be limited by the length of the keystroke sequence required for analysis. In cases where a short keystroke sequence suffices, the technique will be more widely applicable.

For a usage scenario, consider password-protected user accounts. Keystroke dynamics would provide a second line of defense in such an authentication system. In a two-factor authentication system, an attacker would need to also accurately mimic a users tying habits. Note that the second "factor" (i.e., keystroke dynamics) is transparent from a user's perspective—the keystroke-related biometric information is collected passively, and requires no additional actions from a user beyond typing his or her password.

Even in cases where the length of the keystroke sequence must be relatively long in order to achieve the necessary accuracy, keystroke dynamics systems could still be useful. For example, suppose that a user needs to reset their password for a high-security application, such as an online bank account. Most such systems require the user to answer a "secret" security question or multiple such questions. It can be difficult for users to remember the answers to security questions, and the answers themselves (e.g., "mother's maiden name") are often not secret. Replacing these question with a keystroke dynamics system would free the user from the need to remember answers, as the user would simply need to type a sufficient number of characters in the user's usual typing mode.

From the use-case point of view, keystroke dynamics-based systems can be classified into those for which long input sequences are acceptable, and those for which short input sequences are essential. We can also classify keystroke dynamics systems according to whether they are based on fixed-text or free-text. In this paper, we only consider free-text.

2.1 Related Work

Previously, most work in keystroke dynamics was based on fixed-text, but recently more attention has been paid to free-text keystroke analysis. There are two commonly used free-text keystroke datasets, which we refer to as the Buffalo dataset [20] and the Clarkson II dataset [22]. We discuss these datasets in more detail in Sect. 2.2. Yan et al. [20] introduced the Buffalo dataset, which they use to evaluate a Gaussian mixture model (GMM) proposed by Hayreddin et al. [4]. The best EER obtained is 0.01. Their experiments are limited to keystroke data generated using the same keyboard. In our research, we evaluate our models on the entire Buffalo dataset, which includes different keyboards.

Pilsung et al. [9] divide the keyboard into three areas, left, right, and space, which correspond to the keys that are typically typed by the left hand (L), right hand (R), and thumbs (S), respectively. In this way, the time-based features extracted from different adjacent keystroke pairs fall into eight categories, which are denoted as L-L, L-R, R-R, R-L, R-S, S-R, L-S, S-L. Then they compute average time-based histogram over each group and concatenate these values to form a feature vector. In this way, the free-text keystroke sequence is embedded into a vector of fixed length eight, which can then be used in different detection models. However, their method fails to preserve most of the sequential information that is available in keystrokes.

To improve the performance of authentication systems based on free-text, Junhong et al. [10] propose a novel user-adaptive feature extraction method to capture unique typing pattern behind keystroke sequences. The method consists of ranking time-based features, and splitting all of these features into eight categories based on the rank order. Similar to the method proposed in [9], they calculate the average time-based feature of each category as a single feature value and concatenate these features to form a vector. Their experiments show that the method significantly improves performance, as compared with the method in [9]. However, they are still discarding a significant amount of the information available in the raw keystroke dynamics data.

Eduard et al. [17] explore the use of multi-layer perceptrons (MLP) for keystroke based authentication. Their model considers time-based information between different keys separately, and does not aggregate information from the entire keystroke sequence. The performance appears to be relatively poor.

Bernardi et al. [3] propose a feature extraction model to capture user input patterns; additional related work by these authors can be found in [21]. In [3], the authors test the impact of different numbers of layers in various deep learning networks and compared the effectiveness of deep networks with classical machine learning methods. They attain a highest accuracy of 99.9% using an MLP with nine hidden layers. However, their architectures are limited to feed-forward fully-connected layers, and better results require a large number of hidden layers. Also, the dataset used in their research is different from that used in our research, and thus the results are not directly comparable.

Kobojek et al. [11] uses an RNN-based model for classification based on keystroke data. They make use of keystroke sequential data. They achieve a best EER that is relatively high at 13.6%.

Influenced by the work in [11], Xiaofeng et al. [14] divide continuous keystroke dynamics sequences into keystroke subsequences of a fixed length and extracts time-based features from each subsequence. These features are then organized into a fixed-length sequence, and the resulting data is fed to a complex model consisting of a combined CNN and RNN. They consider an overlapping sliding window, and the they use a majority vote system to further improve the accuracy. They best EERs of 2.67 and 6.61% over a pair of open free-text keystroke datasets. In our research, we propose a new architecture that is inspired by the model in [14].

2.2 Datasets

In this paper, we evaluate various models based on two open-source free-text keystroke dynamics datasets. The two datasets we consider are from Clarkson University [22] and SUNY Buffalo [20]. Next, we discuss these datasets.

2.2.1 Buffalo Keystroke Dataset

The Buffalo free-text keystroke dataset was collected by researchers at SUN Buffalo from 148 research subjects. In this dataset, the subjects were asked to finish two typing tasks in a laboratory. For the first task, participants transcribed Steve Jobs' Stanford commencement speech, which was split into three parts. The second task consisted of responses to several free-text questions. The interval between the two sessions was 28 days. Additionally, only 75 of the subjects completed both typing tasks with the same keyboard, while the remaining 73 subjects typed using three different keyboards across three sessions.

The Buffalo dataset includes relatively limited information. Specifically, the key that was pressed, along with timestamps for the key-down time and the key-up events. The average number of keystrokes in the three sessions exceeded 17,000 for each subject. Additionally, some of the participants used keyboards with different key layouts to input text information. This dataset also provides gender information for each subject.

2.2.2 Clarkson II Keystroke Dataset

The Clarkson II keystroke dataset is a popular free-text keystroke dynamics dataset that was collected by researchers at Clarkson University. This dataset includes keystroke timing information for 101 subjects in a completely uncontrolled and natural setting, with the date having been collected over a period of 2.5 years. Compared with other datasets which are controlled to some degree, the participants contribute their data with different computers, different keyboards, different browsers, different software, and even different tasks (e.g., gaming, email, etc.). Models that perform well on this dataset should also perform well in a real-world scenario.

Unfortunately, the Clarkson II dataset only provides very limited features—specifically, the timestamps of key-down and key-up events. The average number of keystrokes for each research subject is about 125,000. However, the number of keystroke events is far from uniform, with some users having contributing only a small number of keystrokes. Therefore, we set a threshold of 20,000 keystrokes, which gives us only 80 subjects.

2.3 Deep Leaning Algorithms

In this research, we apply deep learning methods to the free-text keystroke datasets discussed above. Our best-performing architecture is a novel combination of neural network based techniques. In this section, we briefly discuss the learning techniques that we have employed.

2.3.1 Multilayer Perceptron

Multilayer perceptrons (MLP) [18] are a class of supervised learning algorithms with at least one hidden layer. Any MLP consists of a collection of interconnected artificial neurons, which are loosely modeled after the neurons in the human brain. Nonlinearity is provided by the choice of the activation function in each layer. MLP is related to the classic machine learning technique of support vector machines (SVM).

2.3.2 Convolutional Neural Network

Convolutional neural networks (CNN) [1] are a special class of neural networks that make use of convolutional kernels to efficiently deal with local structure. CNNs are often ideal for applications where local structure dominates, such as image analysis. CNNs with multiple convolutional layers are able to extract the semantic information at different resolutions and have proven to be extremely powerful in computer vision tasks.

2.3.3 Recurrent Neural Network

Recurrent neural networks (RNN) [8] are used to deal with sequential or time-series data. For example, sequential information is essential for the analysis of text and speech. Plain "vanilla" RNNs suffer from vanishing gradients and related pathologies. To overcome these issues, highly specialized RNN architectures have been developed, including long short-term memory (LSTM) [8] and gated recurrent units (GRU) [6]. In practice, LSTMs and GRUs are among the most successful architectures yet developed. In this research, we focus on GRUs, which are faster to train than LSTMs, and perform well in our application.

2.3.4 Cutout

Fully connected neural networks often employ dropouts [19] to reduce overfitting problems. While dropouts work well for models with fully-connected layers, the technique is not suitable for CNNs. Instead, we use cutout regularization [7] with our CNN models. Cutouts are essentially the image-based equivalent of dropouts— we cut out part of the image when training, which forces the CNN to learn from other parts of the image. In addition to helping with overfitting, a model that is able to handle images with such occlusions is likely to be more robust.

3 Feature Engineering

As mentioned in Sect. 2, we consider two open source keystroke datasets. Both the Buffalo and Clarkson II datasets are free-text, and only provide fairly limited information. Therefore, we will have to consider feature engineering as a critical part of our experiments. In this section we consider different categories of features and various types of feature engineering.

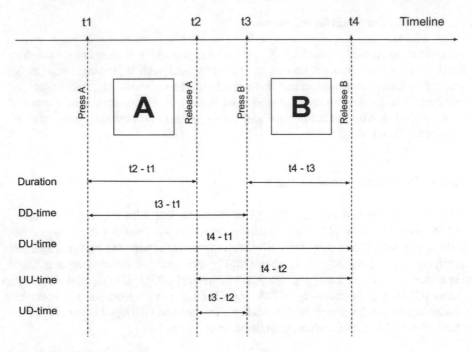

Fig. 1 Five time-based features

3.1 Features

With the development of mobile devices, modern keyboards are no longer limited
to physical keyboards, but also include most virtual devices that allow user input.
Ideally, we would like to consider patterns in user typing behavior with respect
to time-based information and pressure-based features. However, pressure-based
features are not directly available from the datasets used in this research. In the
future, datasets obtained using mobile devices could include such information,
which should enable stronger authentication and identification results.

Again, in this research we necessarily focus on time-based features, because that
is what we have available in our keystroke datasets. The five time-based features
that we consider are illustrated in Fig. 1.

Let A and B represent two consecutive input keys, with press and release representing a key-down and key-up event, respectively. The five time-based features are
duration, down-down time (DD-time), up-down time (UD-time), up-up time (UU-
time), and down-up time (DU-time). Duration is the time that the user holds a key in
the down position, while the other four features are clear from the figure. Note that
for any two consecutive keystroke events, say, A and B, six features can be extracted,
namely, duration-A, duration-B, DD-time, UD-time, UU-time, and DU-time.

3.2   Length of Keystroke Sequence

As mentioned in Sect. 2, we can divide keystroke dynamics-based authentication into four categories depending on the length and consistency of the keystroke sequence. For our free-text keystroke datasets, the data consists of a long keystroke sequence of thousands of characters for each user. In previous research, such long sequences have been split into multiple subsequences, and we do the same here. Each subsequence is viewed as an independent keystroke sequence from the corresponding user. Previous research has shown that short keystroke subsequences decrease accuracy, while the longer keystroke subsequences may incorporate more noise. Therefore, we will need to experiment with different lengths of keystroke subsequence to determine an optimal value.

3.3   Keystroke Dynamics Image

In Sect. 2, we introduced the keystroke datasets used in this paper. As mentioned in the previous section, we divide the entire keystroke sequence into multiple subsequences, and in Sect. 3.1 we discussed the six types of timing features that are available. Thus, for a subsequence of length N, there are $6(N-1)$ features that can be determined from consecutive pairs of keystrokes, where repeated pairs are averaged and treated as a single pair. For example, for a subsequence of length 50, we obtain at most $6 \cdot 49 = 294$ features. We view each keystroke subsequence as an independent input sequence for the corresponding user. Next, we propose a new feature engineering structure to better organize these features.

The features UD-time, DD-time, DU-time, and UU-time are determined by consecutive keystroke events. Therefore, we organize these four features into a transition matrix with four channels, which can be viewed as four $N \times N$ matrices overlaid. This approach is inspired by RGB images, which have a depth of three, due to the R, G, and B channels.

Each row and each column in our four-channel $N \times N$ feature matrix corresponds to a key on the keyboard, and each channel corresponds to one kind of feature. Figure 2 illustrates how we have organized these features into transition matrices. For example, the value at row i and column j in the first channel of the matrix refer to the UD-time between any key presses of i followed by j within the current observation window.

The final feature is duration, which is organized as a diagonal matrix and added to the transition matrix as a fifth channel. Note that if a key or key-pair is pressed more than once, we use the average as the duration for that key or key-pair. In this channel, only diagonal locations have values because the duration feature is only relevant for one key at a time. The final result is that all of the features generated from keystroke subsequence are embedded in a transition matrix with five channels, which we refer to as the keystroke dynamics image (KDI).

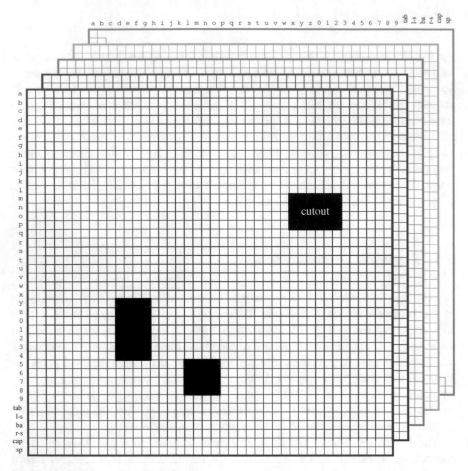

Fig. 2 Keystroke dynamics image for free-text

To prevent the transition matrix from being too sparse, we only consider time-based features for the 42 most common keystrokes. These 42 keys include the 26 English characters (A-Z), the 10 Arabic numerals (0-9), and six meta keys (space, back, left-shift, right-shift, tab, and capital). Therefore, the shape of the transition matrix is $5 \times 42 \times 42$, with the five channels as discussed above.

3.4 Keystroke Dynamics Sequence

Above, we provided details on the time-based image-like feature that we construct, which we refer to as the KDI. In this section, we discuss the application of an

RNN-based neural network to the KDI. Our goal is to use this feature to better take advantage of the inherently sequential nature of the keystroke dynamics data.

A keystroke in a keystroke sequence can be viewed as a word in a sentence. For our two free-text keystroke datasets, the keystroke sequence is different for each input and each user. For this data, we consider various encodings of each keystroke and use this encoding information in the embedding vector. Specifically, we experiment with index encoding and one-hot encoding. The resulting embedding vectors are used to construct a keystroke dynamics sequence, which we abbreviate as KDS. These KDS vectors will be used in our RNN-based neural networks.

3.5 Cutout Regularization

As mentioned in Sect. 2.3.4, we employ a cutout regularization to prevent overfitting in our CNN. By artificially adding occlusions to our image-like data, the network is forced to pay attention to all parts of the image, instead of over-emphasizing some specific parts. We apply cutouts to our novel KDI data structure, which is discussed in Sect. 3.3, and the KDS, which was mentioned in Sect. 3.4. The dark blocks in Fig. 2 illustrate cutouts.

4 Architecture

In this section, we discuss the classification models in more detail. We also discuss hyperparameters tuning for the models considered.

4.1 Multiclass vs Binary Classification

The Buffalo and Clarkson II keystroke datasets are based on 101, and 148 subjects, respectively. Regardless of the dataset, our goal is to verify a user's identity based on features derived from keystroke sequences. While this is a classification problem, we can consider it as either a multiclass problem or multiple binary classification problems. In a practical application, the number of users could be orders of magnitude higher than in either of our datasets. To train a multiclass model on a large number of users would be extremely costly, and each time a new user joins, the entire model would have to be retrained. This is clearly impractical.

To train and test our models, we require positive and negative samples for each user. All the data available for a specific user will be considered as positive samples, while an equivalent number of negative samples are selected at random (and proportionally) from other users' samples. In practice, the number of non-target

Table 2 Best
Hyperparamters of deep
learning models

Parameter	Search space
Training epochs	100, **200**, 500, 1000
Initial learning rate	0.1, **0.01**, **0.001**, 0.0001
Optimizer	**Adam**, SGD, SGD with Momentum
Learning schedule	**StepLR** (**0.1**, 0.3, 0.5), Plateau
Experiments	**50**

users may be very large. In that case, we could draw negative samples from a a fixed
number of non-target users.

4.2   Hyperparameter Tuning

For the deep learning methods used in our experiments, we employ a grid search to
find the best parameters for the initial learning rate, optimizer, number of epochs,
and learning rate schedule. The values shown in Table 2 were tested, and those in
boldface were found to generate the best result. To allow for a direct comparison
of our different models, we use these same hyperparameters for all of our deep
learning models. Note that a learning rate of 0.01 generates the best results for CNN,
MLP, LSTM, GRU, while a learning rate of 0.001 generates best result in our RNN
experiments.

4.3   Implementations

For our keystroke dynamics experiments, we evaluate two kinds of models.
Specifically, a CNN is applied to the our novel KDI image-like features, while a
hybrid model that combines CNN and GRU is applied to the KDS features. The
KDI is presented in Sect. 3.3, while the free-KDS is described in Sect. 3.4.

4.3.1   CNN

The architecture of our CNN is shown in Fig. 16 in the Appendix. The input of this
model is the KDI, and hence we view the transition matrix as an image. Here, a
"stage" includes two `conv2d` layers and a `maxpooling` layer, not counting the
activation function.

In each stage, there are two convolutional layers and a maxpooling layer.
Moreover, a `relu` function is employed after each convolutional layer. Following
these two stages, there are three fully connected layers, and a dropout layer is
added to prevent overfitting. Finally, a sigmoid function is used to compute the final
probability of a positive sample.

4.3.2 CNN-RNN

The architecture of our CNN-RNN is illustrated in Fig. 17 in the Appendix. The input to this model is the KDS mentioned in Sect. 3.4. Note that 32 convolutional kernels shift in the keystroke sequence direction, and thus a sequence matrix with embedding size 32 is generated. This resulting output matrix is fed into a 2-layers GRU network, which is followed by a fully connected layer. Since this is a binary classification model, a sigmoid function is used to compute the probability of a positive sample.

5 Experiment and Result

In this section, we provide experimental results for our free-text binary classification experiments. The results of the various models considered are analyzed and compared. Note that in all of our experiments, we apply 5-folds cross validation and average the performance for each user.

5.1 Metrics

We adopt two metrics to evaluate our results. The first metric is accuracy, which is simply the number of correct classifications divided by the total number of classifications. More formally, accuracy is calculated as

$$\text{accuracy} = \frac{\text{TP} + \text{TN}}{\text{TP} + \text{FP} + \text{TN} + \text{FN}}$$

where TP and TN are true positives and true negatives, while FP and FN are false positives and false negatives.

There are two kinds of classification errors, namely, false positives and false negatives. There is an inherent trade-off between the false positive rate (FPR) and the false negative rate (FNR), in the sense that by changing the threshold that we use for classification, we can lower one but the other will rise. For a metric that is threshold-independent, we compute the equal error rate (EER) which, as the name suggests, is the value for which the FPR and FNR are equal. The EER is obtained by considering thresholds in the range of [0, 1] to find the point where the FPR and FNR are in balance. Figure 3 illustrates a technique for determining the EER.

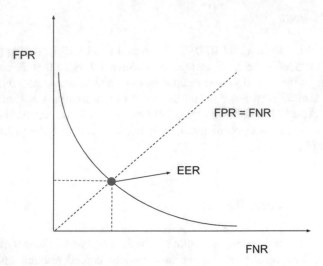

Fig. 3 Equal error rate

5.2   Result of Free-Text Experiments

For our free-text experiments, we focus on the effect of the lengths of keystroke sequences, kernel sizes for the CNN, encoding methods for the keystroke sequence data, and different hyperparameters of the RNN. Additionally, we explore the performance of models with and without cutout regularization.

5.2.1   Length of Keystroke Subsequence

First, we experiment with different lengths of keystroke subsequences. Specifically, we consider lengths of 50, 75, and 100 keystrokes. The results of these experiment are given in Figs. 4 and 5 for the Buffalo and Clarkson II datasets, respectively. From these results, we observe that when the length of a keystroke sequence has minimal impact on the accuracy or EER.

From these results, we observe that when the length of a keystroke sequence is relatively short, there is insufficient information to support strong authentication and, conversely, when the sequence is too long, the additional noise degrades the accuracy. Moreover, the results shows that the CNN-based model is more robust when the length of the keystroke sequence changes, which can be explained by the KDI mitigating the noise inherent in longer sequences. To accelerate the training process, we adopt the length 100 for the keystroke subsequences in all subsequent experiments.

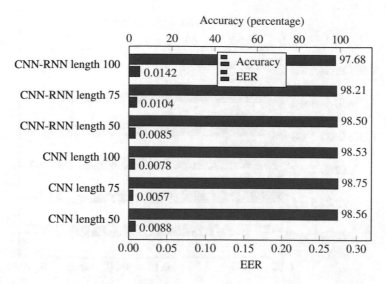

Fig. 4 Keystroke lengths (Buffalo dataset)

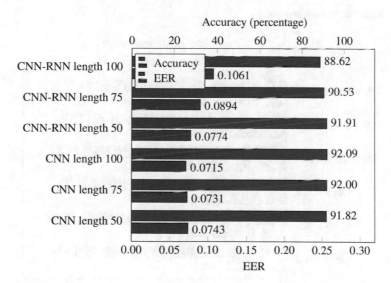

Fig. 5 Keystroke lengths (Clarkson II dataset)

5.2.2 CNN Kernel Sizes

In any CNN, the kernel size is a critical parameter. To determine the optimal kernel size, we experiment with three square kernels (3×3, 5×5, and 7×7) in our basic CNN model. For the CNN part of our hybrid CNN-RNN model, we experiment with

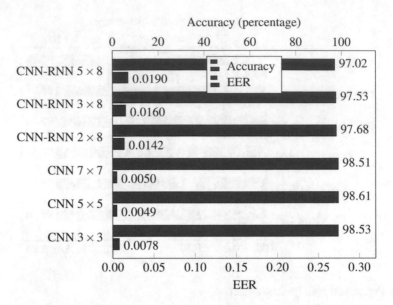

Fig. 6 Kernel size (Buffalo dataset)

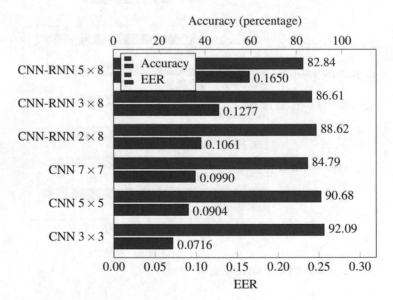

Fig. 7 Kernel size (Clarkson II dataset)

three sizes of rectangle kernels (2×8, 3×8, and 5×8). These experimental results for the basic CNN and hybrid CNN-RNN are given in Figs. 6 and 7.

We note that the kernel size makes no appreciable difference for the basic CNN model on the Buffalo dataset, while the two larger kernels both perform equally well

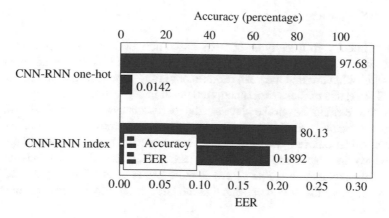

Fig. 8 Embedding methods (Buffalo dataset)

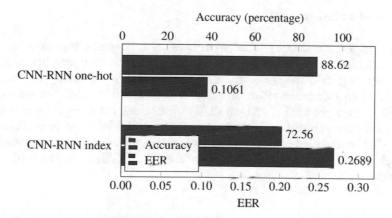

Fig. 9 Embedding methods (Clarkson II dataset)

on the Clarkson II dataset. For the CNN-RNN model, the results are also mixed, with the smaller kernel giving the best results over the two datasets. We adopt 3×3 square kernels for CNN-based models and 2×8 kernels for CNN-RNN based model in subsequent experiments.

5.2.3 Embedding Method

As mentioned above in Sect. 3.4, we consider two embedding methods, namely, index encoding and one-hot encoding. These experimental results are given in Figs. 8 and 9 for the Buffalo and Clarkson II datasets, respectively. From these results, it is clear that one-hot encoding is far superior to index encoding, and hence in subsequent experiments, we use one-hot encoding.

5.2.4 RNN Structure

We experiment with three types of RNN-based networks in our CNN-RNN architecture. Specifically, we consider a plain RNN, GRU, and LSTM. The advantages of GRU and LSTM are that they can capture more long-term information than a plain RNN. The results of these experiments are given in Figs. 10 and 11.

For the Buffalo keystroke dataset, the performances of our three different models are virtually identical, which indicates that the most valuable information is contained in adjacent keystroke pairs. However, for the Clarkson II keystroke dataset, we find that the GRU is more effective than the other two architectures. A plausible explanation is that LSTM is more prone to overfitting, while RNN is simply less powerful. And it appears that the GRU is slightly better at dealing with noisy data.

5.2.5 Cutout Experiments

It is likely that the data extracted from keystroke dynamics sequences is noisy because of the various extraneous factors that can influence typing behavior. We use cutout regularization, since it is useful at preventing overfitting, and since it is believed to reduce the effect of noisy information. The results of our cutout experiments are given in Figs. 12 and 13. We observe that cutout regularization has a significant positive effect on the performance of our models, which is most obvious in the CNN-based model. This is reasonable, since the cutout concept derives from the field of computer vision and our input data (i.e., KDI) is an image-like data structure.

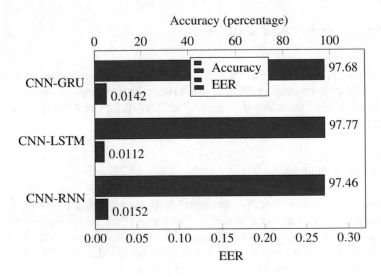

Fig. 10 CNN-RNN (Buffalo dataset)

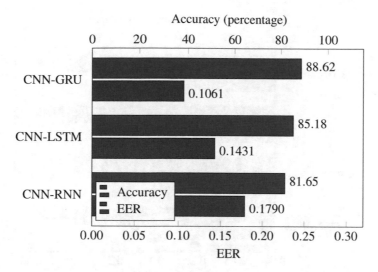

Fig. 11 CNN-RNN (Clarkson II dataset)

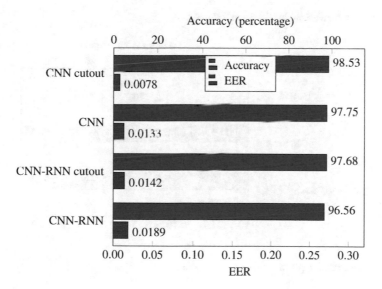

Fig. 12 Cutout regularization (Buffalo dataset)

5.3 Discussion

In our experiments, the performance on the Buffalo dataset is consistently higher than that of the Clarkson II dataset. It is likely the case that the latter dataset contains noisier data, as it was collected over a period of 2.5 years and under far less controlled conditions. We also find that our CNN-based model (KDI + CNN)

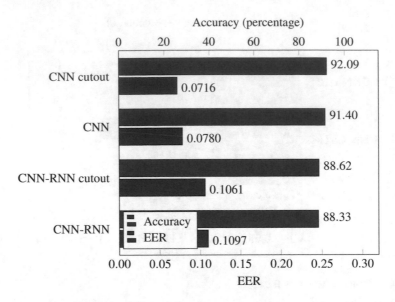

Fig. 13 Cutout regularization (Clarkson II dataset)

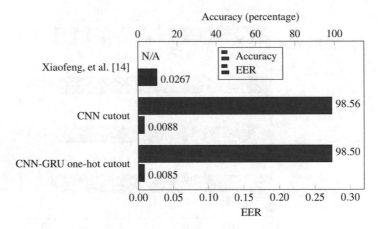

Fig. 14 Comparison to previous work (Buffalo dataset)

consistently generates better results than our RNN-CNN based model (Free-KDS + CNN-RNN). Comparing our results with the previous work in [14], we observe that in terms of EER, our two models both perform better on the Buffalo dataset, but slightly worse on the Clarkson II dataset. These results are summarized in Figs. 14 and 15.

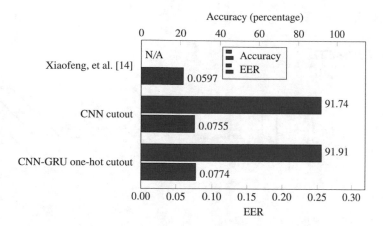

Fig. 15 Comparison to previous work (Clarkson II dataset)

6 Conclusion

This research focused on authentication based on keystroke dynamics derived features in the free-text case. We found that dividing the sequence into a number of fixed-length subsequences was an effective feature engineering strategy. In addition, we developed and analyzed an image-like engineered feature structure that we refer to as KDI, and we compared this to another structure that we refer to as KDS. The KDI was used as the input for our CNN experiments, while the KDS served as the input data for our CNN RNN experiments. In both cases, we applied cutout regularization.

The experimental results reported here show that our pure CNN architecture outperforms our combination of CNN and RNN, and cutout significantly improves the performances of both models. Moreover, our two modeling approaches both outperform previous work on the Buffalo keystroke dataset and yield competitive results for the Clarkson II dataset.

In the realm of future work, we conjecture that generative adversarial networks (GAN) will prove useful in this problem domain. More fundamentally, we believe that improved (and larger) datasets are necessary if we are to make significant further progress on this challenging authentication problem.

Appendix

See Figs. 16 and 17.

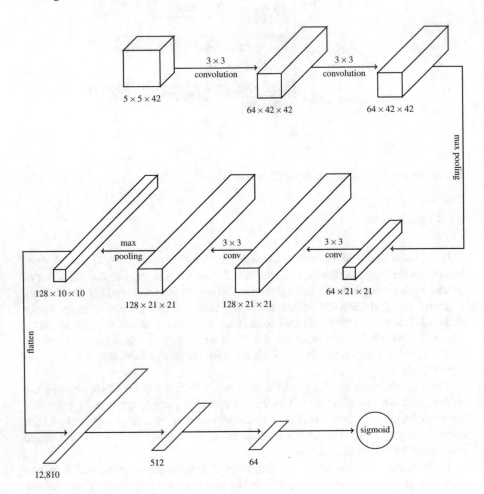

Fig. 16 Architecture of CNN for free-text datasets

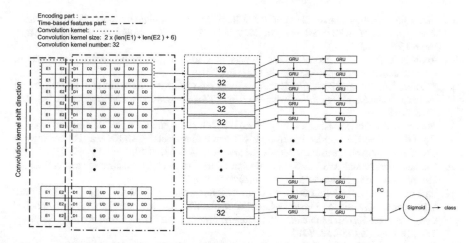

Fig. 17 Architecture of CNN-RNN for free-text datasets

References

1. Saad Albawi, Tareq Abed Mohammed, and Saad Al-Zawi. Understanding of a convolutional neural network. In *2017 International Conference on Engineering and Technology*, ICET, pages 1–6, 2017.
2. Faisal Alshanketi, Issa Traore, and Ahmed Awad Ahmed. Improving performance and usability in mobile keystroke dynamic biometric authentication. In *2016 IEEE Security and Privacy Workshops*, SPW, pages 66–73, 2016.
3. Mario Luca Bernardi, Marta Cimitile, Fabio Martinelli, and Francesco Mercaldo. Keystroke analysis for user identification using deep neural networks. In *2019 International Joint Conference on Neural Networks*, IJCNN, pages 1–8, 2019.
4. Hayreddin Çeker and Shambhu Upadhyaya. Enhanced recognition of keystroke dynamics using gaussian mixture models. In *2015 IEEE Military Communications Conference*, MILCOM, pages 1305–1310, 2015.
5. Hayreddin Çeker and Shambhu Upadhyaya. Sensitivity analysis in keystroke dynamics using convolutional neural networks. In *2017 IEEE Workshop on Information Forensics and Security*, WIFS, pages 1–6, 2017.
6. Junyoung Chung, Caglar Gulcehre, KyungHyun Cho, and Yoshua Bengio. Empirical evaluation of gated recurrent neural networks on sequence modeling. https://arxiv.org/abs/1412.3555, 2014.
7. Terrance DeVries and Graham W Taylor. Improved regularization of convolutional neural networks with cutout. https://arxiv.org/abs/1708.04552, 2017.
8. Sepp Hochreiter and Jürgen Schmidhuber. Long short-term memory. *Neural Computation*, 9(8):1735–1780, 1997.
9. Pilsung Kang and Sungzoon Cho. Keystroke dynamics-based user authentication using long and free text strings from various input devices. *Information Sciences*, 308:72–93, 2015.
10. Junhong Kim, Haedong Kim, and Pilsung Kang. Keystroke dynamics-based user authentication using freely typed text based on user-adaptive feature extraction and novelty detection. *Applied Soft Computing*, 62:1077–1087, 2018.
11. Paweł Kobojek and Khalid Saeed. Application of recurrent neural networks for user verification based on keystroke dynamics. *Journal of Telecommunications and Information Technology*, 2016:80–90, 2016.

12. Gutha Jaya Krishna, Harshal Jaiswal, P. Sai Ravi Teja, and Vadlamani Ravi. Keystroke based user identification with XGBoost. In *2019 IEEE Region 10 Conference*, TENCON, pages 1369–1374, 2019.
13. Andreas Lanitis.   A survey of the effects of aging on biometric identity verification. *International Journal of Biometrics*, 2(1):34–52, 2010.
14. Xiaofeng Lu, Shengfei Zhang, and Shengwei Yi. Free-text keystroke continuous authentication using CNN and RNN. *Journal of Tsinghua University (Science and Technology)*, 58(12):1072–1078, 2018.
15. Andrew Maas, Chris Heather, Chuong (Tom) Do, Relly Brandman, Daphne Koller, and Andrew Ng. Offering verified credentials in massive open online courses: MOOCs and technology to advance learning and learning research. *Ubiquity*, 2014:1–11, 2014.
16. Fabian Monrose and Aviel D. Rubin. Keystroke dynamics as a biometric for authentication. *Future Generation Computer Systems*, 16(4):351–359, 2000.
17. Eduard C. Popovici, Ovidiu G. Guta, Liviu A. Stancu, Stefan C. Arseni, and Octavian Fratu. MLP neural network for keystroke-based user identification system. In *11th International Conference on Telecommunications in Modern Satellite, Cable and Broadcasting Services*, TELSIKS, pages 155–158, 2013.
18. Jürgen Schmidhuber. Deep learning in neural networks: An overview. *Neural Networks*, 61:85–117, 2015.
19. Nitish Srivastava, Geoffrey E. Hinton, Alex Krizhevsky, Ilya Sutskever, and Ruslan Salakhutdinov. Dropout: A simple way to prevent neural networks from overfitting. *Journal of Machine Learning Research*, 15(56):1929–1958, 2014.
20. Y. Sun, Hayreddin Çeker, and Shambhu Upadhyaya. Shared keystroke dataset for continuous authentication. In *2016 IEEE International Workshop on Information Forensics and Security*, WIFS, pages 1–6, 2016.
21. Fabio Di Tommaso, Michele Guerra, Fabio Martinelli, Francesco Mercaldo, Massimo Piedimonte, Giovanni Rosa, and Antonella Santone.   User authentication through keystroke dynamics by means of model checking: A proposal. In *2019 IEEE International Conference on Big Data*, Big Data, pages 6232–6234, 2019.
22. Esra Vural, Jiaju Huang, Daqing Hou, and Stephanie Schuckers.   Clarkson University keystroke dataset. https://citer.clarkson.edu/research-resources/biometric-dataset-collections-2/clarkson-university-keystroke-dataset/.
23. Jatin Yadav, Kavita Pandey, Shashank Gupta, and Richa Sharma. Keystroke dynamics based authentication using fuzzy logic. In *2017 Tenth International Conference on Contemporary Computing*, IC3, pages 1–6, 2017.

Correction to: Artificial Intelligence for Cybersecurity

Mark Stamp, Corrado Aaron Visaggio, Francesco Mercaldo, and Fabio Di Troia

Correction to:
M. Stamp et al. (eds.), *Artificial Intelligence for Cybersecurity*,
Advances in Information Security 54,
https://doi.org/10.1007/978-3-030-97087-1

The original version of the book was inadvertently published with the incorrect title. This has now been amended throughout the book to the correct title "Artificial Intelligence for Cybersecurity".

The updated original version of the book can be found at
https://doi.org/10.1007/978-3-030-97087-1

Printed in the United States
by Baker & Taylor Publisher Services